S0-BMA-513

ADVANCES IN LIPID RESEARCH

Volume 24

Skin Lipids

Advances in Lipid Research

Volume 24

Skin Lipids

Edited by

PETER M. ELIAS
Department of Dermatology
University of California, San Francisco
San Francisco, California

SERIES EDITORS

RICHARD J. HAVEL
Cardiovascular Research Institute
University of California, San Francisco
San Francisco, California

DONALD M. SMALL
Department of Biophysics
Boston University
Boston, Massachusetts

ACADEMIC PRESS, INC.
Harcourt Brace Jovanovich, Publishers
San Diego New York Boston London Sydney Tokyo Toronto

This book is printed on acid-free paper. ♾

Academic Press, Inc.
San Diego, California 92101

United Kingdom Edition published by
Academic Press Limited
24–28 Oval Road, London NW1 7DX

Library of Congress Catalog Number: 63-22330

International Standard Book Number: 0-12-024924-3

PRINTED IN THE UNITED STATES OF AMERICA
91 92 93 94 9 8 7 6 5 4 3 2 1

CONTENTS

Lipid Metabolism in Cultured Keratinocytes

Maria Ponec

Lipid Modulators of Epidermal Proliferation and Differentiation

Walter M. Holleran

X-Ray Diffraction and Electron Paramagnetic Resonance Spectroscopy of Mammalian Stratum Corneum Lipid Domains

Sui Yuen E. Hou, Selwyn J. Rehfeld, and William Z. Plachy

Strategies to Enhance Permeability via Stratum Corneum Lipid Pathways

Russell O. Potts, Vivien H. W. Mak, Richard H. Guy, and Michael L. Francoeur

Lipids in Normal and Pathological Desquamation

Mary L. Williams

Chemistry and Function of Mammalian Sebaceous Lipids

Mary Ellen Stewart and Donald T. Downing

Integumental Lipids of Plants and Animals: Comparative Function and Biochemistry

Neil F. Hadley

Epidermal Vitamin D Metabolism, Function, and Regulation

Sreekumar Pillai and Daniel D. Bikle

PREFACE

This volume of *Advances in Lipid Research* is intended to provide a unique resource, with a comprehensive and current overview of the field of skin lipids. Because of the acknowledged importance of epidermal lipids for cutaneous barrier function, the first three chapters address structural, biochemical, and metabolic aspects of the role of lipids in permeability barrier formation and maintenance. In addition, Chapters Six and Seven describe the lipid biophysics of the intercellular lipid domains in the stratum corneum, and the regulation of percutaneous absorption by these domains, respectively. Chapter Four describes the lipid content and metabolism of cultured keratinocytes, grown under standard immersed conditions and in various *in vitro* systems that attempt to produce an epidermal equivalent, including a competent barrier.

The remaining chapters cover a broad panoply of subjects not directly related to the epidermal barrier. Chapter Eight describes the role of epidermal lipids in the pathogenesis of several disorders of cornification and the insights gained from these "experiments of nature" about the role of specific lipids in normal cohesion and desquamation. Chapter Five discusses the important new field of lipid signaling mechanisms in the epidermis, focusing on the emerging potential role of sphingolipid metabolites in regulating epidermal proliferation and differentiation. A discussion of eicosanoids is specifically not included, however, since this subject has been exhaustively covered in several recent reviews. The ninth chapter provides a comprehensive description of the biochemistry of mammalian sebaceous gland lipids, including speculations about the function of some of these species in normal and diseased skin (e.g., acne). Chapter Ten compares the structure, function, and lipid biochemistry of integumental lipids from plants, invertebrates, and cold-blooded vertebrates to warm-blooded (homeothermic) organisms. Finally, Chapter Eleven reviews the explosion of new information about vitamin D and the skin, including new clinical, pathophysiological, and therapeutic applications. Again, we have specifically chosen not to include chapters on the other major fat-soluble vitamin known to influence epidermal function (i.e., vitamin A), only because so many recent, exhaustive reviews of this subject are already available.

In summary, this volume not only provides abundant current and comprehensive information, but also each of the chapters represents a unique effort in the literature.

PETER M. ELIAS

ADVANCES IN LIPID RESEARCH, VOL. 24

Structural and Lipid Biochemical Correlates of the Epidermal Permeability Barrier

PETER M. ELIAS AND GOPINATHAN K. MENON

Dermatology Service
Veterans Administration Medical Center
San Francisco, California 94121
and
Department of Dermatology
University of California School of Medicine
San Francisco, California 94143

I. Introduction and Historical Perspective

A pivotal point in terrestrial adaptation is prevention of desiccation and maintenance of internal water homeostasis. Mammals have evolved an impressive array of adaptive responses for water conservation, among the most remarkable of which is the development of a cutaneous barrier to water loss. The outermost integumentary tissue, the epidermis, maintains a reserve of germinal cell layers whose proliferation, stratification, and differentiation result in production of the outermost layer, the anucleate stratum corneum. The loose "basketweave" pattern of the stratum corneum in typical histological preparations delayed appreciation of its responsibility for cutaneous barrier function. Physical chemists were the first to show that the stratum corneum is extremely resilient, possessing the permeability properties of a homogeneous film (for review see Scheuplein and Blank, 1971). Later studies revealed the stratum corneum to be composed of interlocking, vertical columns of foreshortened polyhedral cells, with thickened membrane envelopes (MacKenzie, 1969; Christophers, 1971; Menton and Eisen, 1971). More recent work has revealed the unique structural organization of this tissue into a two-compartment system (see below).

The importance of stratum corneum lipids for barrier integrity has been appreciated for several decades. For example, the observation that topical applications of organic solvents produce profound alterations in barrier function is over 40 years old (for review see Scheuplein and Blank, 1971). More recently, the importance of bulk stratum corneum lipids for the barrier has been demonstrated by (1) the inverse relationship between the permeability of the stratum corneum to water and water-soluble molecules at different skin sites (e.g., abdomen versus palms and soles) and the lipid content of the first site (Elias *et al.*, 1981a; Lampe *et al.*, 1983a), (2) the observation that organic solvent-induced perturbations in barrier function occur in direct proportion to the quantities of lipid removed (Grubauer *et al.*, 1989a), (3) the observation that stratum corneum lipid content is deficient or defective in pathological states that are accompanied by compromised barrier function, such as essential fatty acid deficiency (Elias and Brown, 1978), and, finally, (4) that replenishment of stratum corneum lipids, which follows removal by solvents or detergent, parallels the recovery of barrier function (Menon *et al.*, 1985a; Grubauer *et al.*, 1989b).

II. Stratum Corneum Two-Compartment Model

More recently, the concept of the stratum corneum as merely a homogeneous film has been replaced by a model of the stratum corneum consisting of protein-enriched corneocytes embedded in a lipid-enriched, intercellular matrix (Fig. 1) (Elias, 1983), i.e., a continuous lipid phase surrounding a discontinuous protein

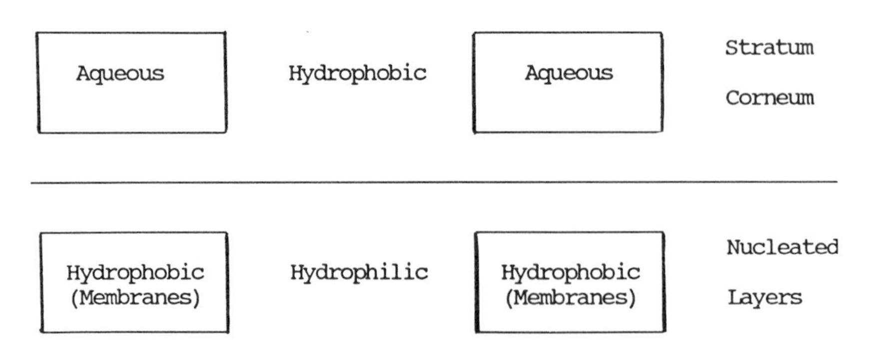

FIG. 1. Depiction of two-compartment model, illustrating the localization of lipid-enriched domains to the stratum corneum interstices. [Modified from Fitzpatrick *et al.* (1987). "Dermatology in General Medicine," 3rd Ed. McGraw-Hill, New York, with permission.]

phase. The evidence for such protein–lipid sequestration is based upon freeze–fracture replication (Elias and Friend, 1975; Elias *et al.*, 1977a,b), histochemical (Elias *et al.*, 1977b), biochemical (Grayson and Elias, 1982), cell fractionation (Grayson and Elias, 1982), cell separatory (Smith *et al.*, 1982), and physical–chemical (Elias *et al.*, 1983; White *et al.*, 1988; Rehfeld *et al.*, 1990) studies (Table I). First, the freeze–fracture method revealed that stacks of intercellular bilayers existed in the intercellular spaces (Fig. 2) (Elias and Friend, 1975; Elias *et al.*, 1977a,b); transmission electron microscopy previously had revealed only empty spaces (Brody, 1964, 1966). Likewise, histochemical stains revealed the membrane domains of the stratum corneum to be enriched in neutral lipids (Fig. 3), but only when these stains were applied to frozen sections (Elias *et al.*, 1977b). Later, it became possible to isolate the peripheral membrane domains as a separate subcellular fraction. Analysis of these preparations showed that (Grayson and Elias, 1982) (1) the bulk of stratum corneum lipids were in these preparations, (2) the lipid composition of these preparations was virtually identical to that of whole stratum corneum, and (3) the freeze–fracture pattern of membrane multilayers, previously described in whole stratum corneum, was duplicated in the membrane preparations. More recently, X-ray diffraction and electron-spin resonance studies have localized all of the bilayer structures, as well as lipid-based thermal phenomena, to these membrane domains (Elias *et al.*, 1983; White *et al.*, 1988; Rehfeld *et al.*, 1990).

The two-compartment arrangement, which is sometimes simplified to a "bricks and mortar" analogy (Elias, 1983), also explains both the ability of cells in the outer stratum corneum to take up water (i.e., lipid-enriched bilayers act as semipermeable membranes) (Middleton, 1968; Imokawa *et al.*, 1986) as well as the differences in rates of percutaneous absorption of topically applied lipophilic versus hydrophilic agents—the latter penetrating at much slower rates, suggesting a separate pathway (Michaels *et al.*, 1975) (see Potts *et al.*, this volume). But the two-compartment lipid-versus-protein model also requires further modification (see below), based upon recent evidence that extracellular proteins, such as desmosomal components

Table I

EVIDENCE FOR LIPID–PROTEIN COMPARTMENTALIZATION IN MAMMALIAN STRATUM CORNEUM (SC)

1. Pulverization destroys osmotically active structures responsible for the water-holding capacity of the SC
2. Freeze–fracture reveals lipid lamellae segregated within the SC interstices
3. Histochemistry displays neutral lipids solely in SC interstices
4. Organic solvents disperse the SC into individual cells
5. Isolated SC membrane sandwiches account for most SC lipids
6. X-Ray diffraction shows ordered lipids in isolated SC membranes
7. Catabolic enzymes colocalize with lipids in SC interstices
8. Electron-spin resonance localizes lipid signals to SC membranes

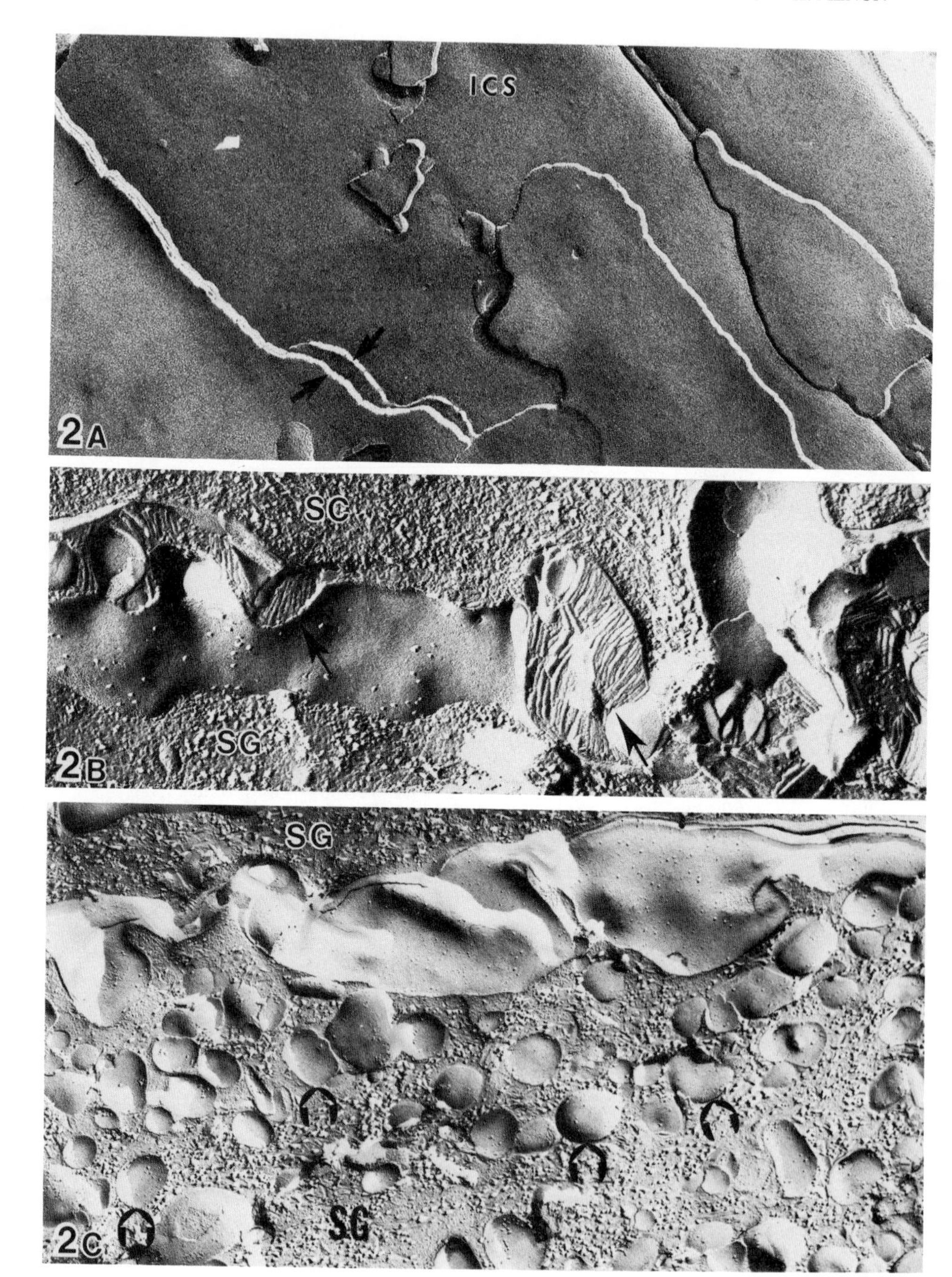
ICS
2A
SC
SG
2B
SG
SG
2C

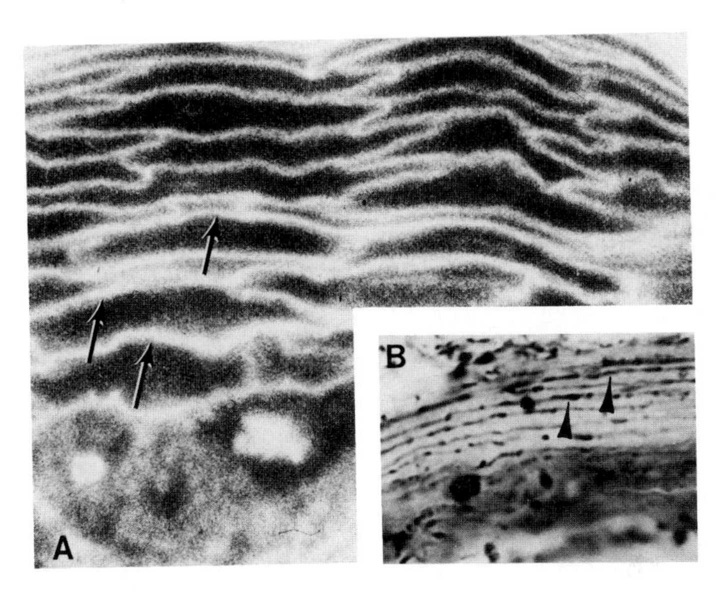

FIG. 3. Frozen sections of neonatal mouse stratum corneum stained with the lipophilic fluorophore, 8-anilino-1-sulfonic acid (A) and oil red O for neutral lipids (B). Note localization of lipid staining to membrane domains (arrows). [Reprinted from Elias *et al.* (1979). *J. Invest. Dermatol.* **73,** 339–348, with permission.]

(Allen and Potten, 1975; Haftek *et al.*, 1986), glycoproteins (Brysk *et al.*, 1988), and abundant enzymatic activity (Menon *et al.*, 1986c) exist within the intercellular spaces. Indeed, even the intercellular lipids are heterogeneous; in various animal species and topographic sites, there are different proportions of nonpolar, sebaceous gland-derived lipids in addition to lipids derived from the epidermis (Nicolaides, 1974) (see Stewart and Downing, this volume).

III. Cellular Basis for Lipid–Protein Sequestration in the Stratum Corneum of Terrestrial Mammals

Since its earliest descriptions (for review see Odland and Holbrook, 1987), hypotheses have abounded about the function of the epidermal lamellar body. These ellipsoidal organelles, measuring about ⅓ x ½ μm, appear initially in the first

FIG. 2. Freeze–fracture replicas of murine epidermis. (A) Note multilamellar stacks (arrows) in the intercellular spaces (ICS). (B) The initially secreted lamellar body contents cross-fracture (arrows), consistent with enrichment in polar lipids (SC, stratum corneum; SG, stratum granulosum). (C) Finally, note abundant lamellar bodies (arrows) in apical cytoplasm of outer granular cell (SG). [Reprinted from Elias *et al.* (1981b). *Lab. Invest.* **44,** 531–540 (A) and Elias *et al.* (1977a). *Anat. Rec.* **189,** 577–593 (B and C), with permission.]

suprabasal cell layer, the stratum spinosum, and they continue to accumulate in the stratum granulosum until they account for up to 25% of the volume of the cytosol (Elias and Friend, 1975). Although the subcellular site of lamellar body generation is not known, cytochemical studies have tentatively traced its origin to elements of the smooth endoplasmic reticulum (Wolff-Schreiner, 1977) or the Golgi apparatus (Wolff and Holubar, 1967; Weinstock and Wilgram, 1970; Chapman and Walsh, 1989).

Many ultrastructural studies have depicted the internal structure of these membrane-enclosed organelles. They are described to contain parallel stacks of lipid-containing disks enclosed by a limiting trilaminar membrane (for review see Odland and Holbrook, 1987). In near-perfect cross-sections, each lamella shows a major electron-dense band (shared by adjacent lamellae) separated by electron-lucent material, divided centrally by a minor electron-dense band (Fig. 4). Yet, despite published information about the fusion of secreted lamellar body disks, as well as the behavior of model liposomes made from stratum corneum lipids (Landmann, 1984, 1988; Abraham *et al.*, 1987), our recent observations suggest that lamellar body contents may actually be composed of bilayers already connected to each other, folded in an accordion-like fashion (Menon *et al.*, 1991b). When the epidermis is permeabilized with acetone, the limiting membranes of lamellar bodies are occasionally disrupted and the folded bilayers appear at different stages of unfurling (Fig. 4).

In the outer granular layer, lamellar bodies are arrayed at the lateral and apical surfaces, where they are poised to undergo rapid exocytosis. Although tracer perfusion studies first demonstrated a potential role for these organelles in the initial formation of the water barrier (Schreiner and Wolff, 1969; Squier, 1973; Elias and Friend, 1975; Elias *et al.*, 1977a), these electron-dense tracers may not reflect the actual diffusion pathway for much smaller molecules, such as water.

Cytochemists provided the next clues about the function of this organelle in the barrier, describing lamellar bodies to be enriched in sugars (Ashrafi and Meyer, 1977) and lipids (Olah and Rohlich, 1966; Breathnach and Wylie, 1966; Schreiner and Wolff, 1969; Elias *et al.*, 1977b; Landmann, 1980; Squier, 1982), thereby generating the initial hypothesis that their contents might be important for epidermal waterproofing (Schreiner and Wolff, 1969). Moreover, tracer perfusion studies demonstrated the role of the lamellar body secretory process in the initial formation of the barrier (Schreiner and Wolff, 1969; Squier, 1973; Elias and Friend, 1975; Elias *et al.*, 1977b). Indeed, the outward egress of water-soluble tracers through the epidermis is blocked at sites of lamellar body secretion, and no other membrane specializations, such as tight junctions, are present at these locations to account for barrier formation (Elias and Friend, 1975).

Biochemical studies on partially purified lamellar body preparations have demonstrated that these organelles are enriched in glycosphingolipids, free sterols, and phospholipids (Fig. 5) (Grayson *et al.*, 1985; Freinkel and Traczyk, 1985; Wertz *et al* 1985). These lipids are the putative source of almost all of the

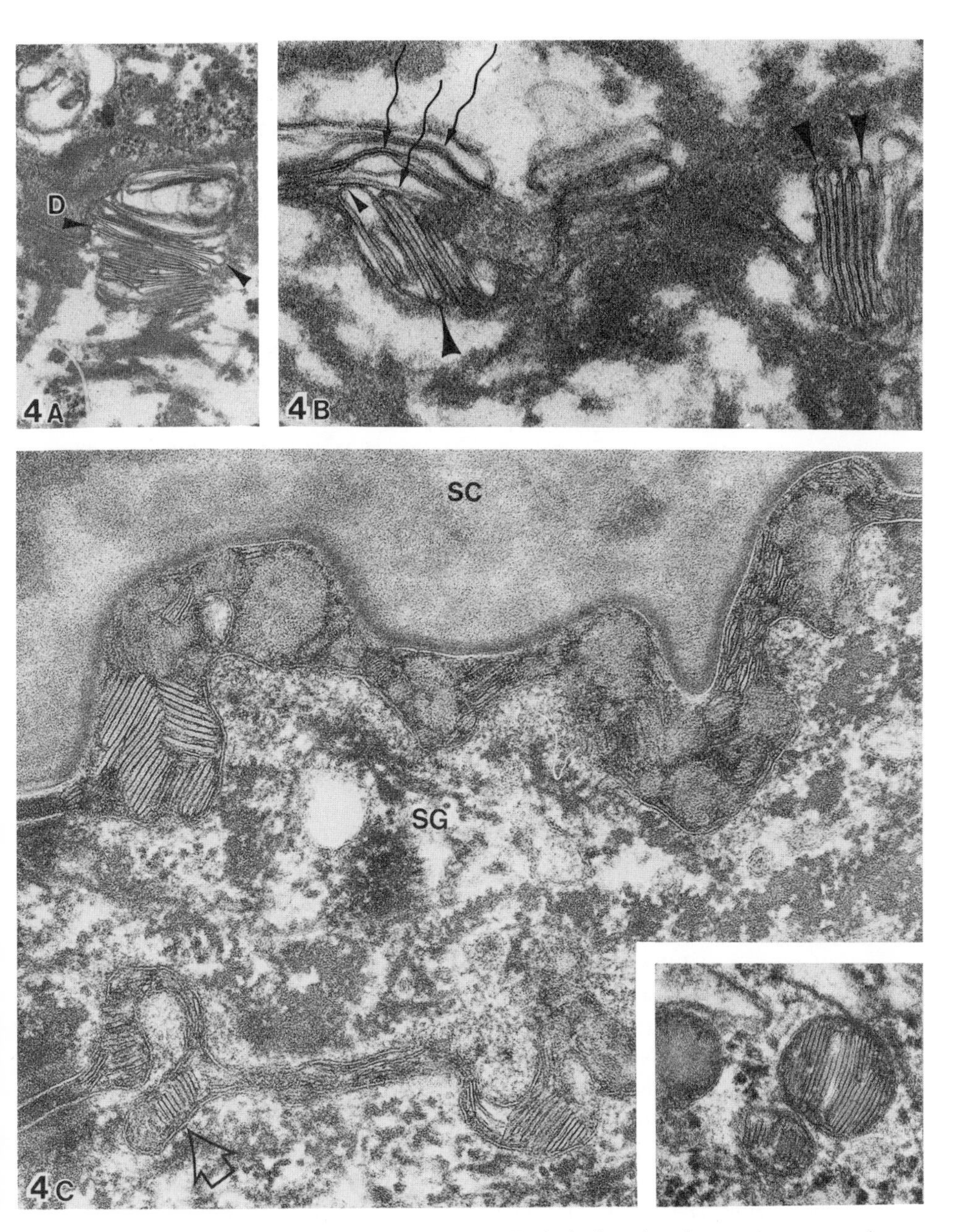

FIG. 4. Electron micrograph of epidermal lamellar body (insert), and secreted contents at the stratum granulosum (SG) and stratum corneum (SC) interface (B). Note that the lamellar body "disks" (D) actually appear to be a continuous sheet within the organelle, which begins to "unfurl" immediately after secretion (arrows) (A; c.f. Fig. 6). [Fig. 4C reprinted from Grayson *et al.* (1983). *Science* **221,** 962–964. Copyright © 1983 by the American Association for the Advancement of Science, with permission.]

Table II
HYDROLYTIC ENZYME CONTENT OF EPIDERMAL LAMELLAR BODIES[a]

Present	Absent or not increased
Lipid catabolic	**Lipid catabolic**
Acid lipase	Steroid sulfatase[b]
Phospholipase A	
Sphingomyelinase	**Others**
Glycosidases	Arylsulfatases A and B[c]
	β-Glucuronidase
Others	
Acid phosphatase	**Protease**
Cathepsins	Plasminogen activator[d]
Carboxypeptidase	

[a]Modified from Grayson *et al.* (1985).
[b]Microsomal.
[c]Typical lysosomal enzyme.
[d]Extracellular.

stratum corneum intercellular lipids (for review see Elias, 1983; Odland and Holbrook, 1987; Williams and Elias, 1987).

However, others described hydrolytic enzymes in lamellar bodies, suggesting alternatively that these organelles might instead modulate desquamation (see below). In addition to lipids, the lamellar body is enriched in certain hydrolytic enzymes, including acid phosphatase, proteases, lipases, nonspecific esterases, and presumably glycosidases (Wolff and Holubar, 1967; Squier and Waterhouse, 1970; Weinstock and Wilgram, 1970; Nemanic *et al.*, 1983; Grayson *et al.*, 1985; Freinkel and Traczyk, 1983, 1985; Wertz *et al.*, 1989). Hence, the lamellar body has been considered to be a type of lysosome (Waterhouse and Squier, 1966; Wolff and Holubar, 1967; Lazarus *et al.*, 1975), and a recent report of proton pumps in its limiting membrane supports this analogy (Chapman and Walsh, 1989). Yet, certain acid hydrolase activities characteristic of lysosomes, such as arylsulfatases A and B and β-glucuronidase, are notably absent (Grayson *et al.*, 1985) (Table II). Moreover, the same enzymes that are concentrated in the outer epidermis and/or in lamellar bodies have been found in high specific activity in the stratum corneum (Nemanic *et al.*, 1983; Elias *et al.*, 1984; Menon *et al.*, 1986b) and further localized to intercellular domains both biochemically and cytochemically (Elias *et al.*, 1984; Menon *et al.*, 1986b). Hence, it is likely that the enzymes present in lamellar bodies fulfill specific functions.

In Fig. 5, we have presented a model that is consistent with current information about the colocalization of selected lipids and catabolic enzymes within the lamellar body, suggesting a dual role for these enzymes in both barrier formation and desquamation. The colocalization of "probarrier" lipids and various lipases (phospholipase A, sphingomyelinase, and acid lipase) and glucosidases to the

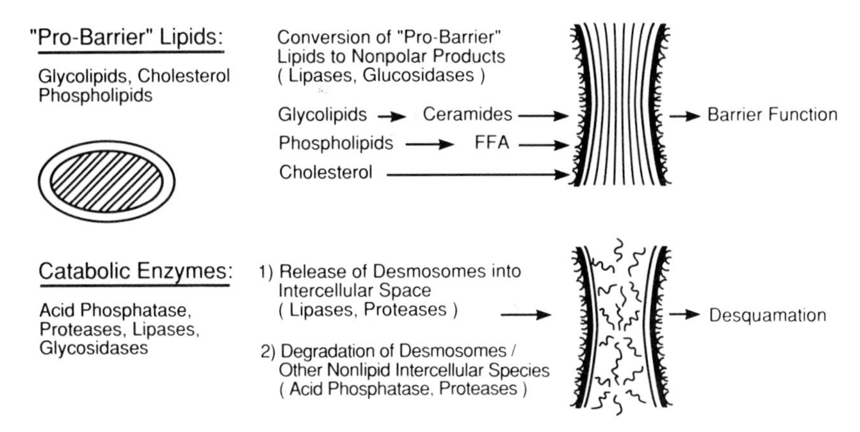

FIG. 5. Speculative program that links available information about lamellar body lipid and hydrolase content to modulations leading to barrier formation and desquamation. The release of desmosomes may be facilitated by the detergent action of fatty acids and/or phospholipases or proteases at sites of desmosomal insertion. [Reprinted from Elias (1987). *In* "Skin Pharmacokinetics" (B. Shroot and H. Schaeffer, eds.), pp. 1–9. Karger, Basel, with permission.]

same tissue compartment may mediate the changes in lipid composition and structure that occur during transit through the stratum corneum (Figs. 5 and 6) (Nemanic *et al.*, 1983; Menon *et al.*, 1986b; Elias *et al.*, 1988; Wertz and Downing, 1989). However, several features of this model are still speculative; e.g., the function of acid phosphatase in the cellular interstices has not been investigated. Moreover, the function of lamellar body-derived proteases is unknown. One possibility would be the activation of other lamellar body-derived enzymes under conditions present in the intercellular spaces. An acidic environment could result either from the deposition of acidic lamellar body contents and/or from the insertion of proton pumps in the plasma membrane in association with lamellar body exocytosis, if these pumps continue to be active in that site (Chapman and Walsh, 1989). These conditions may initiate a sequence that begins initially with

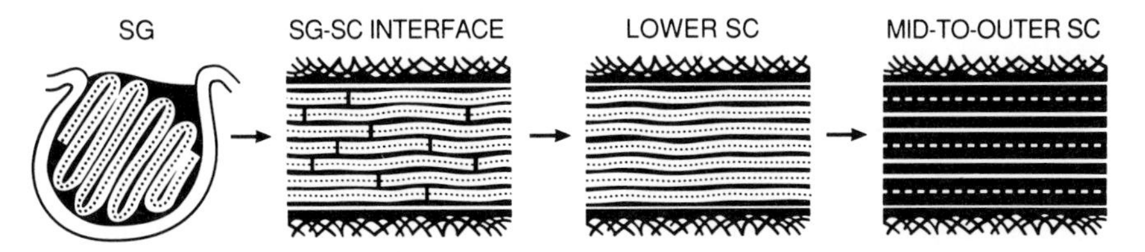

FIG. 6. Diagram that illustrates alterations of membrane structures that follow lamellar body secretion as a result of degradation of polar species by codeposition of hydrolases (SC, stratum corneum; SG, stratum granulosum) (c. f. Fig. 5). [Modified from Elias *et al.* (1988). *J. Invest. Dermatol.* **91,** 3–10, with permission.]

protease activation, followed by conversion of proenzymes to active forms of the lamellar body-derived hydrolases, leading ultimately to the compositional and structural changes known to occur in the intercellular spaces of the lower stratum corneum (Elias *et al.*, 1988).

A second possible function of lamellar body-derived proteases may relate to desmosomal degradation. Although desmosomes cannot form a physiological barrier to water loss (Arnn and Staehelin, 1981), they may contribute to the integrity of this tissue by mediating its cohesiveness. During stratum corneum transit, desmosomes decrease in number (Allen and Potten, 1975), a change that correlates spatially with the gradual loss of cohesiveness of this layer (King *et al.*, 1979). Intercellular proteases appear to mediate desmosomal degradation in plantar stratum corneum, because cell shedding requires an acidic environment and is blocked by serine protease inhibitors (Lundström and Egelrud, 1988, 1990a,b). Moreover, desmosomal proteins, such as desmoglein, are progressively deleted during transit through plantar stratum corneum (Egelrud and Lundström, 1989). However, because plantar stratum corneum is lipid depleted (Elias *et al.*, 1981a; Lampe *et al.*, 1983a), it is possible that in other topographic sites, access of proteases to desmosomes is limited by more extensive, lipid-enriched domains. Yet, very recent studies suggest that proteases may participate in stratum corneum shedding even in nonvolar sites (Egelrud and Lundström, 1990). Thus, although it seems likely that lamellar body-derived proteases contribute to stratum corneum desquamation, and that desmosomes play an important role in stratum corneum integrity, the regulation of these processes remains unknown.

Another inadequately studied potential consequence of lamellar body secretion relates to changes in (1) the intercellular volume and (2) the surface area:volume ratio of the stratum corneum and individual corneocytes, respectively. Massive exocytosis of lamellar bodies results in the deposition of abundant lipid, enzyme protein, and undoubtedly other substances into the stratum corneum interstices. As a result, preliminary studies suggest that this compartment is greatly expanded (5–15% of total volume) in comparison to the volume of the interstices in other epithelia (1–5%) (Elias and Leventhal, 1979). Moreover, the intercellular compartment serves as a selective "sink" for exogenous lipophilic agents, which may result in further expansion of this compartment (Nemanic and Elias, 1980). Finally, the splicing of the limiting membrane of the lamellar body into the plasma membrane of the granular cell should result in a massive expansion of the surface area : volume ratio of individual corneocytes (Elias and Leventhal, 1979). This change may explain the remarkable capacity of corneocytes to absorb up to four times of stratum corneum dry weight in water (for review see Scheuplein and Blank, 1971).

To date, the factors that regulate lamellar body secretion are not known. Recent studies have shown that acute perturbations of the barrier result in lamellar body secretion, accompanied by a striking paucity of these organelles in the cytosol

(Feingold *et al.*, 1990; Menon *et al.*, 1991b). However, by 1–2 hours, abundant nascent lamellar bodies appear in the cytosol. Clearly, secretion must occur under both basal and stimulated conditions, and it is possible that separate factors may regulate each process, as is the case for the surfactant-enriched lamellar bodies of the alveolar type II cell (Chander and Fisher, 1990). In fact, preliminary correlative ultrastructural and confocal microscopic studies suggest that lamellar bodies are organized into a continuous network by components of the cellular cytoskeleton (Cullander *et al.*, 1990). Unfortunately, little work exists in this area, and the control of lamellar body secretion remains a ripe area for investigation.

Although the lamellar body accounts for the delivery and sequestration of the majority of the stratum corneum lipids within the intercellular spaces, other delivery mechanisms may also be operative. For example, cholesterol sulfate, which accounts for up to 5% of total stratum corneum lipids in humans (Williams and Elias, 1981; Lampe *et al.*, 1983a), is not concentrated in lamellar bodies (Grayson *et al.*, 1985). Yet, in the stratum corneum this molecule becomes localized to intercellular domains (Elias *et al.*, 1984). Hence, unless cholesterol sulfate is lost during lamellar body isolation procedures, other mechanisms may account for its delivery to the interstices; e.g., the amphipathic properties of this compound could allow it to move freely across the cell membrane without the requirement of a specific delivery mechanism (Ponec and Williams, 1986). Likewise, steroid sulfatase, the enzyme responsible for desulfation of cholesterol sulfate (for review see Williams and Elias, 1987), is not enriched in lamellar bodies (Table II), yet it is also localized to membrane domains in the stratum corneum (Elias *et al.*, 1984). How this microsomal enzyme reaches the cell periphery is a mystery. Again, it is possible that the enzyme is present in lamellar bodies but is lost or destroyed during isolation. But it also is possible that steroid sulfatase may be transferred from microsomes to the limiting membrane of the lamellar body. This would result in "splicing" of enzyme into the corneocyte periphery during exocytosis, as noted above for the proton pump (Chapman and Walsh, 1989).

IV. Insights from Aves and Marine Mammals (Cetaceans)

A. Aves

Like terrestrial mammals, aves are warm-blooded organisms that face a dry external environment. However, in feathered body regions, avian plumage provides some degree of impediment to water loss, and as a result the epidermis is less waterproof than is epidermis of mammals living under comparable conditions (Webster *et al.*, 1985). Yet, nestlings are initially featherless and often must survive at extremely low ambient humidities (Welty and Baptista, 1988). Hence, avian epidermis must be able to adapt to changing ambient temperatures and humidity.

Like its mammalian counterpart, avian stratum corneum consists of corneocytes embedded in a lipid matrix (Lucas, 1980). However, in comparison to mammals, avian corneocytes are wafer-thin, effete structures (Menon *et al.*, 1986b). Under basal conditions, the mechanism for lipid delivery to the interstices of avian stratum corneum also differs from that in mammals. Rather than delivery by lamellar body secretion, an analogous (Menon *et al.*, 1991a) but larger membrane bound organelle, the multigranular body (MGB), under the usual environmental conditions, does not secrete its lamellar contents, but instead deteriorates and coalesces with its neighbors to form large neutral lipid droplets within the corneocyte cytosol (Menon *et al.*, 1981; Purton, 1988). As the corneocytes become progressively more attenuated, these droplets normally are extruded into the interstices through membrane porosities (Menon *et al.*, 1981; Purton, 1988).

Yet, when zebra finches are xerically stressed, i.e., under conditions of water-deprivation, MGBs appear to be secreted in a manner analogous to mammalian lamellar bodies (Menon *et al.*, 1988). Moreover, secretion of the disklike contents of MGBs gives rise to intercellular bilayer structures in the intercellular spaces, with features similar to those of terrestrial mammals (Menon *et al.*, 1988; cf. Landmann, 1980) (Fig. 7). Simultaneously, epidermal barrier function improves, as shown by significantly lower rates of transepidermal water loss (Menon *et al.*, 1988). With water replenishment or environmental rehumidification, this pattern reverts to basal conditions (Menon *et al.*, 1989b). Interestingly, though secretion of MGB contents gives rise to intercellular membrane structures (and a less permeable barrier), the porous extrusion of MGB-derived lipid droplets, which occurs under basal conditions, does not. The inability of MGB-derived *lipid droplets* (versus MGB-derived *membrane structures*) to provide a significant barrier may be due to hydrolysis of certain key species (e.g., glycosphingolipids and/or phospholipids) (Menon *et al.*, 1986b) by cytosolic hydrolases. This would result in a loss of lamellar structures, with the resultant emergence of a neutral lipid-enriched mixture that is incapable of forming membrane structures. Thus, under basal, hydrated conditions, these polar species would be absent and a more permeable stratum corneum would result (Menon *et al.*, 1988).

B. Marine Mammals (Cetaceans)

Ocean water is slightly hypertonic; hence, glabrous marine mammals, such as whales and dolphins, are exposed to less rigorous barrier requirements than are terrestrial species (Gaskin, 1982; Geraci *et al.*, 1986). Yet, marine mammals must retain as much metabolic water as possible, because exogenous sources are not available. Moreover, because the skin lies outside the subcutaneous fat layer, which effectively insulates the remainder of the organism, cetacean epidermis is exposed to water temperatures as low as 4°C. Furthermore, marine mammals have definite but less well-defined requirements for surface lubrication, solar pro-

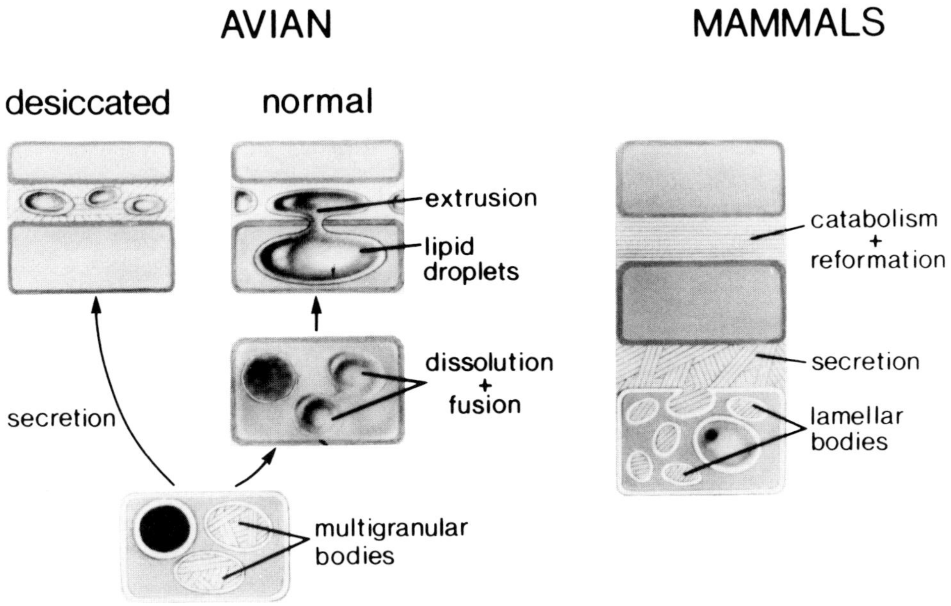

AVIAN MAMMALS

desiccated normal

extrusion
lipid droplets
dissolution + fusion
secretion
multigranular bodies
catabolism + reformation
secretion
lamellar bodies

FIG. 7. Fate of lipid-secretory organelles in avian and mammalian epidermis. Under basal conditions, multigranular bodies are degraded within the cytosol, resulting in the disappearance both of polar lipids and of hydrolytic enzymes. Hence, the resultant extracted lipid displays no lamellar substructure. Under conditions of extreme xeric stress, the fate of multigranular bodies resembles that of terrestrial mammals, i.e., secretion of lamellae into intercellular domains, as occurs in mammals. [Reprinted from Elias *et al.* (1987). *Am. J. Anat.* **180,** 161–177, with permission.]

tection, buoyancy, and antimicrobial activity that may be subserved by epidermal lipids (Gaskin, 1982; Geraci *et al.*, 1986).

Although there have been very few studies on cetacean skin, the epidermis has been shown to contain abundant lamellar bodies (Sokolov *et al.*, 1982; Menon *et al.*, 1986a). Moreover, lamellar body contents are secreted from all suprabasal, nucleated cell layers of cetacean epidermis (Menon *et al.*, 1986a). But, in contrast to terrestrial mammals, lamellar body-derived lipids are not reorganized into the basic unit system of membrane bilayers. Moreover, lamellar body-derived lipids do not appear to be completely catabolized to a more hydrophobic mixture; e.g., glycosphingolipids persist at all levels of the parakeratotic stratum corneum (Menon *et al.*, 1986a). It is possible that the incomplete transformation and metabolism of lamellar body contents reflect the less stringent barrier requirements of cetaceans, and that the partially hydrolyzed, intercellular lipid mixture serves other functions, e.g., lubrication and streamlining. This interpretation is consistent with existing information about the structure and composition of oral mucosal epithelia, which also are exposed to a hydrated environment. These

Table III

POSSIBLE FUNCTION OF NEUTRAL LIPID DROPLETS IN HOMEOTHERMIC SKIN[a]

Animal groups	Cell type	Possible function
Terrestrial mammals	Sebocyte	Natural emollients Antimicrobial activity Pheromones
Cetaceans and Sirenia (Manatees)	Lipokeratinocytes	Thermogenesis Flotation Cryoprotectancy Source of metabolic water
Avians	Sebocytes (uropygial gland)	Feather flexibility Plumage hygiene Antimicrobial activity Pheromones Vitamin D
Avians	Sebokeratinocytes	Permeability barrier Antimicrobial activity Emolliency Ultraviolet filter

[a]Modified from Elias *et al.* (1987), *Am. J. Anat.* **180,** 161–177, with permission.

tissues display abundant lamellar bodies (Squier, 1973; Hayward and Hackermann, 1973; Lavker, 1976; Elias *et al.*, 1977a), but, as in marine mammals, lamellar body-derived lipids form a less effective barrier and appear to be incompletely hydrolyzed; i.e., glycosphingolipids are much more abundant than in the epidermis of the same species (Squier *et al.*, 1986). Moreover, in mucosal epithelia, this compositional profile correlates with a less effective permeability barrier (Squier, 1975).

In contrast to terrestrial mammals, cetaceans also possess large, intracellular lipid droplets at all levels of the epidermis (Sokolov *et al.*, 1982; Geraci *et al.*, 1986; Menon *et al.*, 1986a). And, in contrast to avians, these droplets are not expelled into the intercellular spaces in the stratum corneum. Their tinctorial properties, coupled with the known lipid biochemical composition of cetacean epidermis (Menon *et al.*, 1986a), suggest that they are enriched in neutral lipids, such as triglycerides. Because of their composition and persistence in the epidermis, it is likely that these droplets subserve some other functions of cetacean epidermal lipids described above, e.g., the oxidation of lipid stores to generate calories (Table III) (Gaskin, 1982; Geraci *et al.*, 1986; Elias *et al.*, 1987).

V. Intercellular Membrane Structures in Mammalian Stratum Corneum

As noted above, elucidation of membrane structure in mammalian stratum corneum was impeded by the extensive artifacts produced during processing for light and/or electron microscopy. Typically, in published studies prior to the mid-1970s, the intercellular spaces appear dilated and devoid of membrane structures, or collapsed and lacking in intervening lamellar bilayers (Brody, 1964, 1966). Following the application of freeze–fracture replication to the epidermis (Breathnach *et al.*, 1973), the mid-to-outer stratum corneum interstices in both epidermis and keratinizing mucosal epithelia later were found to be replete with a multilamellar system of broad membrane bilayers (Fig. 2) (Elias and Friend, 1975; Elias *et al.*, 1977a,b). With osmium vapor fixation, these membrane layers comprised a multilayered system of alternating electron-dense and electron-lucent lamellae of approximately equal thickness (Elias and Friend, 1975; Elias *et al.*, 1977a,b). At the stratum granulosum–stratum corneum interface and in the interstices of the lowermost layers of the stratum corneum, a transition can be seen from cross-fractured, lamellar body-derived sheets to successively broader lamellae (Fig. 2) (Elias *et al.*, 1977a; Landmann, 1986).

The sequence of events that leads to the formation of broad intercellular bilayers has been studied ultrastructurally (Elias *et al.*, 1977b, 1988; Landmann, 1984, 1986), biochemically (Elias *et al.*, 1988), and in model vesicles prepared from synthetic and naturally occurring stratum corneum lipids (Landmann, 1984; Wertz *et al.*, 1986; Abraham *et al.*, 1987). As described above, immediately following extrusion, the lamellar body-derived membranes begin to unfold parallel to the plasma membrane. Within the first two layers of the stratum corneum, end-to-end fusion appears to occur, giving rise to broad, uninterrupted lamellae, which undergo further changes in substructure (Fig. 6). From the liposome work, it has been suggested that this fusion process occurs spontaneously, perhaps due to the high radius of curvature at the edges of the disks (Landmann, 1984) and/or calcium-mediated aggregation (Abraham *et al.*, 1987). However, this change in freeze–fracture characteristics also correlates with a sequence of changes in composition (Fig. 5) (Table IV); i.e., from the polar lipid-enriched mixture of glycosphingolipids, phospholipids, and free sterols present in lamellar bodies and at the SG–SC interface to a more nonpolar mixture, enriched in ceramides, free sterols, and free fatty acids, present in the bulk of the stratum corneum (Gray and Yardley, 1975; Elias *et al.*, 1979; Lampe *et al.*, 1983a,b; Bowser *et al.*, 1985; Cox and Squier, 1986). Because the lamellar body also is enriched in proteases, glycosidases, and various types of lipases (Grayson *et al.*, 1985; Freinkel and Traczyk, 1985; Menon *et al.*, 1986c; Elias *et al.*, 1988), deposition and activation of these enzymes presumably account for the change both in composition and structure of the membrane bilayers (Elias *et al.*, 1988). An explanation for the structural changes, more consistent with the compositional changes and enzyme

Table IV

Possible Relationships of Biochemical Modulations and Observed Changes in Stratum Corneum Membrane Structure[a]

Step	Membrane event	Responsible enzyme(s)	Biochemical alteration
1	Unfurling of lamellar body arrays	None	None known
2	Fusion of lamellar body arrays	Phospholipases	Diacylphospholipids → lysolecithin
		Sphingomyelinase	Sphingomyelin → ceramides
		Glycosidases	Glycolipids → ceramides
3	Transformation of elongated disks to broad lamellae	Phospholipases + sphingomyelinase + glycosidases	Degradation of residual polar lipids (e.g., lysolecithin → FFA)[b]
4	Breakup of membrane bilayers → desquamation	Steroid sulfatase	Cholesterol sulfate → cholesterol
		Acid lipase	Triglycerides → FFA
		? Ceramidase	Ceramides → sphingosine base + FFA

[a]After Elias *et al.* (1988).
[b]FFA, free fatty acids.

localization data (Table IV), is that the "unfurled" lamellar body-derived sheets initially fuse end to end (Landmann, 1986; Elias *et al.*, 1988; Menon *et al.*, 1991b), perhaps through the degradation of phospholipids to free fatty acids under acidic conditions by phospholipase A, which is present in abundance in lamellar bodies and in the lower stratum corneum (Berger *et al.*, 1988; Elias *et al.*, 1988). The subsequent transformation of elongated disks into a broad, multilamellar membrane system (see below) may be associated with the further, complete hydrolysis of residual phospholipids and glycosphingolipids, leaving only free fatty acids and ceramides (Lampe *et al.*, 1983b).

Though these membrane bilayers are not seen in routine electron micrographs of the epidermis, elongated membrane bilayers are readily observed in the stratum corneum intercellular spaces in mucosal epithelia (Squier, 1973; Hayward and Hackermann, 1973; Lavker, 1976), in the epidermis of marine mammals (Menon *et al.*, 1986a), and in murine stratum corneum stained with ruthenium tetroxide (Menon *et al.*, 1991b). It is likely that the incomplete hydrolysis of lamellar body-derived lipids, i.e., persistence of relatively polar species such as glycosphingolipids (see above), accounts for the routine visualization of these structures in "moist" epithelia.

Recently, much more detailed information about intercellular membrane structures has resulted from the application of ruthenium tetroxide to the study of stratum corneum membrane structures (Madison *et al.*, 1987). Despite its extreme toxicity to structural proteins, which appear etched away, with improvements in standard fixation procedures this highly reactive and electron-dense substance has revealed finer details of the structural heterogeneity in both electron-dense and electron-lucent lamellae (Hou *et al.*, 1991). The electron-lucent lamellae consist of pairs of continuous bands, alternating with a single fenestrated lamella (Fig. 8A). Each electron-dense lamella is separated by an electron-dense structure of comparable width.

The membrane complex has been variously termed the Landmann (Swartzendruber *et al.*, 1989) or basic (Hou *et al.*, 1991) unit. The lamellar spacing or

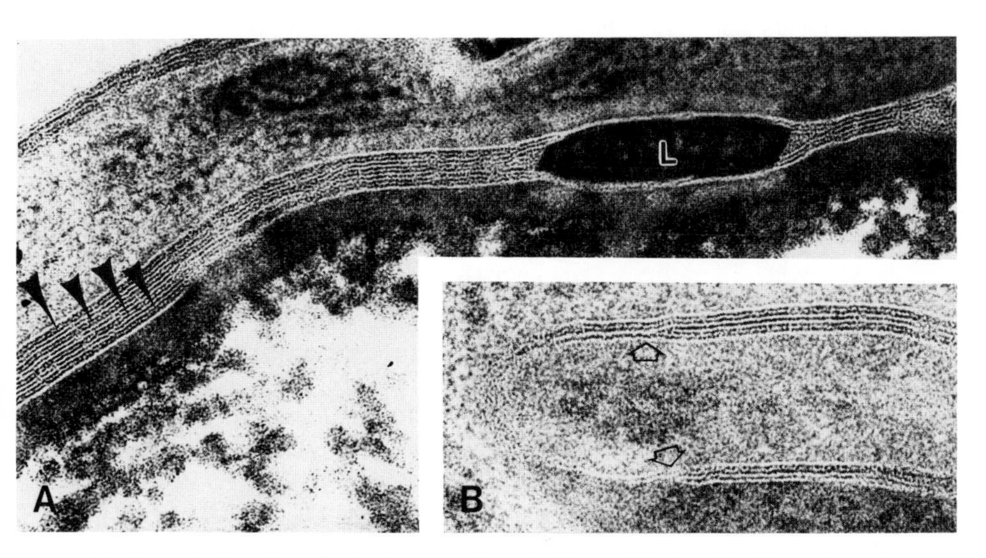

FIG. 8. (A) Overview of ruthenium tetroxide-stained intercellular lamellar bilayers in murine stratum corneum. Note alterations in every third electron-lucent lamella, all of which appear to be fenestrated (arrows), and regular interruptions of the intercellular domains by electron-dense, lenticular dilatations (L); x36,800. (B) Higher magnification of the periphery of ruthenium tetroxide-stained, murine stratum corneum. Note fenestrated, electron-lucent lamellae and attenuation of the numbers of lamellae at lateral margins of cells (arrows). [Reprinted from Hou *et al.* (1991). *J. Invest. Dermatol.* **96,** 215–223, with permission.]

repeat distance of ruthenium tetroxide-fixed lamellae has been analyzed by optical diffraction with computerized reconstruction (Hou *et al.*, 1991). The resultant center-to-center spacing of 12.9 ± 0.2 nm correlates extremely well with independent measurements of unfixed samples by X-ray diffraction (13.1 ± 0.2 nm) (White *et al.*, 1988; Hou *et al.*, 1991), indicating that the ruthenium staining method provides realistic images of intercellular membrane structures (see Hou *et al.*, this volume). Because the repeat distance is more than twice the thickness of typical lipid bilayers, White *et al.* (1988) proposed that each lamellar repeating unit consists of two opposing bilayers (see Hou *et al.*, this volume). Multiples of these units (up to three) occur frequently in murine (and less commonly in human) stratum corneum (Hou *et al.*, 1991). Simplifications of the basic unit, with deletion of one or more lamellae, occur at the lateral surfaces of corneocytes, i.e., at three cell junctures (Fig. 8B). Dilatations of the electron-dense lamellae, corresponding to sites of desmosomal hydrolysis, are visualized with ruthenium staining at all levels of the stratum corneum (Hou *et al.*, 1991). These data, coupled with the known biochemical diversity of these domains, reveal the intercellular domains to be quite heterogeneous (Table V). In fact, the "bricks and mortar" model no longer does justice to this complex region.

The ruthenium staining technique has also provided further information about the membrane leaflet immediately exterior to the cornified envelope. This trilaminar structure survives exhaustive solvent extraction (Fig. 9) (Elias *et al.*, 1977b; Swartzendruber *et al.*, 1987), but is destroyed by saponification (Swartzendruber *et al.*, 1987), which yields a family of very long-chain, ω-hydroxyacid-containing ceramides that are believed to be covalently attached to the cornified envelope (Swartzendruber *et al.*, 1987). Although this leaflet is enriched in ω-hydroxyacid-containing ceramides, it also contains small amounts of free fatty acids and free

Table V
HETEROGENEITY OF STRATUM CORNEUM INTERCELLULAR DOMAINS

Membrane bilayers
Nonpolar domains
Polar domains
Covalently bound envelope
Protein-enriched material
Desmosomal breakdown products
Extracellular glycoprotein(s)
Catabolic enzymes
Others
Sebaceous lipids
Eccrine gland salts
Xenobiotes
Water

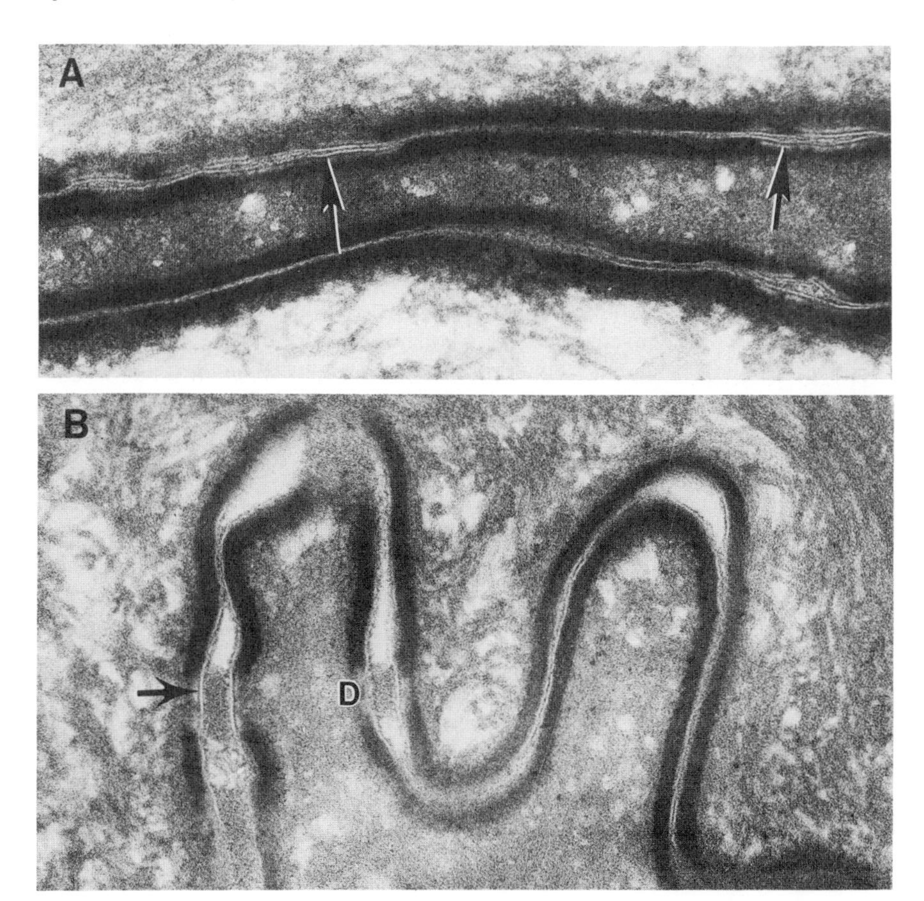

FIG. 9. Electron micrographs of murine stratum corneum after extraction with the organic solvent pyridine. (A and B) Note the loss of intercellular membrane bilayers but persistence of trilaminar structures adjacent to cornified envelope (arrows). This structure presumably correlates with the ceramide-enriched structure, covalently bound to the cornified envelope (Swartzendruber *et al.*, 1987); D, desmosomes. [Reprinted from Elias *et al.* (1977b). *J. Invest. Dermatol.* **69,** 535–546, with permission.]

hydroxyacids (Wertz and Downing, 1987). This leaflet also differs in composition from the intercellular bilayers, because it apparently lacks cholesterol (Kitajima *et al.*, 1985). The persistence of this envelope, after prior solvent extraction has rendered the stratum corneum porous, suggests that it may mediate functions other than permeability; e.g., it has been postulated to function as a scaffold for the deposition and organization of lamellar body-derived, intercellular bilayers (Wertz *et al.*, 1989). Based upon selected ultrastructural and biochemical data, three-dimensional models of these membranes have been proposed, which imply that ceramides are the principal constituents of the intercellular bilayers (Swartzendruber *et al.*, 1987, 1989; Wertz *et al.*, 1989). In light of (1) the demonstrated importance of cholesterol and free fatty acids for barrier homeostasis (see Feingold, this volume) and (2) the presence of approximately equimolar quantities of ceramides, cholesterol, and free fatty acids in these domains, future models will need to be modified to include cholesterol and free fatty acids. Correlation of images obtained with ruthenium tetroxide, biochemical methods, X-ray diffraction methods, and other physical–chemical methods (e.g., ESR and NMR) (see Hou *et al.*, this volume) ultimately should provide an integrated model of the architecture of the stratum corneum intercellular membrane system. Likewise, this correlative approach should yield important new insights about the alterations in membrane structure responsible for altered permeability states and pathological desquamation (see Williams and Potts *et al.*, this volume).

VI. Structural Alterations in Pathological Stratum Corneum

If the intercellular membrane bilayers regulate epidermal barrier function, then perturbations in barrier function should display altered membrane structures. Indeed, both solvent and detergent treatment of stratum corneum lead to depletion of stainable neutral lipids (Menon *et al.*, 1985a; Grubauer *et al.*, 1989a) and loss of intercellular membrane bilayers (Feingold *et al.*, 1990; Menon *et al.*, 1991b). Such acute perturbations of the barrier are followed by an immediate secretion of lamellar bodies (Feingold *et al.*, 1990; Menon *et al.*, 1991b), which leads to restoration of intercellular lipids in 6 to 48 hours (Grubauer *et al.*, 1989a,b). Although the chronology of events that follows barrier perturbation is currently under investigation, available information already indicates that the intercellular domains can be depleted and repleted rapidly in response to acute perturbations of the barrier (Menon *et al.*, 1991b; see also Feingold, this volume), and that perturbations of the normal epidermal calcium gradient are involved (Menon *et al.*, 1985b; Menon and Elias, 1991).

Chronic models of barrier dysfunction, such as essential fatty acid deficiency (EFAD), also are characterized both by depletion of intercellular lipid (Elias and Brown, 1978) and defective intercellular membrane bilayers (Elias and Brown, 1978; Hou *et al.*, 1991). Moreover, the membrane structures in EFAD stratum corneum display a number of alterations in substructure, as observed with ruthenium tetroxide staining (Hou *et al.*, 1991). Finally, the lamellar body secretory system appears to be defective in EFAD epidermis; these organelles display alterations in internal structure and defective secretion into the intercellular spaces (Menon *et al.*, 1989b).

Similar results are observed with repeated applications of lovastatin, a competitive inhibitor of hydroxymethylglutaryl CoA (HMG-CoA) reductase, to intact murine skin. After several daily applications, a defect in barrier function occurs that is accompanied by defective lamellar body contents and secretion (Feingold *et al.*, 1991). Finally, in certain disorders of cornification, such as congenital ichthyosiform erythroderma and psoriasis, which are accompanied by elevated water loss rates, abnormal intercellular membrane structures are observed with ruthenium tetroxide staining (Menon *et al.*, 1989b; Ghadially *et al.*, 1990). These results also point to the requirement of normal intercellular membrane bilayers for barrier function.

VII. Structural–Lipid Biochemical Correlates

The subject of epidermal lipids is discussed in detail in other articles in this volume (see Schurer and Elias), thus the information that follows represents only a brief review of the issues relevant to the aspects of membrane structure discussed in this article. During mammalian epidermal differentiation, characteristic changes in composition occur, consistent with the requirement for cutaneous waterproofing (for reviews see Yardley and Summerly, 1981; Elias, 1983; Williams and Elias, 1987). In porcine, bovine, murine, and human epidermis, these changes include a progressive depletion of phospholipids and glycosphingolipids (Lampe *et al.*, 1983b; Bowser and White, 1985; Long, 1970), with enrichment in ceramides, cholesterol, free fatty acids, and small amounts of other polar (e.g., cholesterol sulfate) and nonpolar species (e.g., hydrocarbons, cholesterol esters, and triglycerides) (Gray and Yardley, 1975; Lampe *et al.*, 1983b) (Table VI).

Table VI

VARIATIONS IN LIPID COMPOSITION DURING EPIDERMAL DIFFERENTIATION AND CORNIFICATION[a]

Composition	Basal/Spinous	Granular	Cornified
Phospholipids	44.5 ± 3.4	25.3 ± 2.6	6.6 ± 2.2
Cholesterol sulfate	2.6 ± 3.4	5.5 ± 1.3	2.0 ± 0.3
Neutral lipids	51.0 ± 4.5	56.5 ± 2.8	66.9 ± 4.8
Free sterols	11.2 ± 1.7	11.5 ± 1.1	18.9 ± 1.5
Free fatty acids	7.0 ± 2.1	9.2 ± 1.5	26.0 ± 5.0
Triglycerides	12.4 ± 2.9	24.7 ± 4.0	Variable
Sterol/wax esters[b]	5.3 ± 1.3	4.7 ± 0.7	7.3 ± 1.2
Squalene	4.9 ± 1.1	4.6 ± 1.0	6.5 ± 2.7
n-Alkanes	3.9 ± 0.3	3.8 ± 0.8	8.2 ± 3.5
Sphingolipids	7.3 ± 1.0	11.7 ± 2.7	24.4 ± 3.8
Glucosylceramides	3.5 ± 0.3	5.8 ± 0.2	Trace
Ceramides	3.8 ± 0.2	8.8 ± 0.2	24.4 ± 3.8
Total	99.1	101.1	99.9

[a]Modified from Lampe *et al.* (1983b). Values given as wt%.

[b]Sterol/wax esters present in approximately equal quantities as determined by acid hydrolysis.

Brody, I. (1966). *Nature (London)* **209,** 472–476.
Brysk, M. M., Rajaraman, S., Penn, P., and Barlow, E. (1988). *Cell Tissue Res.* **253,** 657–663.
Chander, A., and Fisher, A. B. (1990). *Am. J. Physiol.* **258,** L241–253.
Chapman, S. J., and Walsh, A. (1989). *J. Invest. Dermatol.* **93,** 466–470.
Christophers, E. (1971). *Z. Zellforsch. Mikrosk. Anat.* **114,** 441–445.
Cox, P., and Squier, C. A. (1986). *J. Invest. Dermatol.* **87,** 741–744.
Cullander, C., Menon, G. K., Guy, R. H., and Elias, P. M. (1990). *J. Invest. Dermatol.* **94,** 217A.
Egelrud, T., and Lundström, A. (1989). *Acta Derm.-Venereol.* **69,** 470–476.
Egelrud, T., and Lundström, A. (1990). *J. Invest. Dermatol.* **95,** 456–459.
Elias, P. M. (1983). *J. Invest. Dermatol.* **80,** 44–49.
Elias, P. M. (1987). *In* "Skin Pharmacokinetics" (B. Shrout and H. Schaeffer, eds.), pp. 1–9. Karger, Basel.
Elias, P. M., and Brown, B. E. (1978). *Lab. Invest.* **39,** 574–583.
Elias, P. M., and Friend, D. S. (1975). *J. Cell Biol.* **65,** 180–191.
Elias, P. M., and Leventhal, M. E. (1979). *Clin. Res.* **27,** 525A.
Elias, P. M., McNutt, N. S., and Friend, D. (1977a). *Anat. Rec.* **189,** 577–593.
Elias, P. M., Goerke, J., and Friend, D. (1977b). *J. Invest. Dermatol.* **69,** 535–546.
Elias, P. M., Brown, B. E., Fritsch, P. O., Goerke, R. J., Gray, G. M., and White, R. J. (1979). *J. Invest. Dermatol.* **73,** 339–348.
Elias, P. M., Cooper, E. R., Korc, A., and Brown, B. E. (1981a). *J. Invest. Dermatol.* **76,** 297–301.
Elias, P. M., Fritsch, P. O., Lampe, M. A., Williams, M. L., Brown, B. E., Nemanic, M. K., and Grayson, S. (1981b). *Lab. Invest.* **44,** 531–540.
Elias, P. M., Bonar, L., Grayson, S., and Baden, H. P. (1983). *J. Invest. Dermatol.* **80,** 213–214.
Elias, P. M., Williams, M. L., Maloney, M. E., Bonifas, J. A., Brown, B. E., Grayson, S., and Epstein, E. H., Jr. (1984). *J. Clin. Invest.* **74,** 1414–1421.
Elias, P. M., Menon, G. K., Grayson, S., Brown, B. E., and Rehfeld, S. J. (1987). *Am. J. Anat.* **180,** 161–177.
Elias, P. M., Menon, G. K., Grayson, S., and Brown, B. E. (1988). *J. Invest. Dermatol.* **91,** 3–10.
Feingold, K. R., Mao-Qiang, M., Menon, G. K., Cho, S. S., Brown, B. E., and Elias, P. M. (1990). *J. Clin. Invest.* **86,** 1738–1745.
Feingold, K. R., Mao-Quiang, M., Proksch, E., Menon, G. K., Brown, B. E., and Elias, P. M. (1991). (in press).
Fitzpatrick, T. B., Eisen, A. Z., Wolff, K., Freedberg, I. M., and Austen, F. K. (1987). "Dermatology in General Medicine," 3rd Ed. McGraw-Hill, New York.
Freinkel, R. K., and Traczyk, T. N. (1985). *J. Invest. Dermatol.* **85,** 295–298.
Friberg, S., Kayali, I., Beckerman, W., Rhein, L. D., and Simion, A. (1990). *J. Invest. Dermatol.* **94,** 377–380.
Gaskin, D. E. (1982). "The Ecology of Whales and Dolphins," pp. 441–459. Heinemann, London.
Geraci, J. R., St. Aubin, D. J., and Hicks, B. D. (1986). *In* "Research on Dolphins" (M. M. Bryden and R. Harrison, eds.), pp. 1–21. Oxford Univ. Press, London.
Ghadially, R., Menon, G. K., Taylor, N., Williams, M. L., and Elias, P. M. (1990). *Clin Res.* **38,** 223A.
Gray, G. M., and White, R. J. (1978). *J. Invest. Dermatol.* **70,** 336.
Gray, G. M., and Yardley, H. J. (1975). *J. Lipid Res.* **16,** 441–447.
Grayson, S., and Elias, P. M. (1982). *J. Invest. Dermatol.* **78,** 128–135.
Grayson, S., Johnson-Winegar, A. D., and Elias, P. M. (1983). *Science* **221,** 962–964.
Grayson, S., Johnson-Winegar, A. G., Wintroub, B. U., Epstein, E. H., Jr., and Elias, P. M. (1985). *J. Invest. Dermatol.* **85,** 289–295.
Grubauer, G., Feingold, K. R., and Elias, P. M. (1987). *J. Lipid Res.* **28,** 746–752.
Grubauer, G., Elias, P. M., and Feingold, K. R. (1989a). *J. Lipid Res.* **30,** 323–333.
Grubauer, G., Feingold, K. R., and Elias, P. M. (1989b). *J. Lipid Res.* **30,** 89–96.

Haftek, M., Viae, J., Schmitt, D., Gaucherand, M., and Thivolet, J. (1986). *Arch. Dermatol. Res.* **278,** 283–292.

Hamanaka, S., Asagami, C., Suzuki, M., Inagaki, F., and Suzuki, A. (1989). *J. Biochem. (Tokyo)* **105,** 684–690.

Hayward, A. F., and Hackermann, M. (1973). *J. Ultrastruct. Res.* **43,** 205–219.

Holleran, W. M., Feingold, K. R., Man, M.-Q., Gao, W. N., Lee, J. M., and Elias, P. M. (1991a). *J. Lipid Res.* (in press).

Holleran, W. M., Feingold, K. R., Mao-Quiang, M., Menon, G. K., and Elias, P. M. (1991b). *J. Clin. Invest.* (in press).

Hou, S. Y. E., Mitra, A. K., White, S. H., Menon, G. K., Ghadially, R., and Elias, P. M. (1991). *J. Invest. Dermatol.* **96,** 215–223.

Imokawa, G., Akasaki, S., Hattori, M., and Yoshizuka, N. (1986). *J. Invest. Dermatol.* **87,** 758–761.

King, L. S., Barton, S. P., Nicolls, S., and Marks, R. (1979). *Br. J. Dermatol.* **100,** 165–172.

Kitajima, Y., Sekiya, T., Mori, S., Nozawa, Y., and Yaoita, H. (1985). *J. Invest. Dermatol.* **84,** 149–153.

Lampe, M. A., Burlingame, A. L., Whitney, J., Williams, M. L., Brown, B. E., Roitman, E., and Elias, P. M. (1983a). *J. Lipid Res.* **24,** 120–130.

Lampe, M. A., Williams, M. L., and Elias, P. M. (1983b). *J. Lipid Res.* **24,** 131–140.

Landmann, L. (1980). *Eur. J. Cell Biol.* **33,** 258–264.

Landmann, L. (1984). *J. Ultrastruct. Res.* **72,** 245–263.

Landmann, L. (1986). *J. Invest. Dermatol.* **87,** 202–209.

Landmann, L. (1988). *Anat. Embryol.* **178,** 1–13.

Lavker, R. M. (1976). *J. Ultrastruct. Res.* **55,** 79–86.

Lazarus, G. S., Hatcher, V. B., and Levine, N. (1975). *J. Invest. Dermatol.* **65,** 259–271.

Long, V. J. W. (1970). *J. Invest. Dermatol.* **55,** 269–273.

Lucas, A. M. (1980). *In*: "The Skin of Vertebrates" (R. C. Spearman and P. A. Riley, eds.), pp. 33–45. Academic Press, New York.

Lundström, A., and Egelrud, T. (1988). *J. Invest. Dermatol.* **91,** 340–343.

Lundström, A., and Egelrud, T. (1990a). *J. Invest. Dermatol.* **94,** 216–220.

Lundström, A., and Egelrud, T. (1990b). *Arch. Dermatol.* **282,** 234–237.

MacKenzie, I. C. (1969). *Nature (London)* **222,** 881–882.

Madison, K. C., Swartzendruber, D. C., Wertz, P. W., and Downing, D. T. (1987). *J. Invest. Dermatol.* **88,** 714–718.

Menon, G. K., and Elias, P. M. (1991). *Arch. Dermatol.* **127,** 57–63.

Menon, G. K., Aggarwal, S. K., and Lucas, A. M. (1981). *J. Morphol.* **167,** 185–199.

Menon, G. K., Feingold, K. R., Moser, A. H., Brown, B. E., and Elias, P. M. (1985a). *J. Lipid Res.* **26,** 418–427.

Menon, G. K., Grayson, S., and Elias, P. M. (1985b). *J. Invest. Dermatol.* **84,** 508–512.

Menon, G. K., Grayson, S., Brown, B. E., and Elias, P. M. (1986a). *Cell Tissue Res.* **244,** 385–394.

Menon, G. K., Brown, B. E., and Elias, P. M. (1986b). *Tissue Cell* **18,** 71–82.

Menon, G. K., Grayson, S., and Elias, P. M. (1986c). *J. Invest. Dermatol.* **86,** 591–597.

Menon, G. K., Baptista, L. F., and Elias, P. M. (1988). *Ibis* **130,** 503–511.

Menon, G. K., Baptista, L. F., Brown, B. E., and Elias, P. M. (1989a). *Tissue Cell* **21,** 83–92.

Menon, G. K., Hou, S. Y. E., Grayson, S., and Elias, P. M. (1989b). *Clin. Res.* **37,** 233A.

Menon, G. K., Hou, S. Y. U., and Elias, P. M. (1991a). *Tissue Cell* (in press).

Menon, G. K., Feingold, K. R., and Elias, P. M. (1991b). *J. Invest. Dermatol.* (in press).

Menton, D., and Eisen, A. Z. (1971). *J. Ultrastruct. Res.* **35,** 247–264.

Michaels, A. S., Chandrasekaran, S. K., and Shaw, J. E. (1975). *AIChEJ.* **21,** 985–996.

Middleton, J. D. (1968). *Br. J. Dermatol.* **80,** 437–480.

Miller, S. J., Aly, R., Shinefeld, H. R., and Elias, P. M. (1988). *Arch. Dermatol.* **124,** 209–215.

Nemanic, M. K., Whitehead, J. G., and Elias, P. M. (1980). *J. Histochem. Cytochem.* **28,** 573–578.

Nemanic, M. K., Whitehead, J. S., and Elias, P. M. (1983). *J. Histochem. Cytochem.* **31,** 887–897.
Nicolaides, N. (1974). *Science* **186,** 19–26.
Odland, G. P., and Holbrook, K. (1987). *Curr. Probl. Dermatol.* **9,** 29–49.
Olah, I., and Röhlich, P. (1966). *Z. Zellforsch. Mikrosk. Anat.* **73,** 205–219.
Ponec, M., and Williams, M. L. (1986). *Arch. Dermatol. Res.* **279,** 32–36.
Potts, R. O., and Francoeur, M. L. (1990). *Proc. Natl. Acad. Sci. U.S.A.* **87,** 3871–3873.
Purton, M. D. (1988). *J. Anat.* **157,** 43–56.
Rehfeld, S. J., Plachy, W. Z., Hou, S. Y., and Elias, P. M. (1990). *J. Invest. Dermatol.* **95,** 217–223.
Scheuplein, R. J. and Blank, I. H. (1971). *Physiol. Rev.* **51,** 702–747.
Schreiner, E., and Wolff, K. (1969). *Arch. Klin. Exp. Dermatol.* **235,** 78–88.
Smith, W. P., Christiansen, M. S., Nacht, S., and Gans, E. H. (1982). *J. Invest. Dermatol.* **78,** 7–10.
Sokolov, V., Kalashnikova, M., and Rodinov, V. A. (1982). *In* "Morphology and Ecology of Marine Mammals" (K. K. Chepsku and V. A. Sokolov, eds.), pp. 82–101. Wiley, New York.
Squier, C. A. (1973). *J. Ultrastruct. Res.* **43,** 160–177.
Squier, C. A. (1975). *Br. Med. Bull.* **31,** 169–175.
Squier, C. A. (1982). *Arch. Oral Biol.* **27,** 377–382.
Squier, C. A., and Waterhouse, J. P. (1970). *Arch. Oral Biol.* **15,** 153–168.
Squier, C. A., Cox, P. S., Wertz, P. W., and Downing, D. T. (1986). *Arch Oral Biol.* **31,** 741–747.
Swartzendruber, D. C., Wertz, P. W., Madison, K. C., and Downing, D. T. (1987). *J. Invest. Dermatol.* **88,** 709–713.
Swartzendruber, D. C., Wertz, P. W., Kitko, D. J., Madison, K. C., and Downing, D. T. (1989). *J. Invest. Dermatol.* **92,** 251–257.
Webster, M. D., Campbell, G. S., and King, J. R. (1985). *Physiol. Zool.* **58,** 58–70.
Weinstock, M., and Wilgram, G. F. (1970). *J. Ultrastruct. Res.* **30,** 262.
Welty, J. C., and Baptista, L. (1988). "The Life of Birds," 4th Ed. Saunders Coll. Publ., Philadelphia, Pennsylvania.
Wertz, P. W., and Downing, D. T. (1983). *Science* **217,** 1261–1262.
Wertz, P. W., and Downing, D. T. (1982). *J. Lipid Res.* **24,** 759–665.
Wertz, P. W., and Downing, D. T. (1987). *Biochim. Biophys. Acta* **917,** 108–111.
Wertz, P. W., and Downing, D. T. (1989). *Biochim. Biophys. Acta* **1001,** 115–119.
Wertz, P. W., Cho, E. S., and Downing, D. T. (1983). *Biochim. Biophys. Acta* **753,** 350–355.
Wertz, P. W., Downing, D. T., Freinkel, R. K., and Traczyk, T. N. (1985). *J. Invest. Dermatol.* **83,** 193–195.
Wertz, P. W., Abraham, W., Landmann, W., and Downing, D. T. (1986). *J. Invest. Dermatol.* **87,** 582–584.
Wertz, P. W., Swartzendruber, D. C., Kitko, D. J., Madison, K. C., and Downing, D. T. (1989). *J. Invest. Dermatol.* **93,** 169–172.
White, S. H., Mirejovsky, D., and King, G. I. (1988). *Biochemistry* **27,** 3725–3732.
Williams, M. L., and Elias, P. M. (1981). *J. Clin. Invest.* **68,** 1404–1410.
Williams, M. L., and Elias, P. M. (1987). *CRC Crit. Rev. Ther. Drug Carrier Syst.* **3,** 95–122.
Wolff, K., and Holubar, K. (1967). *Arch. Klin. Exp. Dermatol.* **231,** 1.
Wolff-Schreiner, E. (1977). *Int. J. Dermatol.* **16,** 77–102.
Yardley, H. J., and Summerly, R. (1981). *Pharmacol. Ther.* **13,** 357–383.

The Biochemistry and Function of Stratum Corneum Lipids

NANNA Y. SCHURER* AND PETER M. ELIAS†

*Department of Dermatology
Heinrich-Heine Universitat Dusseldorf
D-4000 Dusseldorf, Germany
†Dermatology Service
Veterans Administration Medical Center
San Francisco, California 94121

I. Introduction

In the keratinizing epithelia of terrestrial mammals, an anucleate cornified layer resides above the nucleated layers of the epidermis. Until the 1950s the epidermal barrier to transcutaneous water loss was thought to be located beneath the stratum corneum, within the outermost nucleated layer of the epidermis. This conclusion was based upon the erroneous notion, conveyed in conventional histologic sections, that the stratum corneum is composed of loosely adherent cells, which should readily allow the free diffusion of water (Kligman, 1964). In 1953, Blank performed tape-stripping experiments that proved that the barrier to transcutaneous water loss is located within the stratum corneum. Because the diffusion of water rose sharply after the stratum corneum was removed, Blank concluded that the barrier was located within the innermost layer(s) of stratum corneum (Blank, 1953; Monash and Blank, 1958). Indeed, there is now direct evidence that the stratum compactum layer, the lowest region of the stratum corneum, possesses formidable barrier properties (Bowser and White, 1985). However, as pointed out by Scheuplein and Blank (1971) and Schaefer *et al.* (1982), such tape-stripping experiments do exclude the possibility that the bulk of the stratum corneum is also an effective barrier. In fact, more recent studies

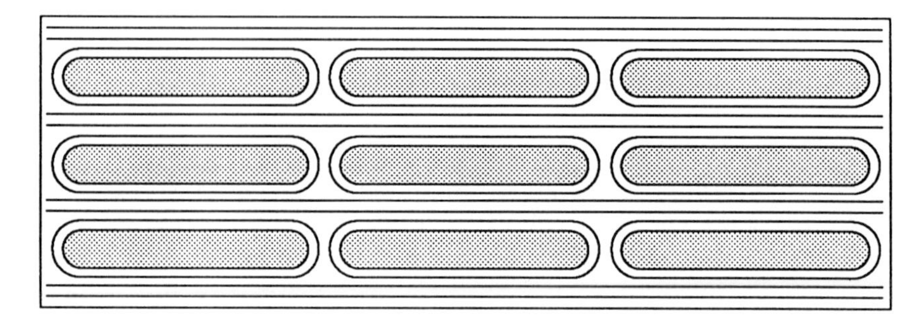

FIG. 1. Schematic diagram of stratum corneum "bricks and mortar" model.

demonstrate a gradual change in transepidermal water loss with increasing numbers of strippings, suggesting that most, if not all, of the cell layers of the stratum corneum may participate in the water barrier (Schaefer *et al.*, 1982).

Biophysical (Michaels *et al.*, 1975; Elias *et al.*, 1983a), morphological (Elias and Friend, 1975), and biochemical (Grayson, and Elias, 1982) data indicate that the stratum corneum forms a continuous sheath of alternating squamae (protein-enriched corneocytes) embedded in an intercellular matrix, enriched in nonpolar lipids deployed as lamellar sheets. This two-compartment model has been analogized to a brick wall, resulting in the "bricks and mortar model" (Fig. 1) (Elias, 1983; Elias *et al.*, 1983b). These intercellular lamellae are thought to mediate transcutaneous water loss, stratum corneum water retention (Imokawa *et al.*, 1986), and possibly desquamation (Elias, 1981, 1983; Williams and Elias, 1987) (see Williams, and Elias and Menon, this volume).

The study of lipids as a class of chemical constituents of the stratum corneum offers a unique opportunity to investigate the functional specialization of this tissue. The daily rate of epidermal lipid synthesis in man is equal to the lipid content times the daily loss of stratum corneum (8% x 0.5–1 g/day = 40–80 mg lipid/day) (Kligman, 1964). Total epidermal lipid constitutes approximately 10–14% of the dry weight of mammalian epidermis (Gray and Yardley, 1975a). However, by themselves, isolated intercellular lipids possess no water-holding capacity (Imokawa *et al.*, 1986). The ability of the intercellular lipids to form lamellar bilayers, in the absence of phospholipids, is dependent upon the amphipathic properties of ceramides, free fatty acids, cholesterol, and perhaps lesser constituents such as cholesterol sulfate and proteolipids. The lamellar bilayers are stabilized in an aqueous environment by van der Waals interactions and hydrogen bonds (Rehfeld *et al.*, 1988, 1990).

Recently, it has been suggested that a major component of the stratum corneum is a ceramide (Abraham and Downing, 1990), consisting of 30 to 34-carbon chain length, *N*-acyl, ω-hydroxyacids covalently bound to the cornified envelope

(Wertz *et al.*, 1987, 1989b). This leaflet may serve as a scaffold for the intercellular bilayers, thereby contributing to both the barrier and the cohesive properties of the stratum corneum. In addition, the stability of the stratum corneum lipid mixture may be enhanced by the presence of large quantities of cholesterol (Cullis and Hope, 1980).

Intercellular hydrolytic enzymes may participate in the regulation of cohesion and desquamation. For example, the ratio of cholesterol sulfate to cholesterol is maintained by enzymatic hydrolysis, which in turn my influence the stability of the intercellular lamellar bilayers (Elias *et al.*, 1984) (see Section III,D,4). Possible mechanisms for the mediation of stratum corneum desquamation include (Epstein, 1985) (1) liquid-crystalline shifts in the intercorneocyte lipids (Rehfeld and Elias, 1982), (2) variations in hydrogen bonding induced by an altered cholesterol sulfate:cholesterol ratio (Rehfeld *et al.*, 1986, 1988), (3) calcium-induced aggregation of stratum corneum intercellular lipids by cholesterol sulfate (Epstein *et al.*, 1981), and/or (4) enzymatic hydrolysis of lipid and nonlipid intercellular constituents, such as desmosomes. Yet, despite these studies, the biochemical and/or physical–chemical mechanisms that regulate the continual, invisible desquamation of the uppermost cornified cells are still unknown (Anton-Lamprecht, 1983; Williams, this volume).

II. Historical Overview

Scientific studies on skin lipids began in the nineteenth century. As early as 1853, researchers were aware that the epidermis was less permeable than the dermis (Homalle, 1853). Liebreich postulated in 1890 that "the orderly formation of mammalian stratum corneum requires the orderly arrangement of lipids" (see also Unna, 1913). Dusham (1918) published that the chemistry of the waxy material on the surface of the insect cuticle "probably serves as a protection against moisture loss." Later, cuticular lipids were shown definitively to be responsible for waterproofing in insects (Ramsay, 1935; Locke, 1965). A frequently cited paper of Kooyman (1932) deserves recognition as the first careful examination of epidermal lipids. By analyzing palmar and plantar epidermis, where sebaceous gland-derived species contribute minimally, he noted a striking decrease in the amount of phospholipid and a corresponding increase in free sterols during cornification (Kooyman, 1932). At that time, however, these shifts in lipid composition were thought to reflect changes in the metabolic activity of the tissue during terminal differentiation, and the link between epidermal lipids and barrier function was not made. Winsor and Burch (1944) first observed that solvent-induced damage to the stratum corneum provoked an increase in water permeability, whereas, in contrast, intraepidermal blisters did not lead to increased transepidermal water loss; i.e., that an intact stratum corneum was required for barrier integrity. This pioneering work has since been confirmed repeatedly (Wheatley and Flesch, 1967; Sweeney and Downing, 1970; Scheuplein and Blank, 1971).

The first detailed studies of human cornified layer lipids from the midabdominal and midscapular regions (Reinertson and Wheatley, 1959; Wheatley *et al.*, 1964) found appreciable amounts of lipid (2.7–9.1%) in both normal and pathological stratum corneum. Based upon their solubility characteristics, two types of lipids were described, one that could be removed with ethanol and another that required more drastic extraction procedures. Wheras the former contained phospholipids, free fatty acids, free sterols, phosphatides, hydrocarbons, sterol esters, and triglycerides, the latter, which comprised approximately 40% of the total lipid, was described as a series of at least six different types of "proteolipids." These authors postulated a physiological role for these proteolipids as cementing and barrier substances in the cornified layer.

Evidence for the presence of lipid substances in close association with keratinized structures was first deduced from X-ray diffraction studies, which suggested that stratum corneum lipids are arranged in a cylindrical array around intercellular keratin filaments (Swanbeck, 1959) (see Hou *et al.*, this volume). However, later freeze–fracture and histochemical studies showed that lipids are restricted to intercellular domains in the stratum corneum (Elias and Friend, 1975; Elias *et al.*, 1977), where they are able to form membrane bilayers despite the absence of phospholipids. This segregation of lipids to intercellular domains was later proved by biochemical (Grayson and Elias, 1982) and physical–chemical (Elias *et al.*, 1983a; Smith *et al.*, 1982; White *et al.*, 1988; Rehfeld *et al.*, 1990) studies. Moreover, topically applied lipid-soluble tracers traverse through the intercellular domains (Nemanic and Elias, 1980). Finally, in essential fatty acid deficiency, a defective barrier is associated with loss of intercellular lipid, and water-soluble tracers gain access to normally inaccessible intercellular domains (Elias and Brown, 1978).

Following the description of so-called proteolipids by Reinerson and Wheatley (1959), the next allusion to a polar lipid fraction was by Nicolaides (1965). However, the definitive assignment of these compounds as sphingolipids first was made by Gray and Yardley (1975a). Gray and White (1978) later separated the ceramides from pig and human epidermis into four fractions, and suggested that despite the paucity of phospholipids in the stratum corneum, the retained sterols, sphingolipids, and free fatty acids could provide sufficient polar groups to account for bilayer formation. Later, they also provided the first definitive analysis of epidermal phospholipids (Gray and Yardley, 1975b; Gray, 1976). Finally, they isolated a species of acylsphingolipids unique to epidermis (Gray and White, 1978), which they hypothesized might be involved in the epidermal water barrier. They suggested that these acylglucosylceramides substitute for phospholipids though their membrane-forming abilities, and supported this hypothesis by preparing stable liposomes from a mixture of stratum corneum lipids enriched in this fraction (Gray *et al.*, 1978).

Of the many excellent reviews that have been written on the subject of skin lipids, only a few can be mentioned here. Several earlier reviews were devoted to studies of either whole skin (Rishmer *et al.*, 1966) or skin surface lipids (Nicolaides, 1974; Yardley, 1983; Downing *et al.*, 1983, 1987). Rothman (1964) reviewed the esterification of sterols in the epidermis, and Yardley (1969) speculated that esterified sterols might be important for barrier function (Yardley, 1969). A still timely review by Yardley and Summerly (1981) describes the lipid composition and metabolism of normal and diseased epidermis. Elias, Williams, and their co-workers have reviewed the implications of stratum corneum lipid segregation to the intercorneocyte spaces (Elias, 1981; Elias *et al.*, 1983b; Williams, 1983; Elias and Williams, 1985) (also see Elias and Menon, this volume). The effects of essential fatty acid deficiency on the skin have been summarized by Budowski (1981) and Sherertz (1986). More recently, Ziboh and colleagues (Ziboh and Chapkin, 1988; Ziboh and Miller, 1990) discussed epidermal linoleic acid and arachidonic acid metabolism, and Melnik (1989) recently reviewed lipids in normal versus diseased human epidermis. Rather than review previously covered material, it is our goal to discuss current views about epidermal lipid biochemistry and, when possible, to relate these concepts to function.

III. Biochemistry and Function

A. The Origin of Stratum Corneum Lipids

The intercellular bilayers, located within the stratum corneum, originate from small ovoid organelles synthesized in the spinous and granular cells (for review see Odland and Holbrook, 1987; Landmann, 1988). These organelles, variously termed lamellar bodies, membrane-coating granules, or Odland bodies, discharge their membranous disks into the intercellular space. Lamellar bodies have been isolated and their contents partially characterized; they contain a selected spectrum of lipids and hydrolytic enzymes (Freinkel and Traczyk, 1985; Wertz *et al.*, 1984; Grayson *et al.*, 1985). Following secretion, the contents of these organelles are reorganized into a system of broad lamellar bilayers, filling the intercellular spaces of the stratum corneum. The lamellar body-derived hydrolases may mediate changes in lipid composition that facilitate membrane fusion and elongation (Elias *et al.*, 1988). It has been hypothesized that these flattened disks reassemble to form the intercellular lamellar sheets of the stratum corneum by an edge-to-edge fusion process that may require calcium (Abraham *et al.*, 1988b; Swartzendruber *et al.*, 1989). Support for this concept has been provided by the demonstration of *in vitro* assembly of lamellar sheets from liposomes composed of lipids similar in composition to those in the stratum corneum (Abraham *et al.*, 1988a). In addition to lipid transformations, lamellar body-derived hydrolases may participate in desquamation by attacking certain intercellular, nonlipid constitutents, e.g., desmosomes, but lipases and glycosidases may also mediate various phases of the desquamation process (Menon *et al.*, 1986a; Elias *et al.*, 1988).

Lamellar bodies are enriched in phospholipids, free sterols, and glycosphingolipids (Grayson *et al.*, 1985), including certain distinctive acylglycosphingolipid species (Wertz *et al.*, 1984). It has been suggested that the latter may be responsible for the formation of the lamellar disks that appear in those organelles (Wertz and Downing, 1982). Immediately following secretion, phospholipids are catabolized to free fatty acids, whereas glycosphingolipids are converted to ceramides (Nemanic and Elias, 1980; Downing *et al.*, 1987; Elias *et al.*, 1988). The hydrolytic enzymes present in the lamellar body seem well situated to mediate the transformation of the relatively polar lipid contents of lamellar bodies to the nonpolar species representative of stratum corneum lipids. However, details of the timing, regulation, and localization of these enzymatic processes, which are crucial to the formation of the lamellar bilayer system that regulates transcutaneous water loss, remain unknown.

B. Human Skin Surface Lipids: Pilosebaceous Versus Epidermal Origin

Human skin surface lipids consist of triglycerides, free fatty acids, wax esters, squalene, cholesterol esters, and cholesterol. The epidermis contributes only a proportion of the total surface lipid (Nicolaides, 1974; Gray and Yardley, 1975a), depending upon the number of sebaceous glands present at the particular site examined. In contrast to the scalp, which has approximately 900 pilosebaceous structures per square centimeter, none exist on the palms and soles (Montagna, 1963). Thus surface lipids collected from the scalp, where lipid production is at least 100 $\mu g/cm^2$ (Ellis and Hendrickson, 1963), represent a reasonable approximation of the lipid composition of the pilosebaceous glands, i.e., 43% triglycerides, 16% free fatty acids, 25% wax esters, 12% squalene, and 2.5% cholesterol esters. In contrast, when surface lipid samples are obtained from the palms or soles, where lipid production is less than $10 \mu g/cm^2$, sebum is virtually absent and surface lipids represent primarily lipids of epidermal origin: variable amounts of triglycerides (see below), 20–25% ceramides, 20% free fatty acids, 15% cholesterol esters, and 20% cholesterol (Table I) (Ebling and Rook, 1979). Pure sebum has no free fatty acids; they derive instead from the lipolytic activity of various species of bacteria on sebum-derived triglycerides and cholesterol esters in the sebaceous gland duct, as well as the action of these organisms on esterified lipids deposited on the skin surface (Nicolaides, 1974).

Table I
APPROXIMATE DISTRIBUTION OF HUMAN SKIN SURFACE LIPIDS DERIVED FROM PILOSEBACEOUS VERSUS EPIDERMAL ORIGIN[a]

Lipid	Origin	
fraction (%)	Pilosebaceous	Epidermal
Ceramides	—	25.0
Cholesterol	—	20.0
Free fatty acids	16.0	20.0
Triglycerides	43.0	10.0 (variable)
Squalene	12.0	—
Wax esters	25.0	—
Cholesterol esters	2.5	15.0
Total	98.5	100.0

[a]Modified from Ebling (1972).

The free fatty acid composition of human sebum changes with sebaceous gland activity (Yamamoto *et al.*, 1990), and endogenous $C_{16:1}$ straight-chain fatty acids correlate with serum testosterone levels. It has been suggested that $C_{16:1}$ may play an important role in comedogenesis (Melnik and Plewig, 1988), as well as indicating the activity of the sebaceous glands in both sexes. Wertz *et al.* (1985) demonstrated that comedonal sphingolipids contain diminished proportions of $C_{18:2}$, which led to the hypothesis that localized essential fatty acid deficiency (EFAD), due to the dilutional effects of increased sebum production, might be an etiologic factor in acne. The characteristic features of hyperkeratosis and decreased barrier function in EFAD, if they existed locally in the follicle, could lead to follicular plugging and, eventually, rupture, i.e., the inflammatory lesion (Downing *et al.*, 1987) (see Stewart and Downing, this volume). As should be apparent from this brief review, descriptions of skin surface lipid composition are relevant for stratum corneum function only when the contribution of sebaceous glands is considered carefully.

C. CHANGES IN EPIDERMAL LIPID COMPOSITION WITH DIFFERENTIATION

The composition of lipids changes markedly during apical migration through successive epidermal layers (Fig. 2). Earlier studies noted a striking shift from polar to neutral lipids during cornification (Kooyman, 1932); Reinertson and Wheatley, 1959). In 1970, Long sliced cow snout epidermis into six horizontal slices and analyzed for phospholipid, triacylglycerol, cholesterol, free fatty acid, and glucose content, noting that neutral lipids (free fatty acids and triglycerides) accumulate in the outermost layers. More recently, using a similar technique, lipid concentrations were compared in 12 consecutive epidermal layers, essentially confirming these observations (Cox and Squier, 1986).

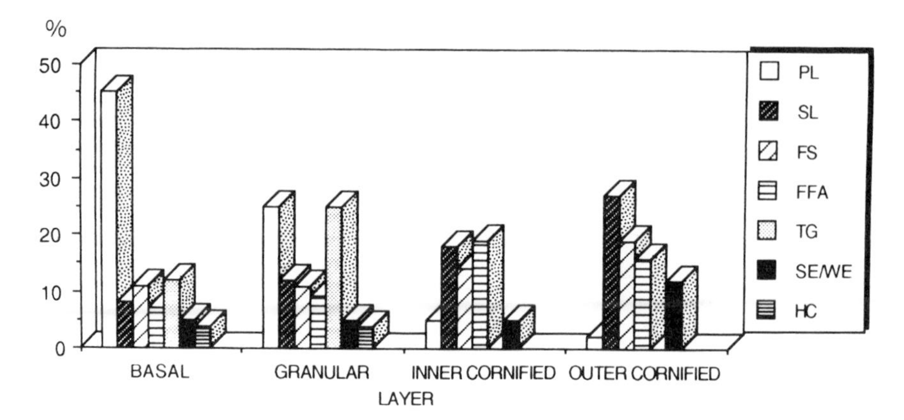

FIG. 2. Changes in lipid distribution in layers of human epidermis during differentiation: PL, phospholipids; SL, sphingolipids; FS, free sterols; FFA, free fatty acids; TG, triglycerides; SE/WE, sterol esters/wax esters; HC, hydrocarbons.

Simultaneously, short-chain fatty acids are replaced by long-chain, highly saturated species during cornification (Carruthers and Heining, 1964; Ansari *et al.*, 1970). Although it has been proposed that the long-chain, saturated species remain after preferential utilization of shorter chain species for energy-requiring processes (Nicolaides, 1974), it also is possible that the long-chain species are generated specifically in response to emerging barrier requirements.

In the basal cell layer of the epidermis, phospholipids account for approximately 50% by weight of the total lipid, with sphingolipids, free sterols, free fatty acids, triglycerides, and sterol esters accounting for the remainder (Gray and Yardley, 1975b; Elias *et al.*, 1979; Lampe *et al.*, 1983b) (Fig. 2). In contrast, about one-third of the lipid remaining in the cornified layer consists of sphingolipids, with the remainder accounted for by free sterols, free fatty acids, and lesser quantities of cholesterol sulfate and nonpolar species, such as triglycerides, sterol esters, and hydrocarbons. The distribution of epidermal lipids in successive epidermal cell layers, shown in Fig. 2, includes a correction for triglyceride content in the stratum corneum, which may vary with pilosebaceous contribution or contamination from subcutaneous fat during tissue preparation.

Cholesterol sulfate is present in all viable epidermal layers, with the highest levels in the stratum granulosum (Lampe *et al.*, 1983b). The activity of the microsomally located enzyme, steroid sulfatase, is also more enriched in the stratum granulosum than in the lower epidermal layers (Elias *et al.*, 1984). In the stratum corneum, enzyme activity and the substrate cholesterol sulfate reside in membrane domains (Williams, 1983; Elias *et al.*, 1984). A further gradient of cholesterol sulfate content appears to exist across the stratum corneum, as one proceeds from the inner to the outer layers (Ranasinghe *et al.*, 1986; Elias *et al.*, 1988). Thus, the progressive desulfation of cholesterol sulfate to cholesterol may be one factor that regulates cell cohesiveness and normal stratum corneum desquamation (Epstein *et al.*, 1981). Alternatively, variations in cholesterol content alone may modulate desquamation (Williams and Elias, 1987). Finally, the ceramide fraction increases with terminal differentiation, as a result of glucosylceramide metabolism by intercellular glycosidases (Nemanic *et al.*, 1983; Downing *et al.*, 1987).

D. Composition of Mammalian Epidermal Lipids

1. *Phospholipids*

As noted above, phospholipids, which are essential for the maintenance of the membrane bilayers in all known cellular organelles, account for less than 5% of the lipids in mammalian stratum corneum. However, as in other tissues, they comprise about 45% of the total lipid in the basal and spinous layer and 25% of the total lipid in the stratum granulosum (Gray and Yardley, 1975b; Elias *et al.*, 1979; Lampe *et al.*, 1983b). The major phospholipid classes in guinea pig epidermal cell membranes are phosphatidylcholine and phosphatidylethanolamine (Miller *et al.*, 1989). The principal phospholipid species found in the nucleated layers of the human epidermis include phosphatidylcholine, -ethanolamine, -serine, and -inositol and sphingomyelin and lysolecithin (Table II). A novel phospholipid species, phosphatidyl-N-acylethanolamine, was identified in porcine granular cells by Gray (1976). A high proportion of the amide-linked fatty acid in this unusual phospholipid is palmitic acid. Whereas ethanolamine lipids account for only 5% of the phospholipids of the basal layer, they persist as other phospholipids disappear, increasing to 25% of all phospholipids in the stratum granulosum, but largely disappearing from the stratum corneum (Gray, 1976; Lampe *et al.*, 1983a). Ethanolamine lipids may represent a storage pool of a compound known to have antiinflammatory properties (Ganley *et al.*, 1958). Phosphatidylserine also remains in small amounts in the lower stratum corneum as the cells terminally differentiate, but it disappears along with all other phospholipids from the outer layers of the stratum corneum, i.e., stratum disjunctum (Lampe *et al.*, 1983b; Bowser and White, 1985; Elias *et al.*, 1988; Yardley, 1990).

These alterations in phospholipid profile may reflect the functional demands that are imposed by terminal differentiation. Phospholipids (particularly phosphatidylinositol) also are stores of arachidonic acid and a subsequent cascade of regulatory eicosanoids (see Holleran, this volume; see also Ziboh and Miller, 1990). Moreover, one phospholipid, sphingomyelin, may serve as high-turnover precursors of sphingolipids (Spector *et al.*, 1980). Phospholipid-derived epidermal fatty acids may serve as substrates for acyltransferases that selectively as-

Table II
PHOSPHOLIPID COMPOSITION OF DIFFERENT CELL POPULATIONS IN PIG EPIDERMIS[a]

	Layer (wt%)		
Phospholipid fraction (%)	Basal/spinous	Granular layer	Cornified
Phosphatidylcholine	47.9	35.3	1.2
Phosphatidylethanolamine	24.2	20.3	1.3
Phosphatidyl-*N*-acylethanolamine	1.6	24.8	?
Lysolecithin	4.9	?	?
Phosphatidylserine	13.2	9.8	Trace
Phosphatidylinositol	11.3	2.5	—

[a]After Gray *et al.* (1980), Lampe *et al.* (1983a), and Elias *et al.* (1988).

sembly these fatty acids within the sphingolipids that accumulate during terminal differentiation (Hedberg *et al.*, 1988). In fact, the transfer of labeled palmitic acid residues to amide-linked fatty acid from phospholipid and/or triglycerides indicates that some, if not all, ceramide is formed by the action of one or more acyl-CoA acyltransferase(s) (Kondoh *et al.*, 1983). Finally, a large proportion of phospholipid-derived fatty acids may not be redistributed to other lipids, but instead may contribute to the true free fatty acid pool (Elias *et al.*, 1988), which comprises a major fraction of the stratum corneum lipid mixture.

2. *Sphingolipids*

a. Biochemistry. Sphingolipids are prominent components of cellular membranes, lipoproteins, and other lipid-rich structures. In porcine human epidermis, their common backbones are the long-chain bases, sphingosine (*trans*-4-sphingenine) and phytosphingosine (4-D-hydroxysphinganine), with lesser amounts of sphinganine (dihydrosphingosine) and other homologues of these compounds (Wertz and Downing, 1983a,b). In contrast, mouse keratinocytes contain only sphingosine as the long-chain base (Madison *et al.*, 1990). Epidermal sphingolipids represent 7.3% of total lipid in the basal layers, increasing to about 15% in the stratum granulosum, 30% in the lower stratum corneum, and reaching 40% in the other stratum corneum (Lampe *et al.*, 1983b). Thus, the transformation of the stratum granulosum into the stratum corneum is accompanied not only by a depletion of phospholipids, but also by an increase in total sphingolipids. Though glycosphingolipids are present in small quantities in the stratum basale and stratum granulosum, where they are localized to the lamellar body (Wertz *et al.*, 1984; Grayson *et al.*, 1985), they are virtually absent in the stratum corneum.

Both glycosphingolipids and ceramides are amphipathic molecules, able to fulfill the structural role of phospholipid in maintaining a stable membrane phase in the intercellular spaces of stratum corneum. Long-chain, saturated fatty acids (>C_{20}) esterified to glycosphingolipids and ceramides have high melting points

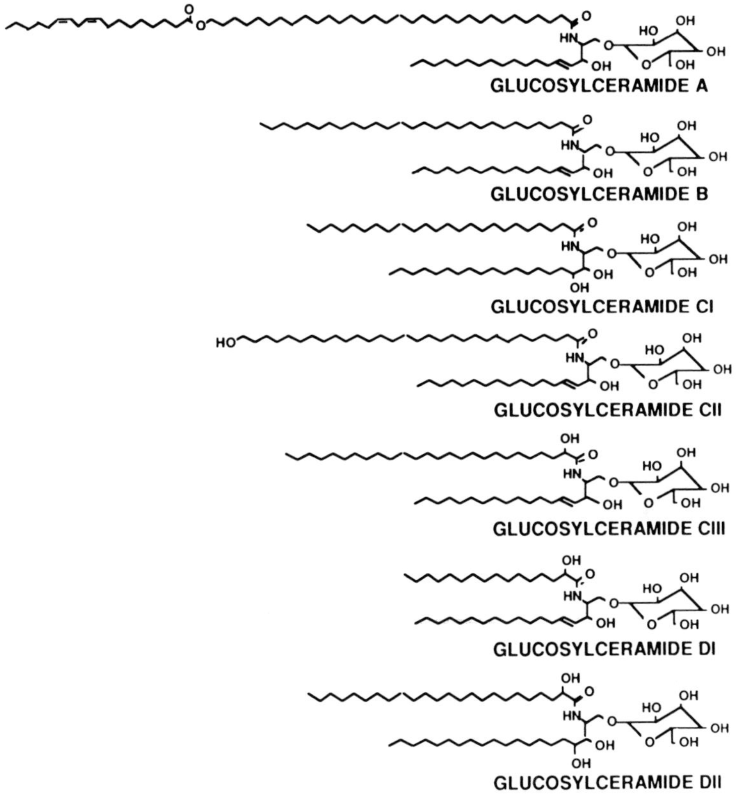

GLUCOSYLCERAMIDE A

GLUCOSYLCERAMIDE B

GLUCOSYLCERAMIDE CI

GLUCOSYLCERAMIDE CII

GLUCOSYLCERAMIDE CIII

GLUCOSYLCERAMIDE DI

GLUCOSYLCERAMIDE DII

FIG. 3. Series of glycosphingolipids isolated from porcine epidermis. (From Wertz and Downing, 1983b.)

and are stable to oxidation, and therefore withstand wide ranges of temperature, UV irradiation, and oxidation processes. The ceramides are of special interest, because this structurally heterogeneous group represents the major polar lipid from which the extracellular membrane structures of the stratum corneum presumably are constructed. Epidermal ceramides represent a unique, heterogeneous group of lipids (Gray and White, 1978; Gray *et al.*, 1978; Wertz *et al.*, 1987). Wertz and colleagues (Wertz and Downing, 1983a; Wertz *et al.*, 1985, 1987; Long *et al.*, 1985) isolated and characterized a series of seven ceramides and an equivalent series of glucoslylceramides from pig and human epidermis (Figs. 3 and 4). These are composed of the long-chain base, sphingosine, with lesser amounts of phytosphingosine, and an amide-linked nonhydroxy or α-hydroxy fatty acid (about 50% αhydroxylated). In the stratum granulosum and stratum

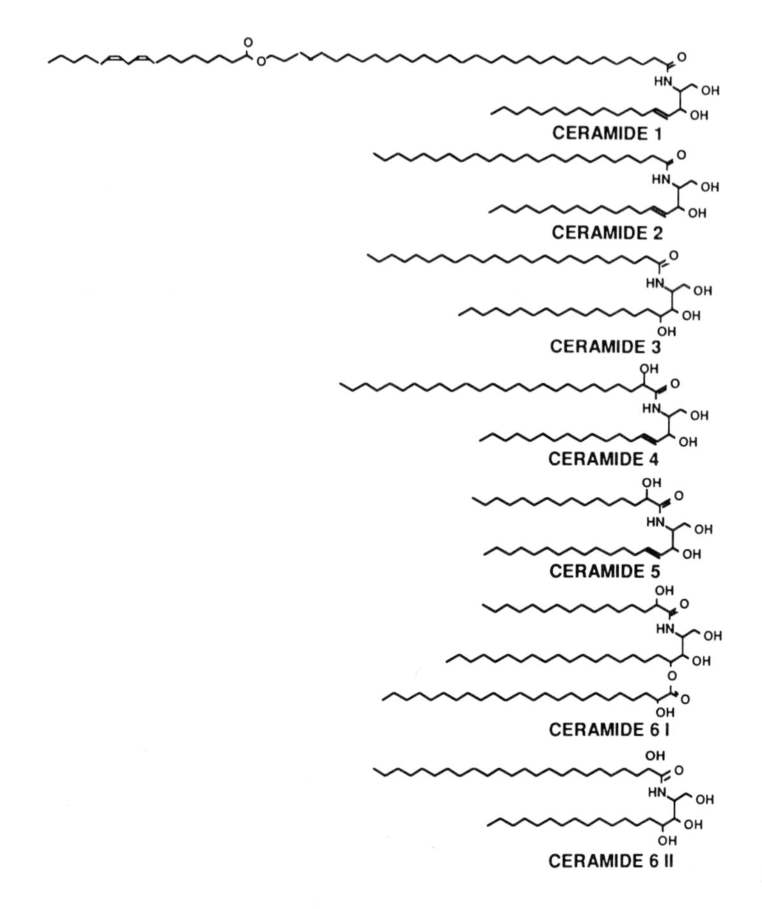

FIG. 4. Series of ceramides isolated from human epidermis. (From Melnik, 1989.)

corneum, long-chain fully saturated fatty acids (C_{22}–C_{26}) predominate in the sphingolipid fraction, whereas other esterified neutral lipid or free fatty acids display a more typical spectrum of fatty acids, with a predominance of $C_{18:1}$ and $C_{16:0}$ (Elias *et al.*, 1979; Lampe *et al.*, 1983a).

One of the most unusual and interesting sphingolipids described to date, and apparently unique to the epidermis, consists of a glycosylated and nonglycosylated sphingosine base with an amide-linked, long-chain, nonhydroxy and α-hydroxy acid, with an additional ω-esterified nonhydroxy acid (primarily $C_{18:2}$). The amide-linked fatty acid ranges from 29 (in humans) to 35 carbon atoms in porcine epidermis, with two hydroxyl groups and two as-yet unlocalized double bonds. These species of acylsphingolipids are present in the epidermis of all terrestrial

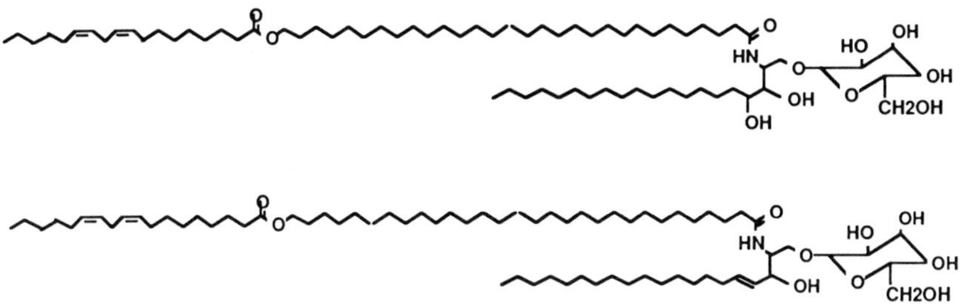

FIG. 5. Acylglucosylceramides from human epidermis. (Modified from Hamanaka *et al.*, 1989.)

mammalians examined to date, including porcine (Gray *et al.*, 1978; Wertz and Downing, 1983b, 1986), murine (Elias *et al.*, 1979; Madison *et al.*, 1990), and human (Wertz and Downing, 1990b) epidermis. Because the predominant ω-linked, esterified fatty acid is linoleic acid, this may indicate a role for these molecules in barrier function (Wertz *et al.*, 1983; Hansen, 1986). As noted above, the esterified acids of the acylsphingolipids contain a high proportion of linoleic acid, ranging from about 35% in mice (Elias *et al.*, 1979) to 50% in humans (Hamanaka *et al.*, 1989; Yamamoto *et al.*, 1990), to 77% in pig (Gray *et al.*, 1978; Bowser and White, 1985; Wertz *et al.*, 1986; Wertz, 1983b).

Hamanaka *et al.* (1989) purified two components from acylglycosylceramide A and glycosylceramide B, named glycolipid I_3, composed of glucosyl-β1-*N*-(ω-*O*-linoleoyl)-triacontanoyl- and -dotriacontanoyl-eicosasphingenine, and glycolipid-II_3, composed of glucosyl-β1-*N*-(ω-*O*-linoleoyl)-triacontanoyl-trihydroxy-eicosasphingenine. A major proportion of the sphingolipid bases of both glycolipids proved to be a novel structure, i.e., trihydroxyeicosasphingenine (Fig. 5).

Acylglucosylceramide is stated to account for up to 50% of the total glycosphingolipids in porcine epidermis (Abraham *et al.*, 1985, 1988b). These acylglucosylceramides are enriched in the lamellar body (Wertz *et al.*, 1984) and perhaps persist within the stratum compactum (Elias *et al.*, 1988), but not in the outer stratum corneum (Lampe *et al.*, 1983b; Elias *et al.*, 1988). However, recent studies suggest that the linoleate moiety in the acylsphingolipids does not seem to be essential for the assembly of lamellar granules (Abraham *et al.*, 1988a). The acylceramide and its structurally related glycosylated precursor acylglucosylceramide have been postulated to carry out membrane-organizing and -stacking functions in the stratum corneum and lamellar body, respectively (Wertz and Downing, 1982; Wertz *et al.*, 1984, 1989). The dimensions of such a molecule would allow the hydroxyacid chains to extend across the hydrocarbon palisades of two adjacent membrane bilayers, with the hydroxyl group anchored in one polar surface and the glucose portion in the other surface. Although these molecules have been hypothesized to span adjacent layers (Wertz and Downing, 1982), their presence is not a prerequisite for a multilamellar arrangement, because marine cetaceans lack acylsphingolipids despite displaying an extensive lamellar body-derived membrane system (Menon *et al.*, 1986a).

b. Metabolism. Long-chain base biosynthesis begins with the condensation of serine and palmitoyl-CoA to form 3-ketosphingenine, a reaction catalyzed by the enzyme, serine palmitoyl transferase (SPT). This enzyme, which is rate limiting for sphingolipid base synthesis in all tissues studied to date, is present in large quantities in both cultured human keratinocytes and epidermis (Holleran *et al.*, 1990, 1991). The subsequent steps in the formation of the ceramide backbone have been studied in other tissues, but not in the epidermis (for review see Merrill and Jones, 1990). A highly efficient NADPH-requiring reductase dehydrogenase, with a subsequent ceramide synthetase reaction, must be involved, because the amounts of sphingosine base present normally are very small (Stoffel *et al.*, 1968); in comparison, large amounts of ceramides accumulate in the outer layers of the epidermis (Lampe *et al.*, 1983b). Ceramides are subsequently converted into glycosphingolipids by specific glycosyltransferases (Kishimoto, 1983; Braun *et al.*, 1970). Glycosphingolipid synthesis has not been studied in the skin.

c. Sphingolipid Catabolism. Human epidermis contains two types of sphingomyelinase, of which the neutral sphingomyelinase appears to be on the outer side of the plasma membrane (Das *et al.*, 1984). It may hydrolyze endogenous sphingomyelin, with the resulting ceramide undergoing further hydrolysis to free sphingosine (Slife *et al.*, 1989). Human epidermis contains ceramidase (*N*-acylsphingosine deacylase), which would cleave the amide-linked fatty acid (Wertz and Downing, 1990b). In the differentiated layers of the epidermis, nearly 1% of the total sphingolipid is in the free form of the sphingoid base (Wertz and Downing, 1989, 1990a). However, little is known about the reutilization of ceramides and/or sphingosine in the epidermis. Recent studies suggest that free sphingosine may regulate cellular differentiation through inhibition of protein kinase C (Hannun *et al.*, 1986; Merrill and Stevens, 1989). Thus, sphingosine in the epidermis could serve as a regulator of epidermal differentiation (Wertz and Downing, 1990a), but the putative mechanism for the epidermis remains speculative.

3. *Neutral Lipids*

Total neutral lipids comprise about half of the total lipids within the whole stratum corneum, increasing to about 60% within the outer stratum corneum. During the process of differentiation, i.e., the shift from stratum granulosum to stratum corneum, the absolute amounts of free sterols, free fatty acids, and certain other nonpolar species increase (*vide supra*).

a. Free Fatty Acids. The increased free fatty acid content of stratum corneum can be ascribed largely to the hydrolysis of phospholipids (Elias *et al.*, 1988; Hedberg *et al.*, 1988) and, perhaps to a lesser extent, to triglycerides (Kondoh *et al.*, 1983). In all epidermal layers the free fatty acid fraction ranges from C_{12} to C_{24}, with C_{16} and C_{18} representing the major species (Ansari *et al.*, 1970; Lampe *et al.*, 1983a,b). The epidermis synthesizes abundant fatty acids, both under basal conditions (Monger *et al.*, 1988) and in response to altered barrier requirements (Grubauer *et al.*, 1987). Moreover, all epidermal layers, including the stratum granulosum, retaining the capacity to synthesize fatty acids (Monger *et al.*, 1988). Yet, circulating fatty acids accumulate in the epidermis under pathological circumstances (Dykes *et al.*, 1978), and both the essential fatty acid, linoleic acid, and arachidonic acid must be acquired from the circulation, the latter because the epidermis lacks the Δ^6-desaturase (Chapkin and Ziboh, 1984). It still remains to be established to what extent fatty acids are synthesized locally versus delivered from the circulation under basal versus perturbed conditions, and which epidermal layer(s) respond(s) to altered requirements for fatty acids.

b. Acylated Fatty Acids. In addition to differentiation-associated modulations in lipid distribution, distinctive changes occur in the fatty acid composition of epidermal phospholipids during differentiation, with longer chain species predominating in the outer layers (Gray and White, 1978; Madison *et al.*, 1986). Most of the esterified fatty acids in the phospholipids of the nucleated cell layers of the epidermis range from C_{12} to C_{24}; $C_{16:1}$ is notably absent, and presumably it is primarily a component of sebum wax esters (Yamamoto *et al.*, 1990). With differentiation, the relative amounts of $C_{16:1}$ and $C_{18:1}$ decrease, whereas the amount of $C_{18:2}$ in epidermal phospholipids proportionally increases. Dietary influences can profoundly alter the fatty acid profile of epidermal phospholipids (for review see Ziboh and Miller, 1990). For example, supplementation with polyunsaturated oils, such as fish or vegetable oils, results in distinctive changes in the unsaturation of epidermal membrane phospholipids (Ziboh *et al.*, 1986). In contrast to phospholipids, short-chain fatty acids ($C_{12:0}$ and $C_{14:0}$) predominate over long-chain species in neutral lipid fractions, including triglycerides and sterol esters (Lampe *et al.*, 1983a). Although there are no significant changes in fatty acid composition of phospholipids and netural lipids within the nucleated layers, the proportion of saturated acids with chain length above C_{20} increases in the stratum corneum neutral lipids (Elias, *et al.*, 1979; Lampe *et al.*, 1983b). Even the small amount of phospholipid (<5%) remaining in the stratum corneum demonstrates a large proportion of long-chain and very long-chain fatty acids (Table III).

The progressive increase in fatty acid chain length has been attributed to the selective utilization of shorter chain species for energy production (Nicolaides, 1983). Alternatively, the emergence of long-chain, saturated fatty acids in the stratum corneum may facilitate the hydrogen bonding of adjacent lipid species,

Table III

RELATIVE DISTRIBUTION OF FREE AND ESTERIFIED FATTY ACIDS IN HUMAN EPIDERMIS[a]

	Fatty acid composition	
Lipid classes	Nucleated layers	Stratum corneum
Phospholipids	C_{12}– C_{14} No $C_{16:1}$	$C_{18:2}$ ↑↑↑ $C_{24:0}$ ↑
Sphingolipids[b]	C_{22}– C_{24}	C_{22}– C_{30}
Free fatty acids	C_{12}– C_{24} C_{16}– C_{18} ↑	C_{12}– C_{24} C_{16}– C_{18} ↑↑
Triglycerides	C_{14}– C_{18} C_{20}– C_{28}	C_{14}– C_{18} —

[a]Modified from Lampe *et al.* (1983b) and Ansarie *et al.* (1970).
[b]Does not reflect ω-hydroxy-*N*-acyl fatty acids of acylsphingolipids, which are C_{33}– C_{37}.

leading to tighter packing of intercellular lipids and formation of a water-repellent barrier. In contrast, the dienoic, unsaturated fatty acid, linoleic acid, has been suggested to function as a molecular rivet between adjacent acylceramides and acylglucosylceramides (Abraham *et al,* 1988b), or it also may facilitate intermolecular bonding (Houtsmuller and van der Beck, 1981). The latter hypothesis is supported by the observation that any *N*-acyl fatty acid with cis-configurated double bonds at the n-6 and n-9 or n-5 and n-8 positions can substitute for linoleic acid (Houtsmuller and van der Beck, 1981). Thus, metabolism of $C_{16:0}$ and $C_{18:0}$ could supply the energy necessary for the later stages of differentiation; $C_{18:2}$ (and possibly $C_{18:1}$) could provide major structural components.

In cultured keratinocytes, the distribution of even-chain fatty acid homologues resembles that of keratinocytes *in vivo* (Wilkinson, 1971), but epidermal fatty acid composition can be altered by exclusion or administration of exogenous lipids (for review see Ziboh and Miller, 1990). Likewise, differences in the esterification of C_{20} polyunsaturated fatty acids and arachidonic acid can be induced in keratinocytes cultured in 10% serum containing arachidonic acid, dihomo-γ-linoleic acid, and eicosapentaenoic acid (4.5, 1.5 and 0.5%, respectively) (Punnonen *et al.*, 1987a). Moreover, cultured keratinocytes, even when grown in serum-containing media, appear to be deficient in linoleic acid (Isseroff *et al.*, 1985). Yet, neither the morphology nor the energy utilization of these cells appears to be altered when these cells are supplemented with linoleate (Boyer and Sato, 1980; Marcelo and Duell, 1989). Likewise, *in vitro* reconstructed epidermis, whether or not provided with linoleic acid supplements, does not generate a permeability barrier with the full competence of the *in vivo* equivalent (Ponec *et al.*, 1988).

c. Hydrocarbons. Significant quantities of hydrocarbons have been noted repeatedly in lipid extracts of mammalian stratum corneum (Gray and Yardley, 1975a; Elias *et al.*, 1979; Lampe *et al.*, 1983a). Because hydrocarbons are critical for waterproofing in both insects (Jackson and Baker, 1970) and plants (Kolattukudy, 1968), a possible role for them in the epidermis must be considered. However, a synthetic mechanism for long-chain *n*-alkane generation in higher vertebrates is not known to exist. Epidermal *n*-alkanes form a homologous series with a bell-shaped distribution from C_{19} to C_{35}, revealing a peak at C_{24}–C_{26} (Lampe *et al.*, 1983a; Williams and Elias, 1984). In addition, these *n*-alkanes comprise equal proportions of odd- and even-chained species. Mammalian tissues, including the epidermis, can catabolize hydrocarbons, the enzymatic activity residing in the microsomal fraction within the P-450 system (Bickers, 1983). Although *n*-alkanes are present in the human epidermis, whether they originate all or in part from sebaceous glands (Nicolaides *et al.*, 1968), epidermis, or environmental contamination (Gloor *et al.*, 1974) has been controversial. The fact that they are present in measurable quantities within the viable epidermis and in plantar stratum corneum, where sebaceous glands are absent, suggests that they are at least not of pilosebaceous origin (Lampe *et al.*, 1983a). However, in a series of time-course experiments, utilizing intradermal injection of C_{14} acetate, Hedberg and co-workers (1988) noted that the hydrocarbon fraction never acquired radioactivity. They concluded that alkanes are not synthesized in the epidermis and most probably derive from the petroleum waxes. Moreover, Bortz and co-workers (1989, 1990) also provided evidence that skin surface *n*-alkanes are of exogenous origin, based on C_{14} dating of hydrocarbons pooled from several normal subjects. On the other hand, *n*-alkanes have been found in significant quantities in the living layer of the epidermis in normal skin (Lampe *et al.*, 1983a) and are known to accumulate in large quantities in the nucleated and cornified epidermal layers of patients with congenital ichthyosiform erythroderma (Williams and Elias, 1984). Whether *n*-alkanes are solely environmental contaminants or are in part generated in the epidermis, as occurs in insects and plants, remains unresolved (Hadley, 1980, 1989; Elias and Williams, 1990).

d. Sterol Esters. The sterol ester fraction consists primarily of iso-branched and antiiso-branched chain fatty acids as well as odd-numbered straight chain and monounsaturated fatty acids (Carruthers and Heining, 1964). The mixture of fatty acids esterified to sterols in mouse epidermis is greatly different from those in the triglyceride fraction, where the esterified fatty acids consist primarily of myristic, palmitic, stearic, oleic, and linoleic acids (Elias *et al.*, 1979). Thus, based on fatty acid profiles and solvent extraction studies (Grubauer *et al.*, 1989a), the sterol esters in rodent stratum corneum appear to derive largely from sebaceous glands. In humans, it is generally accepted that sterol esters are synthesized in the epidermis

(Yardley, 1969), accounting for 6–7% of the lipids in the stratum corneum (Lampe *et al.*, 1983b). The cholesterol-esterifying ability of the guinea pig epidermis does not appear to be dependent upon ATP, coenzyme A, and Mg^{2+}, which suggests that sterol esters are formed by transesterification (Freinkel and Aso, 1968). Esterification has also been demonstrated in homogenates of human callus. However, the function of sterol esters in the epidermis is unknown. If sterol esters are incorporated into the ordered structure of the intercellular membranes, they could contribute to the waterproofing function of the stratum corneum.

e. Free Sterols. Cholesterol is a major constituent of stratum corneum lipids, accounting for 20–25% of total lipid weight (Gray and Yardley, 1975b; Elias *et al.*, 1977; Lampe *et al.*, 1983b). As discussed by (this volume), evidence is accumulating that cholesterol is critical for epidermal barrier function. Moreover, as described by (this volume), cholesterol appears to mediate stratum corneum desquamation, as well. As will be discussed below, most epidermal cholesterol is derived from local synthesis, even under basal conditions, and when the barrier is perturbed this synthetic apparatus is stimulated still further.

Most mammalian cells are equipped with cell surface receptors for the principal plasma cholesterol transport complex, i.e., the low-density lipoprotein (LDL) receptor. This receptor mediates the feedback control of cholesterol synthesis (Goldstein and Brown, 1973) by regulating both hydroxymethylglutaryl-CoA reductase activity as well as transcription of the gene for the LDL receptor (Mazzone *et al.*, 1990). Ponec *et al.* (1985), however, first recognized that confluent keratinocyte cultures lack LDL receptors, whereas in normal human epidermis only basal cells express LDL receptors (Mommaas-Kienhuis *et al.*, 1987). Moreover, circulating cholesterol levels do not affect cutaneous *de novo* cholesterol synthesis (Andersen and Dietschy, 1977), presumably due to the paucity of plasma membrane LDL receptors in more differentiated cell populations (Brannon *et al.*, 1975; Williams *et al.*, 1987b; Mommaas-Kienhuis *et al.*, 1987). Even though the stratum basale is equipped with the LDL receptors to allow the potential uptake of exogenous cholesterol, the basal cells are also an active site of cholesterol synthesis (Feingold *et al.*, 1983; Monger *et al.*, 1988; Proksch *et al.*, 1991b). In fact, sterol synthesis in the skin of mammals accounts for about 30% of the total body sterol synthesis (Feingold *et al.*, 1982), of which 70–80% is localized to the dermis and the remaining 20–30%, to the epidermis (Feingold *et al.*, 1983). Finally, the outer nucleated layers of the epidermis continue to synthesize abundant free sterols (Feingold *et al.*, 1983; Monger *et al.*, 1988; Proksch *et al.*, 1991b).

Cholesterol synthesis and activity of its rate-limiting enzyme, 3-hydroxy-3-methylglutaryl coenzyme A (HMG-CoA) reductase, may be modulated in specific epidermal layers by actue or chronic insults to the permeability barrier. There is a dynamic relationship between the host's water-barrier requirements, transepidermal water loss, and epidermal sterol synthesis (Menon *et al.*, 1985; Feingold *et al.*, 1986; Grubauer *et al.*, 1987, 1989b). Moreover, the time course of synthetic activity closely parallels barrier repair. After solvent-induced disruption of the permeability barrier, both barrier function and the increase in *de novo* cholesterol synthesis return to normal within 24 hours. Furthermore, barrier requirements also modulate both the content and the activation state of HMG-CoA reductase (Proksch *et al.*, 1990). Likewise, in essential fatty acid-deficient mice, increased transepidermal water loss also is accompanied by an increase in cholesterol synthesis and HMG-CoA reductase activity (Feingold *et al.*, 1986; Proksch *et al.*, 1990).

Finally, sterols also serve as important precursors for vitamin D synthesis (Holick *et al.*, 1980), with about 10% of the free sterols in the porcine epidermis comprising 7-dehydrocholesterol (Gray and Yardley, 1975b). However, alterations in systemic vitamin D status do not influence epidermal sterologenesis (Feingold *et al.*, 1987).

4. *Cholesterol Sulfate*

A minor portion of free cholesterol is sulfated by the enzyme, cholesterol sulfotransferase, which transfers a sulfate group from an active sulfate (3′-phosphoadenosine-5′-phosphosulfate) to cholesterol (Epstein *et al.*, 1984). It is one of the few remaining polar lipids of the stratum corneum, accounting for 3–5% of normal human stratum corneum lipids (Williams and Elias, 1981; Lampe *et al.*, 1983b). Due to its amphipathic character and its physicochemical properties, it has been hypothesized to be important for the maintenance of the intercorneocyte lipid bilayers (Williams and Elias, 1981, 1986; Williams, 1983). The enzyme, steroid sulfatase, which hydrolyzes the sulfate esters of 3β-hydroxysterols, including cholesterol sulfate, is localized to the microsomes of the stratum granulosum and accumulates in the peripheral membranes of the stratum corneum (Elias *et al.*, 1984). Because of the enzymatic balance between epidermal synthesis and degradation, cholesterol sulfate content is highest in the stratum granulosum (5%) and lowest in the outer stratum corneum (1.5%) (Lampe *et al.*, 1983b; Ranasinghe *et al.*, 1986; Elias *et al.*, 1988). The progressive decrease of cholesterol sulfate from cohesive to desquamated stratum corneum supports the hypothesis that controlled hydrolysis of cholesterol sulfate by steroid sulfatase may be one factor that destabilizes the intercorneocyte lipid lamellae, thus leading to normal, invisible desquamation (Williams, 1983).

The inherited disorder of cornification, recessive X-linked ichthyosis (RXLI), is associated with steroid sulfatase deficiency, resulting in elevated levels of cholesterol sulfate in the stratum corneum and retention hyperkeratosis (Williams and Elias, 1981). Thus, it has been hypothesized that changes in the levels of cholesterol sulfate mediate the stabilization–destabilization of the intercellular membrane bilayers, thereby regulating normal desquamation (Epstein *et al.*,

1984). This view is supported by work in liposome model systems, where cholesterol sulfate in the presence of calcium promotes the aggregation of stratum corneum lipids (Abraham *et al.*, 1987). On the other hand, physical–chemical studies have shown that cholesterol sulfate does not interact normally with other membrane lipid constitutents; e.g., it fails to form a eutectic mixture with saturated, straight-chain fatty acids (Rehfeld *et al.*, 1986). Moreover, electron-spin resonance studies have shown that the presence of excess cholesterol sulfate, as occurs in RXLI, results in abnormal hydrogen bonding between adjacent lipid species (Rehfeld *et al.*, 1988). Thus, rather than enhancing cohesion, excess cholesterol sulfate might actually have a destabilizing effect on membranes, perhaps leading to the premature cracking and fissuring that characterize RXLI.

Cholesterol sulfate readily enters both fibroblasts and keratinocytes in the absence of LDL receptors, where it inhibits HMG-CoA reductase and sterol synthesis but stimulates fatty acid synthesis (Williams *et al.*, 1985, 1987a,b; Ponec *et al.*, 1985). This finding now provides an alternate explanation for the pathogenesis of abnormal scaling in RXLI; i.e., the reduced cholesterol content of the membranes, due to down-regulation of HMG-CoA reductase by cholesterol sulfate, would alter the physical–chemical properties of the membrane lipids, leading to abnormal desquamation (Williams and Elias, 1987). This view is supported by the occurrence of acquired scaling disorders in association with the administration of hypocholesterolemic drugs.

E. Regional Variations in Human Stratum Corneum Lipid Composition

The lipid composition of human stratum corneum lipids displays striking regional variations that could reflect differences in stratum corneum thickness, turnover, desquamation, and/or permeability (Elias *et al.*, 1981a; Lampe *et al.*, 1983a). However, barrier function is not related to either the thickness of or the number of cell layers in the stratum corneum; instead, an inverse relationship exists between the lipid weight percentage and the permeation of a particular skin site to water (Elias *et al.*, 1981a). However, in addition to total lipid content, significant regional differences occur in the compositional profile of stratum corneum lipids over different skin sites (Lampe *et al.*, 1983a). For example, the proportion of sphingolipids and cholesterol is much higher in palmoplantar stratum corneum than in extensor surfaces of the extremities or abdominal or facial stratum corneum (Lampe *et al.*, 1983a). However, the significance of these differences is not known, particularly because the absolute quantitites of each of these fractions is dependent on the lipid weight percentage of each site. Thus, despite the high proportions of sphingolipids and cholesterol in palmar stratum corneum, when adjusted for the 2% lipid weight of the stratum corneum of this site, their absolute amounts in the intercellular spaces are still much lower than in other more lipid -enriched sites. Moreover, functional interpretations require consideration of not only lipid distribution and weight percentages, but also information about site-related variations in the fatty acid profiles of esterified species, and at present only limited information is available (Lampe *et al.*, 1983a).

The distribution of lipids in the nonkeratinized oral regions, which generally have a higher water permeability than keratinized regions (Squier and Hall, 1985), is different. Nonkeratinized regions such as the buccal epithelia contain no acylglucosylceramides and acylceramides and only very small amounts of ceramides. Yet, glucosylceramides occur in large quantities in nonkeratinizing epithelia. Therefore, a positive correlation appears to exist between the water permeability of nonkeratinizing oral epithelia and the absence of acylglucosylceramides and acylceramides (Wertz *et al.*, 1986). Other nonpolar lipids, such as cholesterol sulfate, cholesterol ester, free sterols, free fatty acids, and triglycerides, also are present, but quantitative comparisons are not available. Keratinized oral epithelium and porcine epidermis reveal a similar lipid pattern (Squier *et al.*, 1986). However, both keratinizing and nonkeratinizing porcine oral epithelia contained more phospholipids than occur in epidermis (Wertz *et al.*, 1986).

F. Variations of Stratum Corneum Lipids in Different Taxa

Mammalian skin surface lipids differ markedly from those of other vertebrates (Nicolaides *et al.*, 1968b). Moreover, there are also distinctive differences among mammals (Gray and Yardley, 1975a). For example only human surface lipids contain appreciable amounts of squalene. Yet, free cholesterol is found in the epidermis of all vertebrates studied to date. Moreover, the stratum corneum of all homeotherms, i.e., mammals and avians, examined to date contains ganglioside sulfates, ceramides, cholesterol, cholesterol sulfate, and free fatty acids, but few phospholipids (Yardley and Summerly, 1981). In contrast, reptilian "barrier layer" lipids lack sphingolipids, but contain abundant phospholipids (Roberts and Lillywhite, 1980). The distinctive nonpolar acylsphingolipids (Gray *et al.*, 1978; Bowser and White, 1985; Wertz and Downing, 1983b, 1986) appear to be unique to mammals but are lacking in marine mammals, i.e., cetaceans (Elias *et al.*, 1987). Moreover, in marine mammals, stratum corneum lipids do not undergo the loss of polarity that characterizes terrestrial adaptation (Menon *et al.*, 1986b; Elias *et al.*, 1987), which may reflect their lesser barrier requirements. The fatty acids of marine mammalian lipids are shorter and more unsaturated than those on equivalent terrestrial species. These comparative studies suggest that it is both the intercellular location and the compositional profile of stratum corneum lipids that subserve barrier function in all terrestrial homeotherms, and that these features may be regulated by similar factors.

IV. Essential Fatty Acid Metabolism

In 1929 Burr and Burr described a characteristic deficiency syndrome, which they attributed to an inability of warm-blooded animals to synthesize certain fatty acids (Burr and Burr, 1929). These so-called essential fatty acids differ from each other according to the position of their first double bond. Linoleic acid ($C_{18:2n\text{-}6}$) and α-linolenic acid ($C_{18:3n\text{-}3}$) are desaturated and cis configurated at the n-6 and n-3 positions (referred to as n6 and n3 positions, respectively). Mammals are incapable of introducing double bonds at these two positions, although fish and plants retain this capacity (Lands, 1981). Whereas fish oils are enriched in the n3 series, but poor in the n6 series of essential fatty acids, plants represent potential sources of both the n3 and n6 series. Holman *et al.* (1982) estimated that the human daily requirement of essential fatty acids is about 0.5% of the total caloric intake. The dietary intake of polyunsaturated fatty acids, including linoleic acid, ordinarily exceeds the minimal requirements for essential fatty acids (Melnik and Plewig, 1989).

Essential fatty acids are required in the epidermis for two purposes: first, maintenance of normal membrane structure (deficiency of linoleic acid results in an increase in transepidermal water loss). Linoleic acid is required to maintain the barrier to transcutaneous water loss, where it is believed to play a direct role in epidermal barrier function (Prottey, 1977; Elias *et al.*, 1980; Houtsmuller and van der Beck, 1981). When linoleic acid is absent in the diet, lamellar bodies are empty or only partially filled (Elias, 1981; Elias and Brown, 1978). As a result, large water-soluble molecules are able to access the stratum corneum interstices from which they are normally excluded.

The second requirement for linoleic acid is as a precursor of arachidonic acid and a variety of its eicosanoid products, including the prostaglandins (via the cyclooxygenase pathway) and the hydroxyacids (via the lipoxygenase pathway) (Duell *et al.*, 1988). Both linoleic acid and arachidonic acid are largely esterified within cell membrane phospholipids. In the epidermis, as in other tissues, arachdonate is liberated by phospholipase A_2, before entering the eicosanoid cascade leading to the biosynthesis of prostaglandins, thromboxanes, and leukotrienes, several of which are important regulators of epidermal inflammation as well as cell division.

Although it has been appreciated for some time that animals deprived of linoleic acid develop a scaling dermatitis and increased transcutaneous water loss (for review see Prottey, 1977), the mechanisms for this effect of essential fatty acid deficiency have been debated. In most tissues, linoleic acid is preferentially converted to arachidonic acid (Rosenthal, 1987). Because arachidonic acid is required for eicosanoid generation (see below), and keratinocyte-derived eicosanoids regulate epidermal cell turnover (Voorhees, 1983), it has been suggested that the dermatitis is due to epidermal hyperproliferation (Ziboh and Chapkin, 1987; Lowe and DeQuoy, 1978). Moreover, arachidonic acid and prostaglandin E_2 both decrease the scaling dermatitis and the hyperproliferation in essential fatty acid deficiency (Ziboh and Hsia, 1972; Lowe and DeQuoy, 1978).

Alternatively, linoleic acid may be important per se for the barrier. When applied topically, prostaglandin E_2 does not correct the abnormality in barrier function (Houtsmuller and van der Beck, 1981; Proksch *et al.*, 1991a). Biochemical changes in essential fatty acid deficiency include a marked increase in oleic acid as well as a marked decrease of linoleic acid in epidermal phospholipids (Ziboh and Chapkin, 1987). Moreover, whereas most of the fatty acid ω-esterified to the *N*-acyl fatty acids of acylglucosylceramides and acylceramides is linoleic acid, in essential fatty acid-deficient epidermis, oleic acid is substituted in the ω-esterified position (Wertz *et al.*, 1983). Both topical and systemic administration of linoleic acid improves epidermal barrier function, even when subsequent conversion to eicosanoids is blocked by coadministration of cyclooxygenase and/or lipoxygenase inhibitors (Prottey, 1976; Elias *et al.*, 1980). Moreover, topical applications of synthetic, nonmetabolizable fatty acids containing n5,8 or n6,9 cis-configurated unsaturated groups correct the barrier abnormality (Houtsmuller and van der Beck, 1981; Elliott *et al.*, 1985). These studies do not, however, exclude the possibility that these analogs may themselves be metabolized to regulatory eicosanoids. Nevertheless, the bulk of evidence suggests that linoleic acid plays a key role in the maintenance of the epidermal permeability barrier.

The Δ^6-desaturase catabolizes the first committed step in the metabolism of linoleic acid. This activity is concentrated in microsomal membranes and requires ATP, Mg^{2+}, O_2, and NADPH (Brenner, 1981). The subsequent step, i.e., desaturation to γ-linolenic acid (6,9,12-octadecatrienoic acid, $C_{18:3n-6}$), is considered the rate-limiting reaction. Further reduction yields a product two carbons longer than $C_{18:3n-6}$), namely dihomo-γ-linolenic acid ($C_{20:3n-6}$). The final step in the conversion of linoleic acid to arachidonic acid ($C_{20:4n-6}$) requires the Δ^5-desaturase. Finally, differentiated keratinocytes contain higher levels of cyclooxygenase and lipoxygenase than do less well-differentiated keratinocytes, i.e., basal cells (Cameron *et al.*, 1990). Although preparations of rat and guinea pig epidermis lack the capacity to transform linoleic acid into arachidonic acid (Chapkin and Ziboh, 1984), cultured human keratinocytes may (Isseroff *et al.*, 1987) or may not (Marcelo and Duell, 1989) possess the ability to transform exogenously supplied linoleic acid into arachidonic acid. Such differences between *in vitro* versus *in vivo* systems mandate caution in interpreting data about epidermal essential fatty acid metabolism.

Many authors have utilized the skin of essential fatty acid-deficient animals as a model to elucidate the functions of essential fatty acids in the skin. Moreover, the possibility of exerting a therapeutic effect on chronic inflammatory skin diseases through manipulation of the diet is a subject of increasing interest (Miller and Ziboh, 1990). Ziboh and Hsia (1972) first demonstrated a significant correlation between the level of linoleic acid in the diet and endogenous prostaglandin E_2 levels. Interestingly, under certain physiological conditions, an enrichment of linoleic acid can occur in epidermal sphingolipids, acylglucosylceramides, and acylceramides (Hansen and Jensen, 1985; Hansen, 1986). The recent observation that orally administered evening primrose oil (enriched in γ-linolenic acid, $C_{18:3n\text{-}6}$) improves atopic eczema suggests that generation of prostaglandins of the E_1 series and/or lipoxygenase products of the n3 series may benefit cutaneous inflammatory and/or hyperproliferative disorders (Melnik and Plewig, 1989). By dietary cross-over feeding with both safflower oil, rich in linoleic acid, and primrose oil, which contains linoleic acid and γ-linolenic acid, inositol phosphates may be generated, which may normalize the hyperproliferative state in cutaneous essential fatty acid deficiency (Tang and Ziboh, 1988). Thus, Tang and Ziboh postulated that epidermal inositol–phospholipid metabolism may play a central role in the pathogenesis of cutaneous hyperproliferative disorders.

V. Fatty Acid Uptake and Binding

Studies on fatty acid uptake and metabolism by isolated mammalian cells have been reviewed in great detail by Rosenthal (1987). In fibroblasts, the cellular uptake of exogenous free fatty acids, complexed to albumin, is nonspecific and quite rapid (Spector *et al.*, 1980).

Unesterified fatty acids are transported in the blood complexed to albumin (Spector, 1975). The liver is the major fatty acid-metabolizing organ in the body and has been extensively studied as an organ of fatty acid synthesis (Mogensen *et al.*, 1987), metabolism (Nugteren *et al.*, 1985; Denning *et al.*, 1983), intracellular transport (Ockner, 1972; Bass, 1988), and uptake (Stremmel and Berk, 1986). Whether uptake into extrahepatic tissues is mediated by (1) passive diffusion, a non-energy-dependent process (Spector *et al.*, 1980; Noy *et al.*, 1986), (2) specific binding to a cell surface albumin receptor, and/or (3) a specific plasma membrane fatty acid-binding protein, i.e., carrier-mediated fatty acid uptake (Stremmel, 1988), remains controversial. According to Spector *et al.* (1980), the capacity of cultured cells to take up exogenous fatty acids is an unsaturable, non-energy-dependent process with excess fatty acid groups stored as triglycerides, which accumulate in cytoplasmic lipid droplets (Spector *et al.*, 1980). However, there is substantial evidence that fatty acid uptake is sodium-, temperature-, and NADPH-dependent in both hepatocytes and intestinal cells (Stremmel *et al.*, 1986). Moreover, uptake also may be partially blocked by inhibitors of oxidative phosphorylation. Furthermore, this putative, membrane fatty acid transport protein is inhibited by trypsin pretreatment of hepatocytes and by pretreatment of cells with specific monoclonal or polyclonal antibodies to isolated rat liver plasma membrane fatty acid-binding protein (Stremmel and Berk, 1986).

Keratinocytes in culture show selectivity in the metabolism and incorporation of exogenous fatty acids into complex lipids (Schürer *et al.*, 1989). It remains to be determined if these observed preferences are dependent upon a plasma membrane fatty acid-binding protein, or whether they reflect the affinity of the enzymes involved in the synthesis of complex lipids. Recent attempts to identify the existence of an intracellular fatty acid-binding protein in cultured human keratinocytes, comparable to the low-molecular-weight protein found in other tissues (Bass, 1988), have been unsuccessful (N. V. Schürer, unpublished observations). The epidermis is a highly active site of fatty acid synthesis (Grubauer *et al.*, 1987; Monger *et al.*, 1988). Yet, arachidonic acid must be first synthesized elsewhere, followed by transport and uptake into the epidermis for esterification (Chapkin and Ziboh, 1984). Likewise, essential fatty acids must be taken up by the epidermis. This suggests that specific or nonspecific mechanisms must exist for fatty acid delivery to the epidermis.

It remains to be determined whether epidermal fatty acid uptake occurs via passive diffusion or a plasma membrane receptor-dependent mechanism, and whether there are separate uptake mechanisms for linoleic, arachidonic, and other fatty acids. The study of fatty acid metabolism in the epidermis is complicated by its capacity to both proliferate and differentiate. With respect to its active fatty acid synthetic capacity, plasma membrane-bound fatty acid-binding proteins may not be expressed in more differentiated epidermal layers, analogous to LDL receptor expression (Mommaas-Kienhuis *et al.*, 1987). Thus, just as a relationship exists between water barrier requirements, cholesterol synthesis, and LDL receptor expression in relation to epidermal level, it is possible that a plasma membrane fatty acid-binding protein is required to further regulate the selective uptake and incorporation of exogenous essential fatty acids, as has been described for arachidonic acid (Ziboh and Chapkin, 1988).

Selective incorporation of arachidonic acid occurs in cultured human skin fibroblasts (Banerjee and Rosenthal, 1985) and keratinocytes (Punnonen *et al.*, 1987b). Rosenthal (1987) and co-workers were able to demonstrate high-affinity uptake for arachidonic acid and other polyunsaturated fatty acids. Selective activation of arachidonic acid by acyl coenzyme A and storage of arachidonic acid in phospholipids may be presumed to serve as the reservoir of arachidonic acid for the synthesis of prostaglandins and leukotrienes (Ziboh, 1989).

References

Abraham, W. A., and Downing, D. T. (1990). *Biochim. Biophys. Acta* **1021,** 119–125.
Abraham, W., Wertz, P. W., and Downing, D. T. (1985). *Clin. Res.* **33,** 621A.

Abraham, W., Wertz, P. W., Landmannn, L., and Downing, D. T. (1987). *J. Invest. Dermatol.* **88,** 212–214.

Abraham, W., Wertz, P. W., and Downing, D. T. (1988a). *Biochim. Biophys. Acta* **939,** 403–408.

Abraham, W., Wertz, P. W., and Downing, D. T. (1988b). *J. Invest. Dermatol.* **90,** 259–262.

Andersen, J. M., and Dietschy, J. M. (1977). *J. Biol. Chem.* **252,** 3652–3659.

Ansari, M. N. A., Nicolaides, N., and Fu, H. C. (1970). *Lipids* **5,** 838–845.

Anton-Lamprecht, I. (1983). *Zentralbl. Haut- Geschlechtskr.* **148,** 911–920.

Banerjee, N., and Rosenthal, M. D. (1985). *Biochim. Biophys. Acta* **835,** 533–541.

Bass, N. M. (1988). *Int. Rev. Cytol.* **111,** 143–184.

Bickers, D. R. (1983). *In* "Biochemistry and Physiology of the Skin" (I. A. Goldsmith, eds.), Vol. 2, pp. 1169–1186. Oxford Univ. Press, New York.

Blank, I. H. (1953). *J. Invest. Dermatol.* **21,** 259–269.

Bortz, J. T., Wertz, P. W., and Downing, D. T. (1989). *J. Invest. Dermatol.* **93,** 723–727.

Bortz, J. T., Wertz, P. W., and Downing, D. T. (1990). *J. Invest. Dermatol.* **94,** 731.

Bowser, P. A., and White, R. J. (1985). *Br. J. Dermatol.* **112,** 1–14.

Brannon, P. G., Goldstein, J. L., and Brown, M. S. (1975). *J. Lipid Res.* 16, 7–11.

Braun, P. E., Morell, P., and Radin, N. S. (1970). *J. Biol. Chem.* **245,** 335–341.

Brenner, R. R. (1981). *Prog. Lipid Res.* **20,** 41–47.

Budowski, P. (1981). *Isr. J. Med. Sci.* **17,** 223–231.

Burr, G. O., and Burr, M. M. (1929). *J. Biol. Chem.* **82,** 345–367.

Cameron, G. S., Baldwin, J. K., Jasheway, D. W., Patrick, K. E., and Fischer, S. M. (1990). *J. Invest. Dermatol.* **94,** 292–296.

Carruthers, C., and Heining, A. (1964). *Cancer Res.* **24,** 1008–1011.

Chapkin, R. S., and Ziboh, V. A. (1984). *Biochim. Biophys. Res. Commun.* **124,** 784–792.

Cox, P., and Squier, C. A. (1986). *J. Invest. Dermatol.* **87,** 741–744.

Cullis, P. R., and Hope, M. J. (1980). *Biochim. Biophys. Acta* **597,** 533–542.

Das, D. V. M., Cook, H. W., and Spence, M. W. (1984). *Biochim. Biophys. Acta* 777, 339–342.

Denning, G. M., Figard, P. H., Kaduce, T. L., and Spector, A. A. (1983). *J. Lipid Res.* **24,** 993–1001.

Downing, D. T., Stewart, M. E., Wertz, P. W., Colton, S. W., and Strauss, J. S. (1983). *Comp. Biochem. Physiol.* **76,** 673–678.

Downing, D. T., Stewart, M. E., Wertz, P. W., Colton, S. W., Abraham, W., and Strauss, J. S. (1987). *J. Invest. Dermatol.* **88,** 2s-6s.

Duell, E. A., Ellis, C. N., and Voorhees, J. J. (1988). *J. Invest. Dermatol.* **91,** 446–450.

Dusham, E. H. (1918). *J. Morphol.* **31,** 563–581.

Dykes, P. J., Marks, R., Davies, M. G., and Reynolds, D. J. (1978). *J. Invest. Dermatol.* **70,** 126–129.

Ebling, F. J., and Rook, A. (1972). *In* "Textbook of Dermatology" (A. Rook, D. S. Wilkinson, and F. J. Ebling, eds.), 2nd ed., pp. 1526–1541. Blackwell, Oxford.

Ebling, F. J., and Rook, A. (1979). *In* "Textbook of Dermatology" (A. Rook, D. S. Wilkinson, and F. J. Ebling, eds.), 3rd Ed., pp. 1694–1695. Blackwell, Oxford.

Elias, P. M. (1981). *Int. J. Dermatol.* **20,** 1–19.

Elias, P. M. (1983). *J. Invest. Dermatol.* **80,** 44s-49s.

Elias, P. M., and Brown, B. E. (1978). *Lab. Invest.* **39,** 574–583.

Elias, P. M., and Friend, D. S. (1975). *J. Cell Biol.* **65,** 180–191.

Elias, P. M., and Williams, M. L. (1985). *Arch. Dermatol.* **121,** 1000–1008.

Elias, P. M., and Williams, M. L. (1990). *J. Invest. Dermatol.* **94,** 730–731.

Elias, P. M., Goerke, J., and Friend, D. S. (1977). *J. Invest. Dermatol.* **69,** 535–546.

Elias, P. M., Brown, B. E., Fritsch, P., Goerke, J., Gray, G. M., and White, R. J. (1979). *J. Invest. Dermatol.* **73,** 339–348.

Elias, P. M., Brown, B. E., and Ziboh, V. A. (1980). *J. Invest. Dermatol.* **74,** 230–233.

Elias, P. M., Cooper, E. R., Korc, A., and Brown, B. E. (1981a). *J. Invest. Dermatol.* **76,** 297–301.

Elias, P. M., Bonar, L., Grayson, S., and Baden, H. P. (1983a). *J. Invest. Dermatol.* **80,** 213–214.

Elias, P. M., Grayson, S., Lampe, M. A., Williams, M. L., and Brown, B. E. (1983b). *In* "The Intercorneocyte Space" (R. Marks and G. Plewig, eds.), pp. 53–68. Springer-Verlag, Berlin.
Elias, P. M., Williams, M. L., Maloney, M. E., Bonifas, J. A., Brown, B. E., Grayson, S., and Epstein, E. H. (1984). *J. Clin. Invest.* **74,** 1414–1421.
Elias, P. M., Menon, G. K., Grayson, S., Brown, B. S., and Rehfeld, S. J. (1987). *Am. J. Anat.* **180,** 161–177.
Elias, P. M., Menon, G. K., Grayson, S., and Brown, B. S. (1988). *J. Invest. Dermatol.* **91,** 3–10.
Elliott, W. J., Sprecher, H., and Needleman, P. (1985). *Biochim. Biophys. Acta* **835,** 158–160.
Ellis, R. A., and Hendrickson, R. C. (1963). *In* "Advances in Biology of the Skin" (W. Montagna *et al.*, eds.), p. 94. Pergamon, Oxford.
Epstein, E. (1985). *Prog. Dermatol.* **19,** 1–7.
Epstein, E., Williams, M. L., and Elias, P. M. (1981). *Arch. Dermatol.* **117,** 761–763.
Epstein, E., Bonifas, J. M., Barber, T. L., and Harber, M. (1984). *J. Invest. Dermatol.* **83,** 332–335.
Feingold, K. R., Wiley, M. H., Moser, A. H., Lau, D. T., Lear, S. R., and Siperstein, M. D. (1982). *J. Lab. Clin. Med.* **100,** 405–410.
Feingold, K. R., Wiley, M. H., MacRae, R., Lear, S. R., Moser, A. H., Zsigmond, G., and Siperstein, M. D. (1983). *Metab., Clin. Exp.* **32,** 75–81.
Feingold, K. R., Brown, B. E., Lear, S. R., Moser, A. H., and Elias, P. M. (1986). *J. Invest. Dermatol.* **87,** 588–591.
Feingold, K. R., Williams, M. L., Pillai, S., Menon, G. K., Holleran, W. P:, Bikle, D., and Elias, P. M. (1987). *Biochim. Biophys. Acta* **930,** 193–200.
Freinkel, R. K., and Aso, K. (1968). *J. Invest. Dermatol.* **50,** 357–361.
Freinkel, R. K., and Traczyk, T. N. (1985). *J. Invest. Dermatol.* **85,** 295–298.
Ganley, O. H., Graessie, O. E., and Robinson, H. J. (1958). *J. Lab. Clin. Med.* **51,** 709–714.
Gloor, M., Josephs, H., and Friedrich, H. C. (1974). *Arch. Dermatol. Forsch.* **250,** 277–284.
Goldstein, J. L., and Brown, M. S. (1973). *Proc. Natl. Acad. Sci. U.S.A.* **70,** 2804–2808.
Gray, G. M., and Yardley, H. J. (1975a). *J. Lipid Res.* **16,** 434–439.
Gray, G. M., and Yardley, H. J. (1975b). *J. Lipid Res.* **16,** 441–447.
Gray, G. M., King, I. A., and Yardley, H. J. (1980). *Br. J. Dermatol.* **103,** 505–515.
Gray, M. G. (1976). *Biochim. Biophys. Acta* **431,** 1–8.
Gray, M. G., and White, R. J. (1978). *J. Invest. Dermatol.* **70,** 336–341.
Gray, M. G., White, R. J., and Majer, J. R. (1978). *Biochim. Biophys. Acta* **528,** 127–137.
Grayson, S., and Elias, P. M. (1982). *J. Invest. Dermatol.* **78,** 128–135.
Grayson, S., Johnson-Winegar, A. G., Wintroub, B. U., Isseroff, R. R., Epstein, E. H., and Elias, P. M. (1985). *J. Invest. Dermatol.* **85,** 289–294.
Grubauer, G., Feingold, K. R., and Elias, P. M. (1987). *J. Lipid Res.* **28,** 746–752.
Grubauer, G., Feingold, K. R., and Elias, P. M. (1989a). *J. Lipid Res.* **30,** 89–96.
Grubauer, G., Elias, P. M., and Feingold, K. R. (1989b). *J. Lipid Res.* **30,** 323–333.
Hadley, N. F. (1980). *Am. Sci.* **68,** 546–553.
Hadley, N. F. (1989). *Prog. Lipid Res.* **28,** 1–33.
Hamanaka, S., Asagami, C., Suzuki, M., Inagaki, F., and Suzuki, A. (1989). *J. Biochem. (Tokyo)* **105,** 684–690.
Hannun, Y. A., Loomis, C. R., Merrill, A. H., and Bell, R. M. (1986). *J. Biol. Chem.* **261,** 12604–12609.
Hansen, H. S. (1986). *Trends Biochem. Sci. (Pers. Ed.)* **11,** 263–265.
Hansen, H. S., and Jensen, B. (1985). *Biochim. Biophys. Acta* **834,** 357–363.
Hedberg, C. L., Wertz, P. W., and Downing, D. T. (1988). *J. Invest. Dermatol.* **91,** 169–174.
Holick, M. F., MacLaughlin, J. A., Clark, M. B., Holick, S. A., Potts, J. T., Anderson, R. R., Blank, I. H., Parrish, J. A., and Elias, P. M. (1980). *Science* **210,** 203–205.
Holleran, W. M., Williams, M. L., Gao, W. N., and Elias, P. M. (1990). *J. Lipid Res.* **31,** 1635–1661.
Holleran, W. M., Feingold, K. R., Mao-Qiang, M., Gao, W. N., Lee, J. M., and Elias, P. M. (1991). *J. Lipid Res.*, in press.

Holman, R. T., Johnson, S. B., and Hatch, T. F. (1982). *Am. J. Clin. Nutr.* **35,** 617–623.
Homalle, A. (1853). *Union Med.* **7,** 462–463.
Houtsmuller, U. M. T., and van der Beck, A. (1981). *Prog. Lipid Res.* 20, 219–224.
Imokawa, G., Akasaki, S., Hutton, M., and Yoshizuka, N. (1986). *J. Invest. Dermatol.* **87,** 758–761.
Isseroff, R. R., Martinez, D. T., and Ziboh, V. A. (1985). *J. Invest. Dermatol.* **85,** 131–134.
Isseroff, R. R., Ziboh, V. A., Chapkin, R. S., and Martinez, D. T. (1987). *J. Lipid Res.* **28,** 1342–1349.
Jackson, L. J., and Baker, G. L. (1970). *Lipids* **5,** 239–246.
Kligman, A. M. (1964). *In* "The Epidermis" (W. Montagna and W. C. Lobitz, Jr., eds.), pp. 387–433. Academic Press, New York.
Kolattukudy, P. E. (1968). *Science* **159,** 498–505.
Kondoh, H., Kanoh, H., and Ono, T. (1983). *Biochim, Biophys. Acta* **753,** 97–106.
Kooyman, D. J. (1932). *Arch. Dermatol. Syph.* **25,** 444–450.
Lampe, M. A., Burlingame, A. L., Whitney, J., Williams, M. L., Brown, B. E., Roitman, E., and Elias, P. M. (1983a). *J. Lipid Res.* **24,** 120–130.
Lampe, M. A., Williams, M. L., and Elias, P. M. (1983b). *J. Lipid Res.* **24,** 131–140.
Landmann, L. (1988). *Anat. Embryol.* **178,** 1–13.
Lands, W. E. M. (1981). *Prog. Lipid Res.* **20,** 875–883.
Locke, M. (1965). *Science* **147,** 295–298.
Long, S. A., Wertz, P. W., Strauss, J. S., and Downing, D. T. (1985). *Arch. Dermatol. Res.* **277,** 285–287.
Lowe, M. J., and DeQuoy, P. R. (1978). *J. Invest. Dermatol.* **70,** 200–203.
Madison, K. C., Wertz, P. W., Strauss, J. S., and Downing, D. T. (1986). *J. Invest. Dermatol.* **87,** 253–259.
Madison, K. C., Swartzendruber, D. C., Wertz, P. W., and Downing, D. T. (1990). *J. Invest. Dermatol.* **95,** 657–664.
Marcelo, C., and Duell, E. (1989). *J. Invest. Dermatol.* **92,** 476A.
Mazzone, T., Basheeruddin, K., Ping, L., and Schick, C. (1990). *J. Biol. Chem.,* **265,** 5145–5149.
Melnik, B. C. (1989). *In* "The Ichthyosis" (H. Traupe, ed.), pp. 15–42.
Melnik, B. C., and Plewig, G. (1988). *Z. Hautkr.* **63,** 591–596.
Menon, G. K., Feingold, K. R., Moser, A. H., Brown, B. E., and Elias, P. M. (1985). *J. Lipid Res.* **26,** 418–427.
Menon, G. K., Grayson, S., and Elias, P. M. (1986a). *J. Invest. Dermatol.* **86,** 591–597.
Menon, G. K., Grayson, S., Brown, B. E., and Elias, P. M. (1986b). *Cell Tissue Res.* **244,** 385–394.
Merrill, A. H., and Jones, D. D. (1990). *Biochim. Biophys. Acta* **1044,** 1–12.
Merrill, A. H., and Stevens, V. (1989). *Biochim. Biophys. Acta* **1010,** 131–139.
Michaels, A. S., Chandrasekaran, S. K., and Shaw, J. E. (1975). *AIChE J.* **21,** 985–995.
Miller, C. C., and Ziboh, V. A. (1990). *J. Invest. Dermatol.* **94,** 353–358.
Miller, C. C., Yamaguchi, R. Y., and Ziboh, V. A. (1989). *Lipids,* **24,** 998–1003.
Mogensen, I. B., Schulenberg, H., Hansen, H. O., Spener, F., and Knudsen, D. (1987). *Biochem. J.* **241,** 189–192.
Mommaas-Kienhuis, A.-M., Grayson, S., Wijsman, M. C., Vermeer, B. J., and Elias, P. M. (1987). *J. Invest. Dermatol.* 89, 513–517.
Monash, S., and Blank. I. H. (1958). *Arch. Dermatol.* **78,** 710–714.
Monger, D. J., Williams, M., Feingold, K. R., Brown, B., and Elias, P. (1988). *J. Lipid Res.* **29,** 603–612.
Montagna, W. (1963). *In* "Advances in Biology of the Skin" (W. Montagna *et al.*, eds.), p. 19. Pergamon, Oxford.
Nemanic, M. K., and Elias, P. M. (1980). *J. Histochem. Cytochem.* **28,** 573–578.
Nemanic, M. K., Whitehead, J. S., and Elias, P. M. (1983). *J. Histochem. Cytochem.* **31,** 887–897.
Nicolaides, N. (1965). *J. Am. Chem. Soc.* **42,** 691–702.
Nicolaides, N. (1974). *Science* **186,** 19–26.

Nicolaides, N. (1983). *Curr. Eye Res.* **2,** 93–98.
Nicolaides, N., Fu, C. H., and Rice, G. R. (1968). *J. Invest. Dermatol.* **51,** 83–89.
Noy, N., Donnelly, T. M., and Zakim, D. (1986). *Biochemistry* **25,** 2013–2021.
Nugteren, D. H., Christ-Hazelhof, E., van der Beek, A., and Houtsmuller, U. M. T. (1985). *N. Eng. J. Med.* **312,** 1210–1216.
Ockner, R. K., Manning, J. A., Poppenhauser, R. B., and Ho, W. K. L. (1972). *Science* **177,** 56–58.
Odland, G. F., and Holbrook, K. (1987). *Curr. Probl. Dermatol.* **9,** 29–49.
Ponec, M., Havekes, K., Kempenaar, J., Lavrijsen, S., Wijsman, M., Boonstra, J., and Vermeer, B. J. (1985). *J. Cell. Physiol.* **125,** 98–106.
Ponec, M., Weerheim, A., Kempenaar, J., Mommaas-Kienhuis, A.-M., and Nugteren, D. H. (1988). *J. Lipid Res.* **29,** 959–961.
Proksch, E., Elias, P. M., and Feingold, K. R. (1990). *J. Clin. Invest.* ***85,*** 874–882.
Proksch, E., Feingold, K. R., and Elias, P. M. (1991a). *J. Clin. Invest.* **87,** 1668–1673.
Proksch, E., Elias, P. M., and Feingold, K. R. (1991b), *Biochim Biophys. Acta* **1083,** 71–79.
Prottey, C. (1976). *Br. J. Dermatol.* **94,** 579–585.
Prottey, C. (1977). *Br. J. Dermatol.* **97,** 29–38.
Punnonen, K., Puustinen, T., and Jansen, C. T. (1987a). *Lipids* **22,** 139–143.
Punnonen, K., Puustinen, T., and Jansen, C. T. (1987b). *J. Invest. Dermatol.* 88, 611–614.
Ramsay, J. A. (1935). *J. Exp. Biol.* **12,** 373–383.
Ranasinghe, A. W., Wertz, P. W., Downing, D. T., and Mackenzie, I. C. (1986). *J. Invest. Dermatol.* **86,** 187–190.
Rehfeld, S. J., and Elias, P. M. (1982). *J. Invest. Dermatol.* **79,** 1–3.
Rehfeld, S. J., Williams, M. L., and Elias, P. M. (1986). *Arch. Dermatol. Res.* **278,** 259–263.
Rehfeld, S. J., Plachy, W. Z., Williams, M. L., and Elias, P. M. (1988). *J. Invest. Dermatol.* **91,** 499–505.
Rehfeld, S. J., Plachy, W. Z., Hou, S. Y., and Elias, P. M. (1990). *J. Invest. Dermatol.* **95,** 217–223.
Reinertson, R. P., and Wheatley, V. R. (1959). *J. Invest. Dermatol.* **32,** 49–59.
Roberts, J. D., and Lillywhite, H. B. (1980). *J. Cell Biol.* **24,** 297–307.
Rosenthal, M. D. (1987). *Prog. Lipid Res.* **26,** 87–124.
Rothman, S. (1964). *In* "The Epidermis" (W. Montagna and W. C. Lobitz, eds.), pp. 1–14. Academic Press, New York.
Rushmer, R. F., Buettner, K. J. K., Short, J. M., and Odland, G. F. (1966). *Science* ***154,*** 343–348.
Schaefer, H., Zesch, A., and Stüttgen, G. (1982). "Skin Permeability." Springer-Verlag, Berlin.
Scheuplein, R. J., and Blank. I. H. (1971). *Physiol. Rev.* **51,** 701–747.
Schürer, N. Y., Monger, D. J., Hincenbergs, M., and Williams, M. L. (1989). *J. Invest. Dermatol.* **92,** 196–202.
Sherertz, E. F. (1986). *In* "Nutrition and the Skin" (R. Alan, ed.), pp. 117–130.
Slife, C. W., Wang, E., Hunter, R., Wang, S., Burgess, C., Liotta, D. C., and Merrill, A. H. (1989). *J. Biol. Chem.* **264,** 10371–10377.
Smith, W. P., Christiansen, M. S., Nacht, S., and Gans, E. H. (1982). *J. Invest. Dermatol.* **78,** 7–10.
Spector, A. A. (1975). *J. Lipid Res.* **216,** 165–179.
Spector, A. A., Denning, G. M., and Stoll, L. L. (1980). *In Vitro* **16,** 932–940.
Squier, C. A., and Hall, B. H. (1985). *J. Invest. Dermatol.* **84,** 174–179.
Squier, C. A., Cox, P. A., Wertz, P. W., and Downing, D. T. (1986). *Arch. Oral Biol.* **31,** 741–747.
Stoffel, W., Le Kim, D., and Sticht, G. (1968). *Hoppe-Seyler's Z. Physiol. Chem.* **349,** 1637–1644.
Stremmel, W. (1988). *J. Clin. Invest.* **81,** 844–852.
Stremmel, W., and Berk, P. D. (1986). *Proc. Natl. Acad. Sci. U.S.A.* **83,** 3086–3090.
Stremmel, W., Strohmayer, G., and Berk, P. D. (1986). *Proc. Natl. Acad. Sci. U.S.A.* **83,** 3584–3588.
Swanbeck, G. (1959). *Acta Derm.-Venereol.* **39,** 43s.
Swartzendruber, D. C., Wertz, P. W., Kitko, D. J., Madison, K. C., and Downing, D. T. (1989). *J. Invest. Dermatol.* **92,** 251–257.
Sweeney, T. M., and Downing, D. T. (1970). *J. Invest. Dermatol.* 55, 135–140.

Tang, W., and Ziboh, V. A. (1988). *Arch. Dermatol. Res.* **280,** 286–292.
Unna, P. G. (1913). "Biochemie der Haut." Fischer, Jena.
Voorhees, J. J. (1983). *Arch. Dermatol.* **119,** 541–547.
Wertz, P. W., and Downing, D.T. (1982). *Science* **217,** 1261–1262.
Wertz, P. W., and Downing, D.T. (1983a). *J. Lipid Res.* **24,** 759–765.
Wertz, P. W., and Downing, D.T. (1983b). *J. Lipid Res.* **24,** 753–758.
Wertz, P. W., and Downing, D.T. (1986). *Biochim. Biophys. Acta* **876,** 469–473.
Wertz, P. W., and Downing, D.T. (1989). *Biochim. Biophys. Acta* **1002,** 213–217.
Wertz, P. W., and Downing, D.T. (1990a). *FEBS Lett.* **268,** 110–112.
Wertz, P. W., and Downing, D.T. (1990b). *J. Invest. Dermatol.* **94,** 159–161.
Wertz, P. W., Cho, E. S., and Downing, D. T. (1983). *Biochim. Biophys. Acta* **753,** 350–355.
Wertz, P. P., Downing, D. T., Freinkel, R. K., and Traczyk, T. N. (1984). *J. Invest. Dermatol.* **83,** 193–195.
Wertz, P. W., Miethke, M. C., Long, S. A., Strauss, K. S., and Downing, D. T. (1985). *J. Invest. Dermatol.* **84,** 410–412.
Wertz, P. W., Stover, P. M., Abraham, W., and Downing, D. T. (1986). *J. Lipid Res.* **27,** 427–435.
Wertz, P. W., Swartzendruber, D. C., Madison, K. C., and Downing, D. T. (1987). *J. Invest. Dermatol.* **89,** 419–425.
Wertz, P. W., Madison, K. C., and Downing, D. T. (1989). *J. Invest. Dermatol.* **92,** 109–111.
Wheatley, V. R., and Flesch, P. (1967). *J. Invest. Dermatol.* **49,** 198–205.
Wheatley, V. R., Flesch, P., Esoda, E. C. J., Coon, W. M., and Mandol, L. (1964). *J. Invest. Dermatol.* **43,** 395–405.
White, S. H., Mirejovsky, D., and King, G. I. (1988). *Biochemisty* **27,** 3725–3732.
Wilkinson, D. I. (1971). *In* "Psoriasis" (E. M. Farber, and A. J. Cox, eds.), pp. 277–285. Stanford Univ. Press, Stanford, California.
Williams, M. L. (1983). *Pediatr. Dermatol.* **1,** 1–24.
Williams, M. L., and Elias, P. M. (1981). *Lipids* **68,** 1404–1410.
Williams, M. L., and Elias, P. M. (1984). *J. Invest. Dermatol.* **74,** 296–300.
Williams, M. L., and Elias, P. M. (1986). *In* "Pathogenesis of Skin Disease" (B. H. Thiers and R. L. Dobson, eds.), pp. 519–551. Churchill-Livingstone, New York.
Williams, M. L., and Elias, P. M. (1987). *CRC Crit. Rev. Ther. Drug Carrier Syst.* **3,** 95–122.
Williams, M. L., Hughes-Fulford, M., and Elias, P. M. (1985). *Biochim. Biophys. Acta* **845,** 349–357.
Williams, M. L., Rutherford, S. L., and Feingold, K. R. (1987a). *J. Lipid Res.* **28,** 955–967.
Williams, M. L., Mommaas-Kienhuis, A.-M., Rutherford, S. L., Grayson, S., Vermeer, B. J., and Elias, P. M. (1987b). *J. Cell. Physiol.* **132,** 428–440.
Winsor, T., and Burch, G. E. (1944). *Arch. Intern. Med.* **74,** 428–436.
Yamamoto, A., Serizawa, S., Ito, M., and Sato, Y. (1990). *J. Dermatol. Sci.* **1,** 269–276.
Yardley, H. J. (1969). *Br. J. Dermatol.* **81,** 29–38.
Yardley, H. J. (1983). *In* "Biochemistry and Physiology of the Skin" (L. A. Goldsmith, ed.), pp. 363–381. Oxford Univ. Press, New York.
Yardley, H. J. (1990). *Br. J. Dermatol.* **81,** 29–38.
Yardley, H. J., and Summerley, R. (1981). *Pharmacol. Ther.* **13,** 357–383.
Ziboh, V. A. (1989). *Arch. Dermatol.* **125,** 241–245.
Ziboh, V. A., and Chapkin, R. S. (1987). *Arch. Dermatol.* **123,** 1686–1690.
Ziboh, V. A., and Chapkin, R. S. (1988). *Prog. Lipid Res.* **27,** 81–105.
Ziboh, V. A., and Hsia, S. L. (1972). *Pure Appl. Chem.* **62,** 1373–1376.
Ziboh, V. A., and Miller, C. C. (1990). *Annu. Rev. Nutr.* **10,** 433–450.
Ziboh, V. A., Cohen, K. A., Ellis, C. N., Miller, C., Hamilton, A., Kragballe, K., Hydrick, C. R., and Voorhees, J. J. (1986). *Arch. Dermatol.* **122,** 1277–1282.

ADVANCES IN LIPID RESEARCH, VOL. 24

The Regulation and Role of Epidermal Lipid Synthesis

KENNETH R. FEINGOLD

Department of Medicine
University of California School of Medicine
San Francisco, California 94143
and
Metabolism Section, Medical Service
Veterans Administration Medical Center
San Francisco, California 94121

I. Introduction

The cutaneous permeability barrier is required for terrestrial life, and a major function of the epidermis is to form this barrier (Blank, 1987; Scheuplein and Blank, 1971). As discussed in detail elsewhere in this volume, this cutaneous barrier resides in the stratum corneum layer of the epidermis (Blank, 1987; Scheuplein and Blank, 1971), primarily in the lipid-enriched intercellular domains (Elias, 1983; Elias *et al.*, 1987). These intercellular lipids are thought to be derived chiefly from the exocytosis of lamellar bodies, which are lipid-enriched organelles formed in stratum spinosum and granulosum cells (Blank, 1987; Elias, 1983; Elias *et al.*, 1987; Elias and Feingold, 1988). The source of the lipids needed for the formation of lamellar bodies is uncertain, but, as will be discussed in detail in this review, it is our hypothesis that these lamellar body lipids are primarily derived from local lipid synthesis in the epidermis. The regulation of epidermal lipid synthesis and its role in barrier maintenance will be addressed in this review.

II. Methodological Considerations

Since the 1950s, it has been well recognized that the skin is an active site of *de novo* lipid synthesis (Srere *et al.*, 1950; Nicolaides and Rothman, 1955; Patterson and Griesemer, 1959). Whereas these earlier studies employed acetate or glucose as the metabolic precursor, more recent studies have employed tritiated water to quantify *de novo* lipid synthesis *in vivo* and *in vitro* under a variety of experimental conditions (Lowenstein *et al.*, 1975; Lakshmanan and Veech, 1977; Dietschy and Spady, 1984). Tritiated water has several important advantages over conventional substrates, such as acetate, glucose, or octanoate, for accurately measuring lipid synthesis (Dietschy and McGarry, 1974; Dietschy and Brown, 1974). First, tritiated water distributes rapidly throughout the entire total body water pool, and, once having reached equilibrium, the specific activity of this substrate remains constant throughout the short period of the experiment. Second, because of the large excess of body water, the specific activity of tritiated water is little affected by the activity of other competing metabolic pathways. Third, the specific activity of the water pool can be determined for individual experimental animals, thus, correcting for differences in pool size. Furthermore, in contrast to many other radiolabels used for measuring lipid synthesis, the specific activity of the precursor pool does not require the prior conversion of the radiolabel to an active form (i.e., acetate to acetyl-CoA, glucose to acetyl-CoA), a process that can fluctuate independent of the lipid synthetic rate. These characteristics make tritiated water the best isotope currently available for the quantitation of *de novo* lipid synthesis. When the rate of sterol synthesis measured in whole skin slices was compared with incorporation rates of either [^{14}C]acetate, [^{14}C]octanoate, or [^{14}C]glucose, the incorporation rates for the latter substrates were less than one-third of the rate of sterol synthesis obtained when tritiated water was used as the radiolabel (Anderson and Dietschy, 1979).

Of course, tritiated water also has certain potential limitations, most notably that a portion of the labeled hydrogen ion incorporated into lipids is derived from NADPH rather than directly from tritiated water (Lakshmanan and Veech, 1977; Dietschy and Spady, 1984). Although the degree of labeling of NADPH by tritiated water could theoretically vary under different conditions and in various organs, to date this has not been proved to be a major liability of tritiated water. Additionally, because of the low labeling efficiency of tritiated water, large quantities (millicuries) must be employed to achieve sufficient incorporation.

In addition to the radioisotope precursor considerations described above, other factors also influence the apparent rates of lipid synthesis. Studies have shown that whereas there is a correlation between *in vitro* and *in vivo* determinations of lipid synthesis using tritiated water, the *in vivo* results are fourfold greater (Jeske and Dietschy, 1980). Presumably, the decreased *in vitro* incorporation is due to tissue injury during preparation, progressive hypoxia during incubation, the absence of crucial cofactors, and/or altered metabolism in the *in vitro* setting. Thus, though the rate of lipid synthesis measured *in vitro* can provide important relative information (i.e., comparisons between different conditions), *in vitro* studies are likely to underestimate significantly actual rates of tissue synthesis. However, *in vitro* studies are very useful in delineating the specific steps in lipid synthesis, the specific requirements for each step, and the factors that regulate synthesis.

Currently, the optimal technique for determining the rate of lipid synthesis is to measure tritiated water incorporation in intact animals. The major liability of *in vivo* studies is the requirement to consider the transport of newly synthesized lipids to and from the tissue under study. In the case of the epidermis, as will be discussed in detail below, this does not represent a significant limitation. Moreover, the confirmation of findings using a variety of different techniques (tritiated water incorporation, other isotopes, *in vivo* and *in vitro* experiments, and direct enzyme measurements) strengthens conclusions and greatly decreases the chances of artifactual observations or invalid interpretation.

III. Lipid Synthesis in the Skin

Using tritiated water in intact animals, Feingold *et al.* observed that the whole skin of rats and primates accounts for 24 and 21%, respectively, of total body cholesterol synthesis (Feingold *et al.*, 1982b, 1983c). Studies by other investigators in rats, primates, and other species have resulted in similar observations (Table I) (Dietschy *et al.*, 1983; Spady and Dietschy, 1983). Additionally, studies have also shown that fatty acid synthesis in the skin of rodents makes an important contribution to total body fatty acid synthesis (Gandemer *et al.*, 1983; Grubauer *et al.*, 1987). It is therefore apparent that the skin is quantitatively an important site of *de novo* lipid synthesis.

A number of earlier investigators compared epidermal and dermal lipid synthesis *in vitro* using [^{14}C]acetate as the radiolabel (Nicolaides and Rothman, 1955; Patterson and Griesemer, 1959; Griesemer and Thomas, 1963; Brooks *et al.*, 1966; Summerly and Woodbury, 1971; Prottey *et al.*, 1972). In general, these studies demonstrated that the epidermis on a weight basis is a very active site of

Table I
CONTRIBUTION OF THE SKIN TO TOTAL BODY STEROL SYNTHESIS[a]

Species	Total body synthesis by the skin (%)
Rat	12
Hamster	19
Guinea pig	18
Rabbit	20
Monkey	26

[a]Data taken from Spady and Dietschy (1983).

sterol and fatty acid synthesis, whereas the dermis, with the pilosebaceous tissues presumably the major source, is an active site of squalene synthesis. In more recent studies in intact animals using tritiated water, we observed that of the sterol synthetic activity in rodent skin, the majority (approximately 75%) occurs within the dermis, whereas the remaining 25% is localized to the epidermis (Feingold *et al.*, 1983a). Again, it is likely that a major portion of the dermal synthesis is localized to the pilosebaceous epithelium. However, administration of isotretinoin, which results in a marked involution of sebaceous glands, did not alter total skin sterologenesis, suggesting that sebum production is not a major contributor to the high rates of sterol synthesis in the dermis (Feingold *et al.*, 1983a). Fatty acid synthesis is also quantitatively greater in the dermis in comparison to the epidermis on a total tissue basis (Grubauer *et al.*, 1987). Because the epidermal layers comprise only a small fraction of total skin mass (approximately 10% or less), on a weight basis the viable epidermis can be considered among the most active sites of lipid synthesis in the entire body. Similar to the results reported in the *in vitro* studies, on a weight basis the epidermis is a much more active site of cholesterol and fatty acid synthesis than is the dermis.

Within the epidermis, all of the viable cell layers (stratum basale, stratum spinosum, and stratum granulosum) synthesize lipids, but the stratum corneum displays very little synthetic capability (Feingold *et al.*, 1983a; Monger *et al.*, 1988). The stratum basale (proliferating layer) is the most active site of sterol synthesis in mature animals (Feingold *et al.*, 1983a). In neonatal animals, the stratum granulosum layer is a much more active site of lipid synthesis, equal to or greater than the stratum basale (Monger *et al.*, 1988), perhaps because of the homeostatic mechanisms required for initial formation of an intact barrier (see below). Thus, one can conclude that the skin, and in particular the epidermis, is one of the most active sites of lipid synthesis in the entire organism.

The remainder of this review will focus on the regulation of epidermal lipid synthesis and the role that epidermal lipid synthesis plays in maintaining the cutaneous permeability barrier. The reader is referred to the excellent comprehensive review by Yardley and Summerly (1981) for additional information regarding earlier studies on cutaneous lipid synthesis.

IV. Systemic Regulation of Epidermal Lipid Synthesis

A large number of studies have examined the effects of systemic factors on epidermal and/or cutaneous lipid synthesis. In the following sections we will review the potential regulatory role of these systemic factors.

A. Effect of Dietary and Circulating Lipids

A number of lines of evidence have demonstrated the accumulation of dietary and circulating lipids in the skin. For example, the epidermis contains a high concentration of essential fatty acids that are required for normal barrier function (Prottey, 1976; Ziboh and Miller, 1990). Moreover, deficiency of essential fatty acids results in both perturbed barrier function and epidermal hyperplasia (McCullough *et al.*, 1978). These essential fatty acids can only be derived from dietary sources because mammals lack the ability to synthesize these lipids *de novo*. Similarly, ingested ω-3 polyunsaturated fatty acids (fish oils), which cannot be synthesized by mammals, also accumulate in the esterified fatty acids found in the epidermis (Miller *et al.*, 1990). In Refsum disease, the plant-derived fatty acid, phytanic acid, accumulates in the skin (Steinberg *et al.*, 1967; Dykes *et al.*, 1978). Finally, the epidermis lacks the enzyme necessary to convert linoleic acid into arachidonic acid (Δ^6-desaturase) and hence must obtain this fatty acid from the circulation (Chapkin and Ziboh, 1984). These observations demonstrate that dietary and circulating fatty acids can be delivered to the skin, but neither the quantitative impact of this delivery nor the effect of these fatty acids on cutaneous lipid metabolism is known except in gross deficiency states such as essential fatty acid deficiency. As will be discussed below, the transport of systemically administered labeled lipids to the epidermis is limited.

Despite the accumulation of some dietary and circulating lipids in the skin, a number of studies have demonstrated that the overall synthesis of cutaneous cholesterol is not influenced by dietary sterol intake (Dietschy and Siperstein, 1967; Feingold *et al.*, 1983c). This is not surprising because intestinally absorbed cholesterol is transported in chylomicrons, and in most species the vast majority of this cholesterol is delivered to the liver rather than to extrahepatic tissues (Havel *et al.*, 1980). Whether dietary sterols affect cholesterol synthesis specifically within the epidermis has not been addressed. The effect of other dietary lipids (for example, fatty acids and triglycerides) on the epidermal synthesis of lipids also has not been extensively investigated.

Studies have also demonstrated that circulating lipoprotein levels do not affect cutaneous sterol synthesis (Anderson and Dietschy, 1976, 1977a,b). Again, although there is a paucity of data specifically addressing the effect of circulating lipoprotein levels on epidermal lipid synthesis, *in vitro* studies in confluent keratinocytes have shown that cholesterol synthesis is not greatly influenced by exogenous cholesterol (Ponec *et al.*, 1983; Williams *et al.*, 1987a,b), likely due to the demonstrated lack of LDL receptors on the cell membranes of these cells (Ponec *et al.*, 1984; Williams *et al.*, 1987a). Moreover, *in vivo* studies have demonstrated that LDL receptor expression is limited to the basal cells in normal murine and human epidermis (Mommaas-Kienhuis *et al.*, 1987). However, in hyperplastic diseases, such as in essential fatty acid-deficient mice and psoriasis, LDL receptors extend into the stratum spinosum and stratum granulosum layers of the epidermis (Mommaas-Kienhuis *et al.*, 1987). This finding correlates both with the presence of proliferating cells in suprabasal layers and an altered differentiation schedule in these conditions (Grove, 1979). Alternatively, both in essential fatty acid-deficient mice and in psoriasis, an altered permeability barrier could create an increased need for cholesterol and other lipids (see below), which could up-regulate the number of LDL receptors. Although the above data suggest that epidermal lipid synthesis is not influenced greatly by dietary or circulating lipids, more definitive studies specifically addressing this issue are needed.

B. Vitamin D Homeostasis

It is well known that the free sterol, 7-dehydrocholesterol, is converted within the epidermis by a complex series of steps to vitamin D_3 following UV irridation of the skin (see Pillai and Bikle, this volume). However, vitamin D deficiency does not affect epidermal or dermal sterol synthesis (Feingold *et al.*, 1987). Moreover, the administration of physiological or supraphysiological concentration of vitamin D_3 or 1,25-$(OH)_2D_3$ to vitamin D-deficient animals did not alter either epidermal or dermal sterol synthesis (Feingold *et al.*, 1987).

C. Starvation and Glucose Feeding

Adding glucose to whole skin slices or epidermal samples *in vitro* increases fatty acid and sterol synthesis (Gaylor, 1961; Greisemer and Thomas, 1963; Ziboh *et al.*, 1970; Hsia *et al.*, 1970; Wheatley *et al.*, 1971). Moreover, the ingestion of glucose by intact humans or animals also increases lipid synthesis when epidermal samples subsequently are studied *in vitro* (Hsia *et al.*, 1966). In contrast, starvation decreases cutaneous and epidermal lipid synthesis (Hsia *et al.*, 1966, 1970; Ziboh and Hsia, 1969), a decrease comparable to observations in other organs such as liver and intestine, where starvation also inhibits lipid synthesis (Dietschy and Siperstein, 1967; Anderson and Dietschy, 1977a). However, not all investigators have observed a decrease in cutaneous cholesterol synthesis with starvation (Dietschy and Siperstein, 1967; Anderson and Dietschy, 1977a). The mechanism accounting for this glucose stimulation of lipid synthesis is uncertain, but it may be that the metabolism of glucose provides the substrate needed for lipid synthesis (acetyl-CoA) or necessary cofactors (NADPH derived from the pentose phosphate shunt pathway).

D. Effect of Diabetes/Insulin

In vitro and *in vivo* studies, using [^{14}C]acetate as the radiolabel for measuring lipid synthesis, have shown that cutaneous fatty acid and sterol synthesis are decreased in diabetic humans and animals (Van Bruggen *et al.*, 1954; Wong and Van Bruggen, 1960; Hsia *et al.*, 1966; Ziboh and Hsia, 1969). Conversely, insulin administration either *in vitro* or *in vivo* stimulates cutaneous lipid synthesis in diabetics but produces only a modest effect in normals (Hsia *et al.*, 1966; Ziboh and Hsia, 1969; Wilkinson, 1973). However, when tritiated water is employed to measure lipid synthesis *in vivo*, these alterations in cutaneous cholesterol synthesis in diabetic animals have not been found (Feingold *et al.*, 1982a). The metabolic status of diabetes is an important variable, and in these studies could have been a crucial difference. In many of the studies using [^{14}C]acetate, samples were taken during severe diabetes or during diabetic ketoacidosis, a condition of marked catabolism. In contrast, the tritiated water study was carried out in nonketotic animals, who were not severely catabolic (Feingold *et al.*, 1982a). Additionally, the diabetic state could effect acetate metabolism or pool sizes and thereby alter acetate incorporation into lipid (see Section III). Further work will be needed to clarify this issue.

E. Effect of Other Hormones

The skin of male rats synthesizes 73% more cholesterol than does the skin of female rats (Feingold *et al.*, 1983b). This sex difference in sterologenesis can be attributed to sex steroid hormones, because testosterone stimulates, whereas estrogens inhibit, cutaneous cholesterol synthesis (Feingold *et al.*, 1983b). In immature male and female rats a sex difference in skin cholesterol synthesis is not observed (Feingold *et al.*, 1983b). The dermis is the site that primarily accounts for the sex difference in sterologenesis (Feingold *et al.*, 1983a). However, this sex difference is not simply attributable to differences in sebaceous gland activity, because hairless mice also show this difference and the administration of isotretinoin, which causes the involution of sebaceous glands, does not ameliorate these differences (Feingold *et al.*, 1983a). Moreover, sterologenesis is also increased in the epidermis of male animals (1.9-fold increase in males) (Feingold *et al.*, 1983a). These results indicate that sex hormones influence both dermal and epidermal lipid synthesis, and that these differences are independent of sebaceous gland activity.

Thyroid hormone has also been shown to affect epidermal lipid synthesis (Rosenberg *et al.*, 1986). In hypothyroid rats the synthesis of phospholipids, sterol esters, free fatty acids, triglycerides, and free sterols is decreased in the epidermis (Rosenberg *et al.*, 1986). Similarly, the synthesis of lipids in human keratinocytes in culture is also decreased when cells are grown in a medium that does not contain thyroid hormone (Rosenberg *et al.*, 1986). These findings demonstrate that thyroid hormone, which effects lipid metabolism in many tissues (Dayton *et al.*, 1954; Kritchevsky, 1960), also regulates lipid metabolism in the epidermis.

Though these data indicate that epidermal lipid synthesis may be influenced by systemic factors, it is also apparent that the physiological regulation of epidermal lipid synthesis is unlikely to be mediated primarily by systemic events. Systemic factors affect epidermal lipid synthesis most dramatically when there are major perturbations (hypothyroidism, diabetes, or starvation).

V. Local Regulation of Epidermal Lipid Synthesis—Role of the Barrier

A. Effect on Lipid Synthesis

1. *Cholesterol Synthesis*

As discussed in detail by Elias and Menon elsewhere in this volume, the stratum corneum permeability barrier is primarily composed of extracellular lipids, chiefly cholesterol, fatty acids, and sphingolipids (Yardley and Summerly, 1981; Elias, 1983; Elias and Feingold, 1988). Thus, one of the factors that might regulate epidermal lipid synthesis would be the status of the permeability barrier (see Table II). Our initial studies in intact animals indicated that the incorporation of tritiated water into sterols is regulated by alterations in the permeability barrier of the skin (Menon *et al.*, 1985). When the permeability barrier is disrupted by topical treatment with either detergents such as sodium dodecyl sulfate (SDS) or lipid solvents (acetone), epidermal sterol synthesis increases two- to threefold, whereas, in contrast, dermal sterologenesis remains unchanged (Menon *et al.*, 1985). Moreover, epidermal lipid synthetic activity parallels barrier status with an almost immediate increase in sterologenesis followed by a rapid decrease so that by 12 to 24 hours, both the rate of synthesis and barrier function have returned to base line (Menon *et al.*, 1985). Additionally, the extent of the increase in sterol synthesis directly correlates with the extent of damage to the barrier, determined by measuring transepidermal water loss (Menon *et al.*, 1985). Most importantly, the increase in epidermal sterol synthesis following detergent or solvent treatment can be completely inhibited if the treated skin sites are occluded with a vapor-impermeable, polyethylene or latex film (Menon *et al.*, 1985). Thus, artificial correc-

Table II

SUMMARY OF THE EFFECTS OF BARRIER DISRUPTION ON EPIDERMAL LIPID SYNTHESIS

Model/condition[a]	Lipid synthesis		
	Cholesterol	Fatty acid	Sphingolipid
Acute disruption			
Acetone	↑	↑	↑ (delayed)
SDS	↑	↑	↑ (delayed)
Acetone + impermeable membrane	Normal	Normal	Normal
Acetone + permeable membrane	↑	↑	Not determined
Chronic disruption			
EFAD	↑	↑	↑
EFAD + impermeable membrane	Normal	Normal	Normal

[a]Abbreviations: SDS, sodium dodecyl sulfate; EFAD, essential fatty acid deficiency.

tion of the barrier abnormality prevents the expected stimulation of sterologenesis, provoked by either detergents or solvents. In contrast, occlusion with a water vapor-permeable membrane (GoreTex) did not prevent the increase in sterol synthesis (Grubauer *et al.*, 1989b). These results show that acute disturbances in barrier function stimulate epidermal sterol synthesis and support the concept that barrier requirements regulate epidermal sterologenesis.

Whereas the studies described above utilized acute perturbations in barrier function, in a chronic model of barrier impairment, i.e., essential fatty acid deficiency induced by dietary restriction of linoleic acid, epidermal sterol synthesis also was increased twofold in comparison to controls (Feingold *et al.*, 1986). The relationship to the alteration in barrier function is underscored by the fact that dermal synthesis was similar in both normal and essential fatty acid-deficient animals (Feingold *et al.*, 1986). Moreover, when the barrier abnormality was corrected by either oral or topical linoleic acid administration, the rates of epidermal sterologenesis decreased in parallel with the normalization of barrier function (Feingold *et al.*, 1986). Most importantly, when the permeability barrier defect in the essential fatty acid-deficient animals was artificially corrected by occlusion with a vapor-impermeable membrane, epidermal sterol synthesis was reduced to normal levels despite the presence of the ongoing, underlying deficiency state (Feingold *et al.*, 1986). Thus, in both acute and chronic models of barrier disruption, epidermal sterol synthesis is stimulated in response to altered barrier requirements.

2. *Fatty Acid Synthesis*

Further studies have addressed the effects of barrier disruption on the epidermal versus dermal synthesis of other lipid classes. As is the case with sterols, incorporation of tritiated water into fatty acids was also increased in the epidermis, but not in the dermis of intact animals following acute barrier disruption with either lipid solvents (acetone) or detergents (SDS) (Grubauer *et al.*, 1987). Similar to our observations regarding sterol synthesis, the increase in fatty acid synthesis occurred almost immediately after barrier disruption and returned toward normal in parallel with barrier recovery (Grubauer *et al.*, 1987). Moreover, epidermal fatty acid synthesis also increased in the chronic model of barrier disruption, essential fatty acid deficiency (Grubauer *et al.*, 1987). Finally and most pertinently, occlusion with a water vapor-impermeable membrane prevented the characteristic increase in epidermal fatty acid synthesis that occurs after barrier disruption, whereas occlusion with a water vapor-permeable membrane had no effect (Grubauer *et al.*, 1987).

3. *Sphingolipid Synthesis*

As detailed by Schurer and Elias (this volume), in addition to sterols and fatty acids, the stratum corneum is enriched in sphingolipids, and this class of lipids is hypothesized to play an important structural role in the barrier (Wertz and Downing, 1982; Wertz *et al.*, 1984). Therefore, recent studies have examined the relationship of barrier function to sphingolipid synthesis. As with sterol and fatty acid syntheses, sphingolipid synthesis was also increased following barrier disruption by acetone or SDS, or in essential fatty acid deficiency (Holleran *et al.*, 1991a). However, in contrast to sterols and fatty acids, wherein an almost immediate increase occurs, the increase in sphingolipid synthesis was delayed, being first observed 6–7 hours following barrier disruption (Holleran *et al.*, 1991a). Similar to observations regarding sterol and fatty acid synthesis, artificial restoration of the barrier with an impermeable membrane restored the rates of sphingolipid synthesis to normal (Holleran *et al.*, 1991a). Together, these observations indicate that barrier requirements globally regulate epidermal lipogenesis, resulting in an increase in each of the principal lipid classes present in the stratum corneum.

4. *Potential Criticisms*

A potential criticism of these findings is that the increase in epidermal lipid synthesis represents a nonspecific injury rather than a specific response to barrier disruption. Against this argument are the following observations: (1) the increase in epidermal lipid synthesis is seen in a number of different models of acute and chronic barrier disruption; (2) the increased synthesis is localized to the epidermis, with no consistent effect on the dermis; (3) the time course of the increased synthesis parallels the disturbance in the barrier; (4) the extent of the increase in epidermal lipid synthesis correlates with the degree of barrier perturbation; (5) the characteristic increase in epidermal lipid synthesis in all of the models can be prevented by artificial restoration of barrier function with a water vapor-impermeable membrane, whereas, conversely, a water vapor-permeable membrane does not block the increase; and finally, (6) in essential fatty acid-deficient animals, when barrier function is restored by treatment with linoleic acid, lipid synthesis is restored to normal in the epidermis. Together, these observations suggest a specific role for barrier function in regulating epidermal lipid synthesis.

Another potential problem in these *in vivo* studies relates to whether the newly synthesized lipid, which localizes to the epidermis, is derived from epidermal lipid synthesis or is delivered to the epidermis via transport from extracutaneous sites. The latter seems unlikely because the transport of lipid to the epidermis is limited (Lipkin *et al.*, 1968). Moreover, following the systemic administration of either labeled cholesterol or stearic acid, the absolute quantities of labeled cholesterol or fatty acid found in the epidermis are relatively small in comparison to the rates of epidermal lipid synthesis (Menon *et al.*, 1985; Grubauer *et al.*, 1987). Furthermore, the quantity of labeled lipids delivered to epidermis with a perturbed barrier is similar to that observed in control epidermis (Menon *et al.*, 1985; Grubauer *et al.*, 1987). Finally, as will be discussed in the next section, barrier disruption stimulates the activity of the enzymes involved in lipid synthesis. Thus, the increased amount of labeled lipid in the epidermis following perturbation of the barrier is not likely to be due to lipid transported from extracutaneous sites.

B. Effect of Barrier Function on Lipid Synthetic Enzymes

1. HMG-CoA Reductase

Recent studies have focused on the changes in specific enzymes that account for the barrier-induced increase in lipid synthesis (see Table III). In most mammalian systems, the enzyme hydroxymethylglutaryl-CoA (HMG-CoA) reductase, which catalyzes the conversion of HMG-CoA to mevalonic acid, is the rate-limiting step in cholesterol biosynthesis (Rodwell *et al.*, 1976). To determine if the conversion of HMG-CoA to mevalonate also is the regulatory step for cholesterol biosynthesis in the epidermis, we compared the *in vitro* incorporation of either [^{14}C]acetate or [^{3}H]mevalonate into sterols. Because acetone treatment resulted in an increase in [^{14}C]acetate incorporation into cholesterol in the epidermis in comparison to the control site, whereas the incorporation of [^{3}H]mevalonate into cholesterol was unchanged, barrier function evidently stimulates cholesterol synthesis at a step prior to mevalonate formation (Proksch *et al.*, 1990).

To ascertain whether barrier function regulates sterol synthesis by affecting HMG-CoA reductase, we measured enzyme activity in several acute models (acetone, SDS, or tape stripping) of barrier disruption and in the chronic, essential fatty acid-deficiency model. In each model, barrier disruption is associated with an increase in HMG-CoA reductase activity (Proksch *et al.*, 1990). The degree of the increase in HMG-CoA reductase activity correlates directly with the extent of

Table III

SUMMARY OF THE EFFECTS OF BARRIER DISRUPTION ON EPIDERMAL LIPID SYNTHESISZING ENZYMES[a]

Model/condition	HMG-CoA reductase activity	HMG-CoA reductase activation state	Serine palmitoyl transferase activity
Acute disruption			
Acetone	↑	↑	↑ (delayed)
SDS	↑	↑	↑ (delayed)
Tape stripping	↑	↑	↑ (delayed)
+ Impermeable membrane (all models)	Normal	Reduced toward normal	Normal
Chronic disruption			
EFAD	↑	↑	↑
EFAD + impermeable membrane	Normal	Normal	Normal

[a]Abbreviations: SDS, sodium dodecyl sulfate; HMG-CoA, hydroxymethylglutaryl coenzyme A; EFAD, essential fatty acid deficiency.

the disturbance in barrier function and returns toward normal as barrier function recovers (Proksch *et al.*, 1990). Moreover, the time course of the increase in HMG-CoA reductase activity after barrier disruption is similar to the time course of the increase in cholesterol synthesis (Proksch *et al.*, 1990). Acetone disruption of the barrier, which is associated with a rapid repair of barrier function and a rapid return toward normal in cholesterol synthesis, leads to a transient increase in HMG-CoA reductase activity. By contrast, SDS treatment provokes a sustained increase in HMG-CoA reductase activity (Proksch *et al* 1990), consistent with both the previously demonstrated slower repair rate and more prolonged increase in cholesterol synthesis (Menon *et al.*, 1985). Most importantly, when barrier function in each of the acute and chronic models is artificially restored by application of a water vapor-impermeable membrane, the increase in epidermal HMG-CoA reductase activity is inhibited (Proksch *et al.*, 1990). These data indicate that increases in HMG-CoA reductase activity account for the increase in epidermal cholesterol synthesis that follows barrier disruption.

HMG-CoA reductase activity is modulated by a reversible phosphorylation–dephosphorylation process, with the phosphorylated form of the enzyme being inactive and the dephosphorylated form active (Gibson and Ingelbritsen, 1978; Beg *et al.*, 1978). Under basal conditions, approximately 50% of epidermal HMG-CoA reductase is in the active form (Proksch *et al.*, 1990), a finding similar to that in other mammalian tissues, with the exception of the liver, in which only 15–30% of the enzyme is active (Brown *et al.*, 1979; Feingold *et al.*, 1983d). Acetone disruption of the barrier produces a marked increase in the activation state of HMG-CoA reductase (Proksch *et al.*, 1990). This increase in activation state occurs almost immediately after acetone treatment, a response that is much

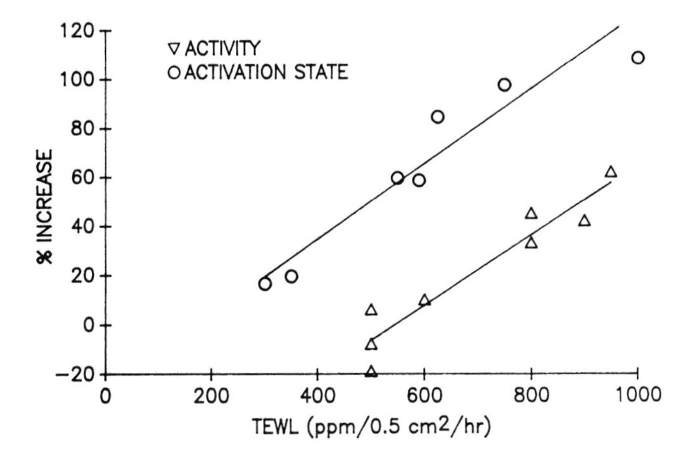

FIG. 1. Correlation between transepidermal water loss (TEWL) and increases in HMG-CoA reductase activity and activation state. The correlation coefficients for reductase activity and activation state are $r = 0.95, p < 0.01$ and $r = 0.96, p < 0.01$, respectively. [Reprinted from *J. Clin. Invest.* **85,** 874 (1990), with permission.]

more rapid than the increase in total enzyme activity (i.e., enzyme content) that occurs following barrier disruption (earliest increase seen at 1.5 hours; see above) (Proksch *et al.*, 1990). Additionally, though the increase in enzyme content returns to normal by 7 hours, the proportion of activated enzyme remains elevated at 15 hours and beyond (Proksch *et al.*, 1990). Moreover, as with enzyme content, the extent of the increase in activation state of HMG-CoA reductase also correlates directly with the degree of disruption of the barrier (Proksch *et al.*, 1990) (Fig. 1). Furthermore, other conditions that disturb barrier function acutely, such as topical SDS treatment and tape stripping, also provoke an increase in the activation state of HMG-CoA reductase, and, again, occlusion with a water vapor-impermeable membrane largely prevents the expected increase in HMG-CoA reductase activation state (Proksch *et al.*, 1990). Finally, in the chronic model of barrier disruption, essential fatty acid deficiency, a marked increase in the proportion of enzyme in the activated state is observed (Proksch *et al.*, 1990).

However, though increases in HMG-CoA reductase content occur only after profound barrier disruption (transepidermal water loss is >550 ppm/0.5 cm^2/hour), changes in activation state occur in association with lesser degrees of barrier disruption (transepidermal water loss is >300 ppm/0.5 cm^2/hour) (Proksch *et al.*, 1990) (Fig. 1). Thus, after extensive barrier disruption, increases in epidermal cholesterol synthesis result from changes both in the amount of enzyme protein and in the activation state. In contrast, with less extensive barrier disruption, changes in cholesterol synthesis are due primarily to alterations in enzyme activation state. Recent studies have suggested that dephosphorylated HMG-CoA reductase is degraded more slowly than the phosphorylated enzyme, and therefore it is possible that these rapid changes in phosphorylation state could also influence the quantity of enzyme present in the epidermis (Parker *et al.*, 1989).

2. *Serine Palmitoyl Transferase*

In all mammalian tissues studied to date, the rate-limiting enzyme of sphingolipid synthesis is serine palmitoyl transferase, which catalyzes the initial and first committed step in sphingolipid synthesis (Merrill and Jones, 1990). Serine palmitoyl transferase is a microsomal enzyme that catalyzes the condensation of palmitoyl-CoA and serine to form 3-ketodihydrosphingosine. Very high levels of serine palmitoyl transferase activity have been found in both cultured keratinocytes and murine epidermis (Holleran *et al.*, 1990).

Very recent studies have demonstrated that the activity of serine palmitoyl transferase increases following disruption of the barrier with either acetone, SDS, or tape stripping, and in essential fatty acid deficiency (Holleran *et al.*, 1991b). Moreover, the time course of the increase in serine palmitoyl transferase activity parallels the increase in sphingolipid synthesis described above (Holleran *et al.*, 1991b). Serine palmitoyl transferase activity is maximally increased 6 hours after barrier disruption (Holleran *et al.*, 1991b). Furthermore, the increase in serine palmitoyl transferase activity can be prevented by occlusion with a vapor-impermeable membrane (Holleran *et al.*, 1991b). These results suggest that increases in the activity of serine palmitoyl transferase account for the changes in sphingolipid synthesis following barrier disruption.

C. Localization of the Increase in Cholesterol Synthesis Within the Epidermis

As discussed above, in adult mice, the major contributor to epidermal cholesterol synthesis is the lower epidermis. However, the stratum granulosum also possesses considerable lipid biosynthetic capacity (Feingold *et al.*, 1983a; Monger *et al.*, 1988), despite the fact that most other synthetic activity is absent or in decline. Depending on the separation method, approximately 25% of total epidermal cholesterol synthesis is localized to the upper epidermis (stratum granulosum and stratum corneum), with approximately 75% in the lower epidermis (stratum basale and stratum spinosum) (Feingold *et al.*, 1983a). When the distribution of epidermal HMG-CoA reductase is compared to the localization data for cholesterol synthesis, enzyme activity and synthetic activity are strikingly similar (Proksch *et al.*, 1991a). Moreover, the activation state of HMG-CoA reductase in the upper epidermis is 46%, which is similar to results seen in the entire epidermis (Proksch *et al.*, 1991a). Because the activation state of HMG-CoA reductase can only be accurately assessed in very rapidly processed tissues, it has not been possible to determine the activation state in the lower epidermis, which requires a

Table IV

SUMMARY OF THE EFFECT OF BARRIER DISRUPTION ON HMG-COA REDUCTASE ACTIVITY IN UPPER AND LOWER EPIDERMIS[a]

Model/condition	HMG-CoA reductase activity	
	Lower epidermix (SS/SB)	Upper epidermis (SC/SG)
Acute disruption		
Acetone	↑	Normal
SDS	↑	Normal
+ Impermeable membrane (both models)	Normal	Normal
Chronic disruption		
EFAD	↑ (but due to hyperplasia)	↑
EFAD + impermeable membrane	Normal	Normal
Long-term occlusion		
Normal animal	Normal	↓

[a]Abbreviations: HMG-CoA, hydroxymethylglutaryl coenzyme A; SDS, sodium dodecyl sulfate; EFAD, essential fatty acid deficiency; SC, stratum corneum; SG, stratum granulosum; SS, stratum spinosum; SB, stratum basale.

prolonged *in vitro* incubation step for isolation. However, given the results in the entire epidermis and in the upper epidermis, one can surmise that the basal activation state in the lower epidermis is also in the range of 40–50% of HMG-CoA reductase in the active form (see Table IV).

Acute disruption of the barrier with either topical acetone or SDS treatment results in an increase of HMG-CoA reductase activity in the lower epidermis that is due to an increase in enzyme content (Proksch *et al.*, 1991a). In contrast, enzyme content is unchanged in the upper epidermis (Proksch *et al.*, 1991a). However, the activation state of HMG-CoA reductase in the upper epidermis increases significantly after acetone treatment (Proksch *et al.*, 1991a). As noted above, our isolation technique does not allow determination of the activation state of HMG-CoA reductase in the lower epidermis, but based on the increased activation state in the total epidermis and our results in the upper epidermis, it is highly likely that barrier function also influences the activation state in the lower epidermis. Moreover, occlusion with an impermeable membrane attenuates both the increase in total enzyme content in the lower epidermis and the increase in activation state in the upper epidermis (Proksch *et al.*, 1991a). These results suggest that acute perturbations in barrier function stimulate an increase in the quantity of HMG-CoA reductase in the lower epidermis as well as modulating the activation state in the upper epidermis, and presumably in the lower epidermis as well. The absence of an increase in enzyme quantity in the upper epidermis may be due to the limited capacity of the stratum granulosum for new protein synthesis. The increase in

HMG-CoA reductase activity in the upper epidermis could provide a source of lipid for the reformation of the lipid permeability barrier (see below), whereas the increase in enzyme activity in the lower epidermis could provide (1) cholesterol and other products derived from mevalonate, which are required for cell proliferation (see below), and/or (2) replenish the lipid pool that is depleted immediately after barrier disruption. Whether lipids can be transported from the lower epidermis to the upper epidermis is unknown.

Chronic disruption of the barrier (essential fatty acid deficiency) results in an increase in HMG-CoA reductase activity in both the upper and lower epidermis (Proksch *et al.*, 1991a). However, on a per milligram protein basis, the increase in enzyme activity is restricted to the upper epidermis, whereas the increase in total activity in the lower epidermis can be attributed to the tissue hyperplasia that accompanies essential fatty acid deficiency rather than an actual increase in enzyme specific activity (Proksch *et al.*, 1991a). Artificial restoration of barrier function in essential fatty acid-deficient animals results in a normalization of HMG-CoA reductase activity in the upper epidermis and a decrease in HMG-CoA reductase activity on a surface area basis in the lower epidermis, which can be attributed to the partial reversal of epidermal hyperplasia by occlusion (Proksch *et al.*, 1991a). Thus, though HMG-CoA reductase activity is increased in both the upper and lower epidermis of essential fatty acid-deficient animals, different mechanisms account for these changes. In the upper epidermis, stimulation of HMG-CoA reductase activity can be attributed to barrier requirements, whereas in the lower epidermis, these changes primarily reflect epidermal hyperplasia.

Finally, because long-term occlusion with an impermeable membrane reduces HMG-CoA reductase activity in the upper epidermis of the skin of normal mice, with little effect on enzyme activity in the lower epidermis, the activity of HMG-CoA reductase evidently can be modulated in the upper epidermis by barrier requirements, even in normal animals (Proksch *et al.*, 1991a).

VI. Importance of Epidermal Lipid Synthesis for Barrier Repair

A. Relationship of Lipid Synthesis to Barrier Repair

Topical treatment with organic solvents or detergents extracts lipids from the stratum corneum, thereby resulting in the disruption of barrier function (Scheuplein and Blank, 1971; Grubauer *et al.*, 1989a). Almost immediately following barrier disruption, epidermal lamellar bodies, small secretory organelles enriched in lipids present in the outer nucleated cell layers of the epidermis (see Elias and Menon, this volume), are secreted into the intercellular spaces (Feingold *et al.*, 1990). Over time, the content of lipids in the stratum corneum is restored to normal levels in parallel with the return of barrier function (Grubauer *et al.*, 1989a;

Table V
SUMMARY OF THE EFFECTS OF OCCLUSION OR LIPID SYNTHESIS INHIBITORS[a]

Model/condition	Lipid synthesis	Stratum corneum lipid content	Barrier function
Following acetone disruption			
Impermeable membrane	Decreased	Decreased	No recovery
Permeable membrane	Increased	Normal	Normal recovery
Lovastatin	Decreased CHOL	Decreased CHOL	Delayed recovery
β-Chloroalanine	Decreased SL	Decreased SL	Delayed recovery, particularly the late phase
Chronic barrier dysfunction			
Repeated lovastatin — intact skin	Increased fatty acids	Increased fatty acids	Abnormal barrier
EFAD	Increased	Decreased[b]	Abnormal barrier

[a]Abbreviations: CHOL, cholesterol; SL, sphingolipid; EFAD, essential fatty acid deficiency.
[b]Intercellular lipid (Elias and Brown, 1978).

Feingold *et al.*, 1990). As discussed above, an increase in epidermal lipid synthesis is associated with this restoration of stratum corneum lipids and function. If this increase in epidermal lipid synthesis is inhibited by occlusion with a water vapor-impermeable membrane, the usual restoration of lipids in the stratum corneum and barrier recovery is prevented (Grubauer *et al.*, 1989b). In acetone-treated animals covered with a water vapor-impermeable membrane for 48 hours, transepidermal water loss does not improve, whereas in acetone-treated animals exposed to air, transepidermal water loss returns to base line by 24–48 hours (Grubauer *et al.*, 1989b). In parallel with these changes in function, the lipid content of the stratum corneum, as assessed by histochemical staining or biochemical lipid analysis, does not return to normal in animals covered with a water vapor-impermeable wrap (Grubauer *et al.*, 1989b). When acetone-treated animals are covered with a water vapor-permeable membrane, which does not inhibit epidermal lipid synthesis, both the return of lipid to the stratum corneum and the recovery of barrier function occur normally (Grubauer *et al.*, 1989b). These observations provide a direct link between local epidermal lipid synthesis, lipid reaccumulation, and barrier repair (see Table V).

B. EFFECT OF PHARMACOLOGIC INTERVENTIONS ON BARRIER REPAIR

To further link modulations in the synthesis of epidermal lipids with barrier repair, we also have utilized specific inhibitors of cholesterol and sphingolipid synthesis. Following the disruption of the barrier with topical acetone treatment, the normal rapid recovery of barrier function is impaired in animals treated topically with lovastatin (Feingold *et al.*, 1990), a competitive inhibitor of the enzyme HMG-CoA reductase (Alberts *et al.*, 1980). A single topical application of lovastatin leads to a marked inhibition in epidermal and dermal cholesterol synthesis, with inhibition of barrier recovery by as much as 50% at early time points (Feingold *et al.*, 1990). With time, the lovastatin-induced inhibition of cholesterol synthesis decreases and the inhibition of barrier recovery lessens such that by 6 hours barrier recovery is reduced by approximately 20% (Feingold *et al.*, 1990). Both histochemical staining with filipin, a macrolide antibiotic that fluoresces after binding to the 3β-OH group of free sterols, and direct biochemical analysis demonstrate that the return of free sterols is impaired; in contrast, other lipid classes return at normal rates to the stratum corneum in the lovastatin-treated animals (Feingold *et al.*, 1990). That this inhibition of barrier recovery by lovastatin is due to a specific block in cholesterol synthesis and not due to nonspecific toxicity is shown by the reversal of this inhibition by simultaneous topical treatment with either mevalonate, the immediate product of HMG-CoA reductase, or cholesterol, the final end product of the sterol biosynthetic pathway (Feingold *et al.*, 1990). These results provide direct evidence that epidermal cholesterol synthesis plays an important role in barrier repair. Additionally, they indicate that cholesterol is an essential component of the stratum corneum lipid membrane bilayers that are required for normal barrier function.

Repeated daily topical applications of lovastatin also produce a marked perturbation in barrier function resulting in transepidermal water losses greater than 500 ppm/0.5 cm^2/hour (Feingold *et al.*, 1991). As discussed above (see also Elias and Menon, this volume), lovastatin treatment leads to an initial inhibition of HMG-CoA reductase activity that results in a selective decrease in cholesterol synthesis (Feingold *et al.*, 1990). However, with time, homeostatic mechanisms lead to an increase in the quantity of HMG-CoA reductase in the epidermis such that, despite the presence of the inhibitor, cholesterol synthesis is restored to normal (Feingold *et al.*, 1991). This compensatory response to lovastatin-induced inhibition of cholesterol synthesis has been reported in a number of other experimental systems both *in vitro* and *in vivo* (Brown *et al.*, 1978; Kita *et al.*, 1980; Li *et al.*, 1988). As expected, given the normalization of cholesterol synthesis, the barrier defect in the chronically lovastatin-treated animals is not associated with a decrease in stratum corneum cholesterol content (Feingold *et al.*, 1991).

The mechanism accounting for the barrier perturbation in animals treated topically with lovastatin for several days is not known with certainty (Fig. 2), but our studies have demonstrated that accompanying the up-regulation of epidermal HMG-CoA reductase is an increase in epidermal fatty acid synthesis (Feingold *et al.*, 1991). This stimulation of fatty acid synthesis in response to lovastatin and other HMG-CoA reductase inhibitors also has been observed in other experimental models, but the explanation for this increase is unknown (Endo *et al.*, 1977; Williams *et al.*, 1987b; Menon *et al.*, 1989a,b; Mosley *et al.*, 1989). This increase

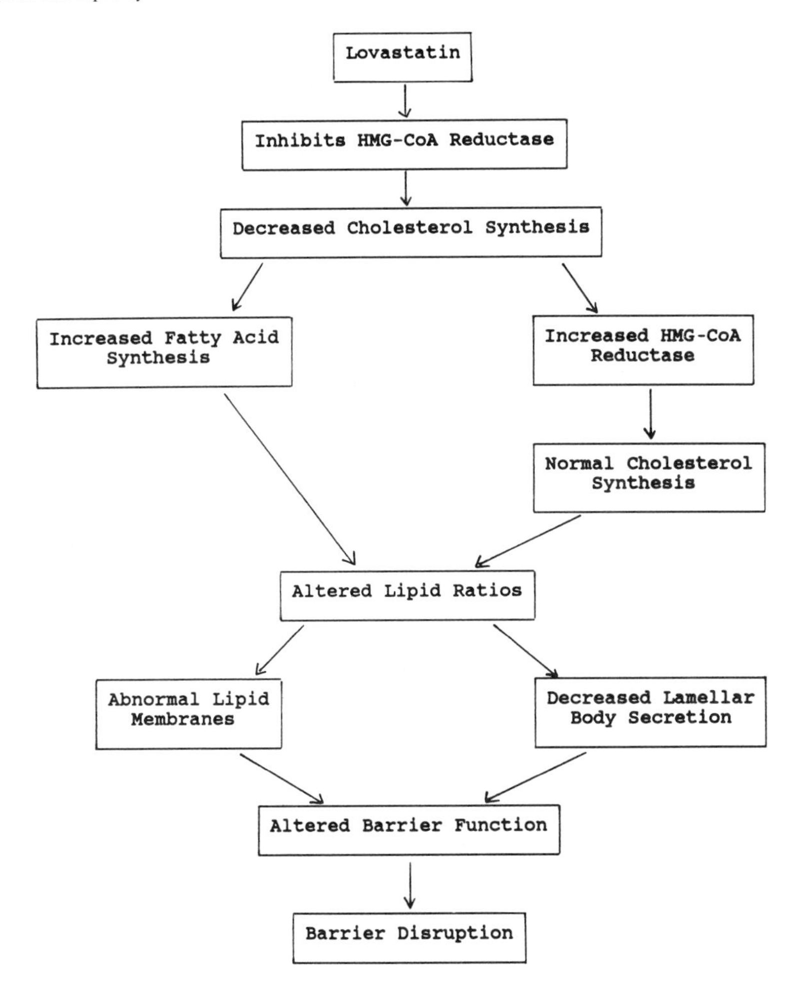

FIG. 2. Speculative diagram of the mechanism by which chronic application of topical lovastatin disrupts barrier function.

in epidermal fatty acid synthesis results in an altered epidermal fatty acid:free sterol ratio that coincides with the appearance of abnormal barrier function (Feingold *et al.*, 1991). These altered lipid ratios could directly affect barrier function by resulting in abnormal intercellular membranes, a phenomenon that has been demonstrated in model lipid systems (McKersie *et al.*, 1989). Alternatively, the abnormal lipid ratios could affect the structure of lamellar bodies, resulting in defective secretion and function. We have observed that in animals chronically treated with topical lovastatin, the lamellar bodies appear abnormal and some lamellar bodies are retained within the corneocyte cytosol, suggesting impaired secretion (Feingold *et al.*, 1991). This impaired secretion would result in a maldistribution of lipid (i.e., less intercellular and more intracellular lipid in the stratum corneum), which could adversely effect barrier function. Regardless of the precise mechanism, it is apparent that chronic lovastatin treatment impairs barrier function by altering lipid ratios rather than by simply inhibiting cholesterol synthesis and causing cholesterol deficiency.

Analogous inhibitor studies have been carried out recently using topically applied β-chloroalanine, a suicide inhibitor of serine palmitoyl transferase (Medlock and Merrill, 1988), the key enzyme in sphingolipid synthesis (Merrill and Jones, 1990). Inhibition of sphingolipid synthesis also slows the recovery of barrier function after acetone treatment, but in contrast to the effects of topical lovastatin, this inhibition of recovery occurs only during the late phases of repair (i.e., 12–24 hours) (Holleran *et al.*, 1991b). Associated with the decrease in barrier recovery is a decreased quantity of sphingolipids in the stratum corneum of β-chloroalanine-treated animals (Holleran *et al.*, 1991b). Moreover, as with lovastatin, this inhibition of barrier repair can be reversed by simultaneous topical applications of ceramides, the major sphingolipid product of serine palmitoyl transfer present in the stratum corneum (Holleran *et al.*, 1991b). The reversal of the β-chloroalanine effect with topical ceramides not only demonstrates the specificity of the inhibitor but also indicates that the inhibition of barrier repair is not a nonspecific toxic effect. Why inhibition of sphingolipid synthesis predominantly affects the late phase of barrier repair is unclear, but it may reflect the presence of adequate reservoirs of stored sphingolipids, which suffice for the synthesis of new lamellar bodies and the formation of new intercellular membranes. Alternatively, it is possible that sphingolipids are not required for the early stages of barrier recovery. These results demonstrate that sphingolipid synthesis also is required for barrier repair. Together with the lovastatin data these pharmacologic interventions provide independent verification that epidermal lipid synthesis plays a key role in barrier homeostasis.

VII. Signals Initiating Barrier Repair

As noted above, if after acute disruption of the barrier with acetone, animals are covered with a water vapor-impermeable membrane, the increase in epidermal lipid synthesis is inhibited, the return of stratum corneum lipids is prevented, and barrier recovery does not occur (Grubauer *et al.*, 1989b). In contrast, if comparably treated animals are covered with a water vapor-permeable membrane, epidermal lipid synthesis is increased, stratum corneum lipid content returns, and barrier function recovers (Grubauer *et al.*, 1989b). These results suggest that transcutaneous water loss may be an important signal initiating the repair of the barrier. However, if following acute barrier perturbation with acetone, the disrupted site is immersed in either an isoosmolar sucrose solution or an isotonic sodium chloride solution, both of which should decrease net water transit, barrier recovery proceeds normally (S. H. Lee *et al.*, unpublished observations). Similarly, varying the osmolarity of the immersion solution from hypotonic (distilled water) to hypertonic (560 mOsm/kg), which would affect the rate and direction of net water flux, also does not alter barrier repair (S. H. Lee *et al.*, unpublished observations). These observations indicate that water transit per se is unlikely to be the crucial signal that stimulates barrier recovery. Very recent preliminary studies by our laboratory have demonstrated that immersion in calcium and/or potassium solutions inhibits barrier recovery, suggesting that these ions may play an important role in signaling barrier repair (S. H. Lee *et al.*, unpublished observations).

VIII. Local Regulation of Epidermal Lipid Synthesis—Potential Role of Cell Proliferation

The epidermis is a rapidly replicating tissue, continually replacing cells that are lost during desquamation. Numerous studies in other tissues have demonstrated that lipid synthesis is required for cell replication (Siperstein, 1984). Thus, one can speculate that, in addition to barrier homeostasis, epidermal lipid synthesis also is required for cell division. However, to date, no studies have examined the influence of inhibition of epidermal lipid synthesis on epidermal cell replication. Conversely, the effect of stimulating or inhibiting cell division on epidermal lipid synthesis has also not been extensively examined. One would predict in a tissue such as the epidermis, where most of the lipid appears to be made *in situ*, that an increase in cell division would stimulate epidermal lipid synthesis. However, the relationship between epidermal lipid synthesis and proliferation may be more complex. Recent studies by our group have extended the observations of other investigators (Fisher and Maibach, 1972a,b) and have demonstrated that acute (acetone or tape stripping) and chronic (essential fatty acid deficiency or chronic lovastatin administration) perturbations of barrier function are associated with increases in epidermal DNA synthesis (Proksch *et al.*, 1991b). The extent of the increase in epidermal DNA synthesis correlates with the degree of barrier disruption (Proksch *et al.*, 1991b). Moreover, artificial restoration of barrier function with a vapor-impermeable membrane in these models inhibits the increase in DNA synthesis (Proksch *et al.*, 1991b). These results indicate that barrier function, in addition to regulating epidermal lipid synthesis, also influences epidermal cell replication. It is possible that the increases in both epidermal lipid synthesis and cell division in response to barrier perturbations represent parallel responses aimed at barrier repair. How and if these responses are coordinated and the role of cell division in barrier repair remain to be elucidated. An increase in cell proliferation potentially could provide additional cells, which in turn could contribute additional lipids required for total restoration of the barrier.

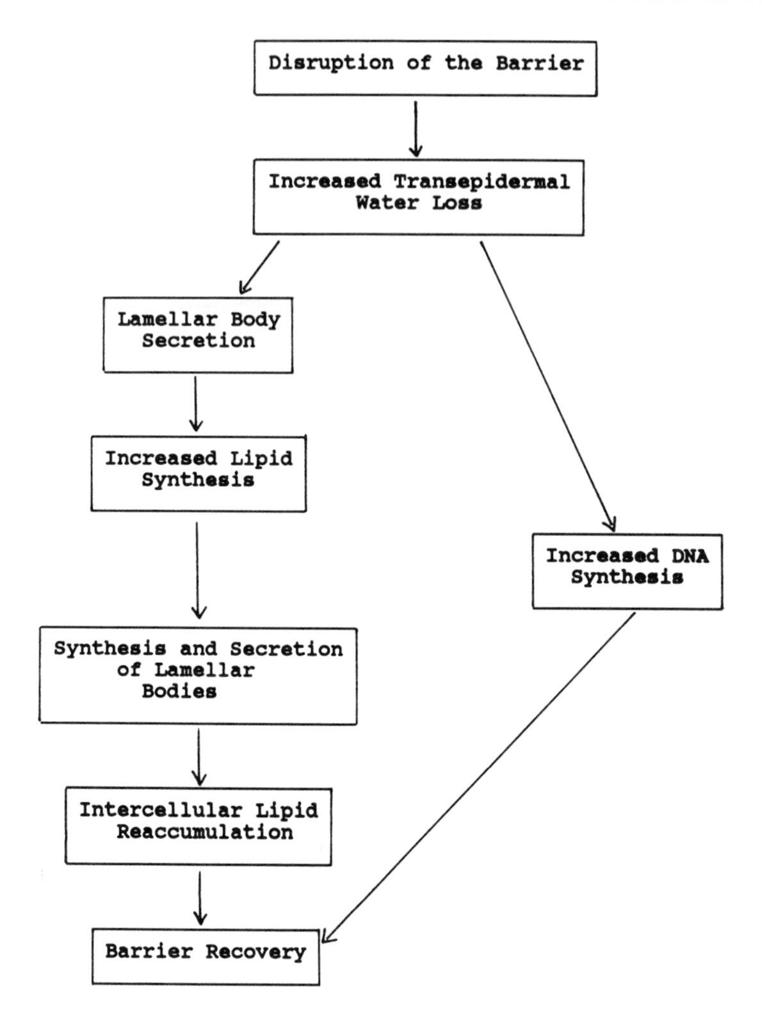

FIG. 3. Speculative diagram of the effect of barrier disruption on epidermal metabolism.

IX. Summary

One of the key functions of the epidermis is to form a barrier between the organism and the outside world. As shown in Fig. 3, disruptions of the barrier result in a cascade of events that ultimately leads to barrier repair. The initial signal that initiates this repair response is unknown. The exocytosis of preformed lipid-enriched lamellar bodies is the first step in this response, which is followed by an increase in lipid synthesis in the epidermis. Our studies demonstrate that this increase in epidermal lipid synthesis is required for the synthesis of new lamellar bodies and repair of the barrier. Inhibition of epidermal lipid synthesis by artificial membranes or drugs impairs barrier recovery by preventing the reformation of lamellar bodies and the continued secretion of lipid. Whether the stimulation of lipid synthesis is primarily regulated by disturbances in barrier function or secondarily by decreases in the lipid content of the cells due to the utilization of lipid for the formation of lamellar bodies is unknown. Additionally, the precise mechanisms by which lipid synthesis is increased (enzyme activation, transcriptional regulation, etc.) remain to be elucidated. The secretion of lipid-containing lamellar bodies results in the reaccumulation of lipid in the intercellular spaces of the stratum corneum and the recovery of normal barrier function. Epidermal lipid synthesis also is probably required to provide lipid for new cell membrane formation to allow for the increase in epidermal cell proliferation, which is stimulated following barrier disruption. Additionally, epidermal lipid synthesis may provide regulatory molecules or crucial substrates that are required for DNA synthesis. Thus, epidermal lipid synthesis plays a key role in the major biological functions of the epidermis, the cutaneous permeability barrier, and cell proliferation.

References

Alberts, A. W., Chen, J., Kuron, G., Hunt, V., Huff, I., Hoffman, C., *et al.* (1980). *Proc. Natl. Acad. Sci. U.S.A.* **77,** 3957–3961.

Anderson, J. M., and Dietschy, J. M. (1976). *Science* **193,** 903–905.

Anderson, J, M., and Dietschy, J. M. (1977a). *J. Biol. Chem.* **252,** 3646–3651.

Anderson, J. M., and Dietschy, J. M. (1977b). *J. Biol. Chem.* **252,** 3652–3659.

Anderson, J. M., and Dietschy, J. M. (1979). *J. Lipid Res.* **20,** 740–752.

Beg, Z. H., Shonik, J. A., and Brewer, H. B., Jr. (1978). *Proc. Natl. Acad. Sci. U.S.A.* **75,** 3678–3682.

Blank, I. H. (1987). *In* "Dermatology in General Medicine" (T. B. Fitzpatrick, A. Z. Eisen, K. Wolff, I. M. Freedberg, and K. R. Austin, eds.), Chap. 28. McGraw-Hill, New York.

Brooks, S. C., Godefroi, V. C., and Simpson, W. L. (1966). *J. Lipid Res.* **7,** 95–102.

Brown, M. S., Faust, J. R., Goldstein, J. L., Kaneko, I., and Endo, A. (1978). *J. Biol. Chem.* **253,** 1121–1128.

Brown, M. S., Goldstein, J. L., and Dietschy, J. M. (1979). *J. Biol. Chem.* **254,** 5144–5149.

Chapkin, R. S., and Ziboh, V. A. (1984). *Biochem. Biophys. Res. Commun.* **124,** 784–792.

Dayton, S., Dayton, J., Drimmer, F., and Kendall, F. E. (1954). *Circulation* **10,** 595.

Dietschy, J. M., and Brown, M. S. (1974). *J. Lipid Res.* **15,** 508–516.

Dietschy, J. M., and McGarry, J. D. (1974). *J. Biol. Chem.* **249,** 52–58.

Dietschy, J. M., and Siperstein, M. D. (1967). *J. Lipid Res.* **8,** 97–104.

Dietschy, J. M., and Spady, D. K. (1984). *J. Lipid Res.* **25,** 1469–1476.

Dietschy, J. M., Spady, D. K., and Stange, E. F. (1983). *Biochem. Soc. Trans.* **11,** 639–641.

Dykes, P. J., Marks, R., Davies, M. G., and Reynolds, D. J. (1978). *J. Invest. Dermatol.* **70,** 126–129.

Elias, P. M. (1983). *J. Invest. Dermatol.* **80,** 44–49.

Elias, P. M., and Brown, B. E. (1978). *Lab. Invest.* **39,** 574–583.

Elias, P. M., and Feingold, K. R. (1988). *Ann. N.Y. Acad. Sci.* **548,** 4–13.

Elias, P. M., Feingold, K. R., Menon, G. K., Grayson, S., Williams, M. L., and Grubauer, G. (1987). *In* "Pharmacology and the Skin" (H. Schaeffer and B. Shroot, eds.), pp. 1–8. Karger, Basel.

Endo, A., Tsujita, Y., Kuroda, M., and Tanzawa, K. (1977). *Eur. J. Biochem.* **77,** 31–36.

Feingold, K. R., Wiley, M. H., MacRae, G., Moser, A. H., Lear, S. R., and Siperstein, M. D. (1982a). *Diabetes* **31,** 388–395.

Feingold, K. R., Wiley, M. H., Moser, A. H., Lear, S. R., and Siperstein, M. D. (1982b). *J. Lab. Clin. Med.* **100,** 405–410.

Feingold, K. R., Brown, B. E., Lear, S. R., Moser, A. H., and Elias, P. M. (1983a). *J. Invest. Dermatol.* **81,** 365–369.

Feingold, K. R., MacRae, G., Moser, A. H., Wu, J., Siperstein, M. D., and Wiley, M. H. (1983b). *Endocrinology (Baltimore)* **112,** 96–103.

Feingold, K. R., Wiley, M. H., MacRae, G., Lear, S. R., Moser, A. H., Zsigmond, G., and Siperstein, M. D. (1983c). *Metab. Clin. Exp.* **32,** 75–81.

Feingold, K. R., Wiley, M. H., Moser, A. H., and Siperstein, M. D. (1983d). *Arch. Biochem. Biophys.* **226,** 231–241.

Feingold, K. R., Brown, B. E., Lear, S. R., Moser, A. H., and Elias, P. M. (1986). *J. Invest. Dermatol.* **87,** 588–591.

Feingold, K. R., Williams, M. L., Pillai, S., Menon, G. K., Halloran, B. P., Bikle, D. D., and Elias, P. M. (1987). *Biochim. Biophys. Acta* **930,** 193–200.

Feingold, K. R., Mao-Quiang, M., Menon, G. K., Cho, S. S., Brown, B. E., and Elias, P. M. (1990). *J. Clin. Invest.* **86,** 1738–1745.

Feingold, K. R., Mao-Quiang, M., Proksch, E., Menon, G. K., Brown, B. E., and Elias, P. M. (1991). *J. Invest. Dermatol.* **96,** 201–209.

Fisher, L. B., and Maibach, H. I. (1972a). *Br. J. Dermatol.* **86,** 593–600.

Fisher, L. B., and Maibach, H. I. (1972b). *J. Invest. Dermatol.* **59,** 106–108.

Gandemer, G., Durand, G., and Pascal, G. (1983). *Lipids* **18,** 223–228.

Gaylor, J. L. (1961). *Proc. Soc. Exp. Biol. Med.* **106,** 576–579.

Gibson, D. M., and Ingelbritsen, T. S. (1978). *Life Sci.* **23,** 2649–2664.

Griesemer, R. D., and Thomas, R. W. (1963). *J. Invest. Dermatol.* **41,** 235–238.

Grove, G. L. (1979). *Int. J. Dermatol.* **18,** 111–122.

Grubauer, G., Feingold, K. R., and Elias, P. M. (1987). *J. Lipid Res.* **28,** 746–752.

Grubauer, G., Feingold, K. R., and Elias, P. M. (1989a). *J. Lipid Res.* **30,** 89–96.

Grubauer, G., Elias, P. M., and Feingold, K. R. (1989b). *J. Lipid Res.* **30,** 323–334.

Havel, R. J., Goldstein, J. L., and Brown, M. S. (1980). *In* "Metabolic Control and Disease" (P. K. Bondy and L. E. Rosenberg, eds.), pp. 393–494. Saunders, Philadelphia, Pennsylvania.

Holleran, W. M., Williams, M. L., Gao, W. N., and Elias, P. M. (1990). *J. Lipid Res.* **31,** 1655–1661.

Holleran, W. M., Feingold, K. R., Mao-Quiang, M., Gao, W. N., Lee, J. M., and Elias, P. M. (1991a). *J. Lipid Res.* (in press).

Holleran, W. M., Mao-Quiang, M., Gao, W. N., Menon, G. K., Elias, P. M., and Feingold, K. R. (1991b). *J. Clin. Invest.* (in press).

Hsia, S. L., Dreize, M. A., and Marquez, M. D. (1966). *J. Invest. Dermatol.* **47,** 443–448.

Hsia, S. L., Fulton, J. E., Jr., Fulghum, D., and Buch, M. M. (1970). *Proc. Soc. Exp. Biol. Med.* **135,** 285–291.

Jeske, D. J., and Dietschy, J. M. (1980). *J. Lipid Res.* **21,** 364–376.

Kita, T., Brown, M. S., and Goldstein, J. L. (1980). *J. Clin. Invest.* **66,** 1094–1100.

Kritchevsky, D. (1960). *Metab., Clin. Exp.* **9,** 884–994.

Lakshmanan, M. R., and Veech, R. L. (1977). *J. Biol. Chem.* **252,** 4667–4673.

Li, C. A., Tanaka, R. D., Callaway, K., Fogelman, A. M., and Edward, P. A. (1988). *J. Lipid Res.* **29,** 781–796.

Lipkin, G., Wheatley, V. R., Woo, T. H., and March, C. (1968). *J. Invest. Dermatol.* **50,** 456–458.
Lowenstein, J. M., Brunengraber, H., and Wadke, M. (1975). *Methods Enzymol.* **35,** 279–287.
McCullough, J. L., Schreiber, S. H., and Ziboh, V. A. (1978). *J. Invest. Dermatol.* **70,** 318–320.
McKersie, B. D., Crowe, J. H., and Crowe, L. M. (1989). *Biochim. Biophys. Acta* **982,** 156–160.
Medlock, K. A., and Merrill, A. H., Jr. (1988). *Biochemistry* **27,** 7079–7084.
Menon, G. K., Feingold, K. R., Moser, A. H., Brown, B. E., and Elias, P. M. (1985). *J. Lipid Res.* **26,** 418–427.
Menon, G. K., Placzak, D., Hincenbergs, M., and Williams, M. L. (1989a). *J. Invest. Dermatol.* **92,** 480A.
Menon, G. K., Placzak, D., Hincenbergs, M., and Williams, M. L. (1989b). *Clin. Res.* **37,** 620A.
Merrill, A. H., Jr., and Jones, D. D. (1990). *Biochim. Biophys. Acta* **1044,** 1–12.
Miller, C. C., Ziboh, V. A., Wong, T., and Fletcher, M. P. (1990). *J. Nutr.* **120,** 36–44.
Mommaas-Kienhuis, A. M., Grayson, S., Wijsman, M. C., Vermeer, B. J., and Elias, P. M. (1987). *J. Invest. Dermatol.* **89,** 513–517.
Monger, D. J., Williams, M. L., Feingold, K. R., Brown, B. E., and Elias, P. M (1988). *J. Lipid Res.* **29,** 603–612.
Mosley, S. T., Kalinowski, S. S., Schafer, B. L., and Tanaka, R. D. (1989). *J. Lipid Res.* **30,** 1411–1420.
Nicolaides, N., and Rothman, S. (1955). *J. Invest. Dermatol.* **24,** 125–129.
Parker, R. A., Miller, S. J., and Gibson, D. M. (1989). *J. Biol. Chem.* **264,** 4877–4887.
Patterson, J. F., and Griesemer, R. D. (1959). *J. Invest. Dermatol.* **33,** 281–285.
Ponec, M., Havekes, L., Kempenaar, J., and Vermeer, B. J. (1983). *J. Invest. Dermatol.* **81,** 125–130.
Ponec, M., Havekes, L., Kempenaar, J., Levrijsen, S., and Vermeer, B. J. (1984). *J. Invest. Dermatol.* **83,** 436–440.
Proksch, E., Elias, P. M., and Feingold, K. R. (1990). *J. Clin. Invest.* **85,** 874–882.
Proksch, E., Elias, P. M., and Feingold, K. R. (1991a). *Biochim. Biophys. Acta* (to be published).
Proksch, E., Feingold, K. R., Mao-Quiang, M., and Elias, P. M. (1991b). *J. Clin. Invest.* **87,** 1668–1673.
Prottey, C. (1976). *Br. J. Dermatol.* **94,** 549–587.
Prottey, C., Hartop, P. J., and Ferguson, T. F. M. (1972). *Br. J. Dermatol.* **87,** 586–607.
Rodwell, V. W., Nordstrom, K. C., and Mitschelen, J. H. (1976). *Adv. Lipid Res.* **14,** 1–74.
Rosenberg, R. M., Isseroff, R. R., Ziboh, V. A., and Huntley, A. C. (1986). *J. Invest. Dermatol.* **86,** 244–248.
Scheuplein, R. J., and Blank, I. H. (1971). *Physiol. Rev.* **51,** 701–747.
Siperstein, M. D. (1984). *J. Lipid Res.* **25,** 1462–1468.
Spady, D. K., and Dietschy, J. M. (1983). *J. Lipid Res.* **24,** 303–315.
Srere, P. A., Chaikoff, I. L., Tretman, S. S., and Burstein, L. S. (1950). *J. Biol. Chem.* **182,** 629–634.
Steinberg, D., Herndon, J. H., Uhlendorf, B. W., Mize, C. E., Avigan, J., and Milne, G. W. A. (1967). *Science* **156,** 1740–1742.
Summerly, R., and Woodbury, S. (1971). *Br. J. Dermatol.* **85,** 424–431.
Van Bruggen, J. T., Yamada, P., Hutchens, T. T., and West, E. S. (1954). *J. Biol. Chem.* **209,** 635–640.
Wertz, P. W., and Downing, D. T. (1982). *Science* **217,** 1261–1262.
Wertz, P. P., Downing, D. T., Frienkel, R. K., and Traczyk, T. N. (1984). *J. Invest. Dermatol.* **83,** 193–195.
Wheatley, V. R., Hodgins, L. T., Coon, W. M., Kumarasiri, M., Berenzweig, H., and Feinstein, J. M. (1971). *J. Lipid Res.* **12,** 347–360.
Wilkinson, D. I. (1973). *J. Invest. Dermatol.* **60,** 188–192.
Williams, M. L., Mommas-Kienhuis, A. M., Rutherford, S. I., Grayson, S., Vermeer, B. J., and Elias, P. M. (1987a). *J. Cell. Physiol.* **132,** 428–440.

Williams, M. L., Rutherford, S. L., and Feingold, K. R. (1987b). *J. Lipid Res.* **28,** 955–967.
Wong, R. K. L., and Van Bruggen, J. T. (1960). *J. Biol. Chem.* **235,** 26–29.
Yardley, H. J., and Summerly, R. (1981). *Pharmacol. Ther.* **13,** 357–383.
Ziboh, V. A., and Hsia, S. L. (1969). *Arch. Biochem. Biophys.* **131,** 153–162.
Ziboh, V. A., and Miller, C. C. (1990). *Annu. Rev. Nutr.* **10,** 433–450.
Ziboh, V. A., Dreize, M. A., and Hsia, S. L. (1970). *J. Lipid Res.* **11,** 346–354.

Lipid Metabolism in Cultured Keratinocytes

MARIA PONEC

Department of Dermatology
University Hospital Leiden
2300 RC Leiden, The Netherlands

I. Introduction

Epidermal differentiation is a process during which keratinocytes undergo marked morphologic and biochemical changes: they leave the stratum basale and move upward through the stratum spinosum and granulosum and end up in the uppermost layer—the stratum corneum—consisting of anucleate corneocytes surrounded by extracellular lipids organized in multilamellar structures. Their composition and organization into membrane bilayers have been shown to play an essential role in the barrier function of the stratum corneum (Elias, 1983; Elias and Friend, 1975). The finding that the composition of the stratum corneum lipids markedly differs from that found in the lower layers of the epidermis (Gray and Yardley, 1975; Long, 1970; Elias *et al.*, 1979; Yardley and Summerly, 1981; Lampe *et al.*, 1983a,b, Wertz *et al.*, 1987) raises two important questions: (1) How is such lipogenesis related to the stage of epidermal differentiation? (2) How is lipogenesis regulated in relation to epidermal function?

Recently developed *in vitro* systems for the cultivation of human epidermal keratinocytes offer an attractive alternative to using the human skin. The extent of

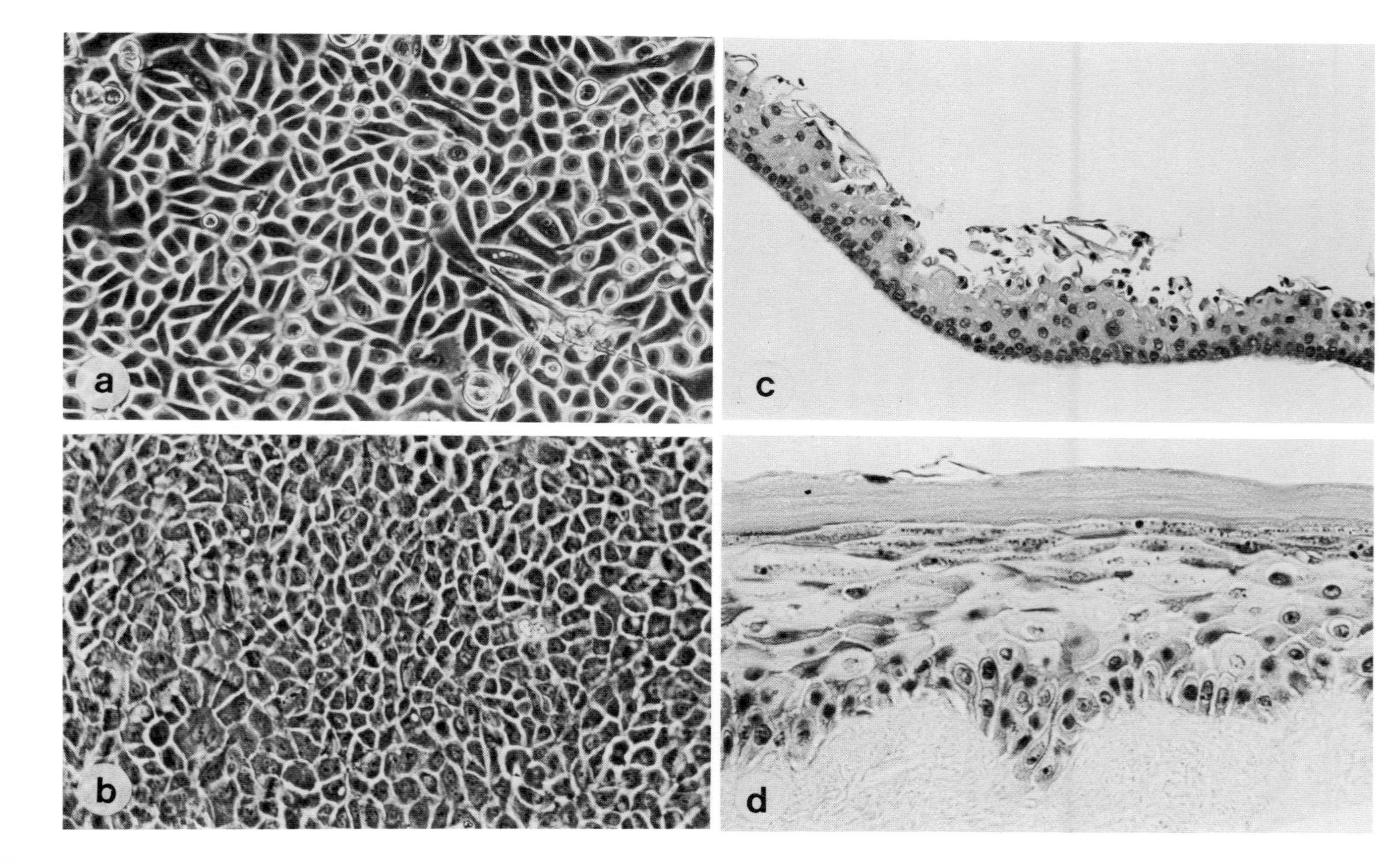
a
c
b
d

keratinocyte differentiation can be modulated by changing culture conditions, making this approach particularly versatile. When keratinocytes are grown under conventional culture conditions (i.e., on a plastic substratum and immersed in a culture medium), the extend of keratinocyte differentiation is regulated by extracellular calcium concentration. When cultured at low calcium concentrations (<0.1 m*M*), keratinocytes grow as a monolayer (Fig. 1a) and exhibit a low capacity to differentiate (Hennings *et al.*, 1980; Watt and Green, 1982). In contrast, when grown at physiological calcium concentrations (1–2 m*M*), keratinocytes display a markedly higher capacity to differentiate (Fig. 1b). Under the latter conditions keratinocytes form a multilayered epithelium consisting of a basal and from two to five suprabasal layers (Fig. 1c). This regulatory role of calcium likely occurs both *in vitro* as well as *in vivo* (Hennings *et al.*, 1980; Watt and Green, 1982; Malmqvist *et al.*, 1984; Menon *et al.*, 1986).

Despite the fact that the use of such cultures has facilitated progress in understanding the biology of keratinocytes, their differentiation under submerged conditions is still incomplete, according to both morphological (Fig. 1b and c) and biochemical criteria (see, e.g., Holbrook and Hennings, 1983; Watt, 1989). A considerably greater degree of tissue organization can be obtained by culturing keratinocytes at an air–medium interface using a substrate of either (1) deepidermized dermis (Régnier *et al.*, 1981; Ponec *et al.*, 1988), (2) fibroblast-populated collagen lattices (Bell *et al.*, 1983; Mackenzie and Fusenig, 1983; Lillie *et al.*, 1980; Asselineau *et al.*, 1985, 1989), or (3) Millipore filters (Bernstam *et al.*, 1986; Williams *et al.*, 1988a; Madison *et al.*, 1988). Under such more physiological conditions, wherein the cells are nourished by diffusion of nutrients through the underlying substrate, the extent of epidermal stratification and cornification approaches that found *in vivo* (Fig. 1d) and the reconstructed "epidermis" displays many of the morphological (Régnier *et al.*, 1981; Asselineau *et al.*, 1985, 1989; Ponec *et al.*, 1988) and biochemical markers (Asselineau *et al.*, 1985, 1989; Régnier *et al.*, 1986; Ponec *et al.*, 1988; Madison *et al.*, 1988; Bohnert *et al.*, 1986) and functional characteristics (Ponec *et al.*, 1990; Régnier *et al.*, 1990) of the native tissue from which it has been derived.

Epidermal lipogenesis, both *in vivo* and *in vitro*, depends on the balance between cellular requirements and access to the exogenous (extracellular) lipid supply. *In vivo* serum lipids carried by dermal blood microcapillaries, after diffusion first through the endothelial wall, next through the uppermost layers of the dermis, and finally through the basement membrane, may potentially influence

---

FIG. 1. Phase-contrast photomicrograph of normal human keratinocytes grown to confluence in submerged culture system either at (a) low calcium or (b and c) normal calcium concentrations; hematoxylin and eosin staining of keratinocyte cultures grown at normal calcium concentration (c) to confluence under submerged conditions and (d) for 10 days at the air–liquid interface. For details, see Appendix (Section VIII).

lipogenesis by the cells of the stratum basale. Subsequently, after diffusion through this cell layer, lipogenesis in cells located in the upper layers of the epidermis may be affected. The amount of a particular lipid that reaches the basal cell layer depends on its physicochemical properties and on its molecular size. For example, it has been demonstrated with low-density lipoproteins (LDLs) that about one-fifth of the LDL content in blood reaches the basal cell layer (Vermeer *et al.*, 1979; Vessby *et al.*, 1987). It can be assumed that amounts reaching the suprabasal cell layers will be even lower.

The purpose of this review is to demonstrate the usefulness and limitations of various keratinocyte culture systems as models for the study of the regulation of epidermal lipid metabolism in relation to differentiation. The following issues will be discussed: (1) differentiation-linked changes in the LDL receptor expression and cholesterol synthesis, (2) differentiation-linked changes in the rate of lipogenesis, (3) differentiation-linked modulations of lipid composition, (4) changes in the plasma membrane composition and physicochemical properties in relation to keratinocyte differentiation, and (5) retinoid-induced modulations of lipid composition in cultured keratinocytes. In the Appendix (Section VIII), various methods relevant to studies of lipid metabolism in cultured keratinocytes are presented in further detail.

II. Changes in LDL Receptor Expression and in Regulation of Cholesterol Synthesis in Relation to Keratinocyte Differentiation

An important aspect of cellular lipid metabolism is the network of processes that involve the cholesterol biosynthetic pathway. Cultured mammalian cells derive their cholesterol, a key component of plasma membrane, in two ways: (1) by uptake from lipoproteins present in the serum in the culture medium and (2) by *de novo* synthesis within the cells (Goldstein and Brown, 1974). Cultured cells usually utilize any available exogenous lipoprotein and their endogenously synthesized cholesterol.

In the last decade abundant information has accumulated concerning the uptake of lipoproteins and the role of membrane receptors in this process (for review see Goldstein and Brown, 1984). The specific serum lipoprotein particle, which normally supplies cholesterol to cells, is the LDL, which is taken into the cell by means of specific LDL receptor-mediated endocytosis. Following LDL binding to the high-affinity receptors located on the plasma membrane, the receptor–ligand complexes cluster into coated pits and are subsequently internalized. Within endosomes, the LDL particle and its receptor dissociate, and a fraction of receptor molecules is recycled to the cell membrane. After the fusion of endosomes with lysosomes, the degradation of LDLs takes place. Released cholesterol affects three cellular processes: (1) suppression of *de novo* sterologenesis through inhibition of the rate-limiting enzyme, 3-hydroxy-3-methylglutaryl coenzyme A (HMG-CoA) reductase; (2) increase in the storage of cholesterol as cholesteryl esters by activation of the enzyme acyl-CoA:cholesterol acyltransferase (ACAT); and (3) suppression of LDL receptor synthesis (Goldstein and Brown, 1984).

In order to study differentiation-linked changes in LDL receptor expression and cholesterol synthesis in cultured keratinocytes, the following approaches have been used to manipulate the degree of keratinocyte differentiation: (1) modulation of extracellular calcium concentration, (2) modulation of cell density, and (3) the use of transformed keratinocytes with defects in their terminal differentiation program. Transformed cells, originating from various squamous cell carcinomas (SCCs) of the head and neck area, demonstrate distinctive phenotypic variations in their ability to form cross-linked cornified envelopes (Rheinwald and Beckett, 1980, 1981), in keratin polypeptide profiles (Kim *et al.*, 1985; Kopan *et al.*, 1987), in cell surface proteins (Brysk *et al.*, 1984), and in plasma membrane epidermal growth factor (EGF) receptor expression (Boonstra *et al.*, 1985). When cultured under low-calcium (<0.1 m*M*) conditions, both normal and neoplastic cell lines form only a monolayer. In contrast, at physiologic calcium concentrations (>1.0m*M*), normal keratinocytes form a basal and from two to five suprabasal layers, whereas SCC cell lines stratify to no more than two to three cell layers (Ponec *et al.*, 1989b). The extent of differentiation of keratinocytes grown in normal calcium medium also changes during the time they spend in culture: it is low in sparse cultures and gradually increases, reaching a maximum in postconfluent cultures.

All studies to date clearly show that both LDL receptor expression and the extent of LDL-regulated sterologenesis are inversely related to the capacity of keratinocytes to differentiate. Namely, LDL receptors are expressed in normal human keratinocytes grown under differentiation-retarding conditions; i.e., in low-calcium media or in preconfluent cultures in normal-calcium media. Moreover, as in other cell types, under these conditions LDL is taken up by the cells, where it is degraded, and *de novo* cholesterol synthesis is inhibited. Induction of keratinocyte differentiation by exposure of cells to physiological calcium concentration results in a rapid decrease in LDL receptor binding, resulting in the loss of LDL-dependent cholesterol synthesis (Ponec *et al.*, 1985). However, even at physiological calcium concentrations in which LDL-independent sterologenesis occurs, the rate of cholesterol synthesis remains very high, suggesting a high requirement of cholesterol in differentiating keratinocytes (Ponec *et al.*, 1983, 1985; Williams *et al.*, 1987a, 1988b). Differentiation-linked modulations of LDL receptor activity at the plasma membrane level are a reflection of the modulations in LDL receptor synthesis: induction of keratinocyte differentiation is accompanied by a two- to threefold decrease in LDL receptor mRNA expression and by about a 40-fold decrease in the LDL receptor protein expression (Fig. 2) (te Pas *et al.*, 1991b).

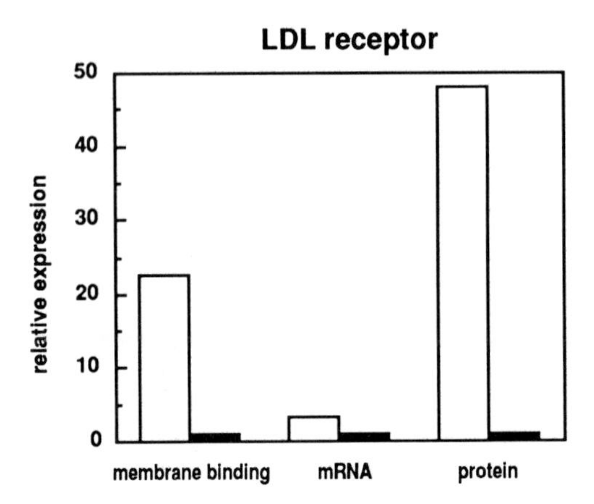

FIG. 2. Differentiation-linked change in the expression of the LDL receptor at membrane, mRNA, and protein levels in normal human keratinocytes cultured at low (0.06 m*M*) (□) and normal (1.6m*M*) (■) calcium levels. The results are normalized, taking the expression in cells grown at normal calcium concentration as a reference. For details, see Appendix (Section VIII).

When keratinocytes are grown at physiological calcium concentration they generate functional LDL receptors only at the early stages of growth, at which time the extent of keratinocyte differentiation is low, i.e., in preconfluent cultures, and they progressively loose this characteristic over time. In confluent and postconfluent cultures the rate of cholesterol synthesis is independent of extracellular lipoproteins (Ponec *et al.*, 1983; Williams *et al.*, 1988b). The morphological observations in which receptor binding and internalization were studied in normal keratinocytes paralleled the biochemical observations. By immunofluorescence, keratinocytes cultured at a physiological calcium concentration showed LDL binding only in subconfluent cultures in the outermost (nonstratifying) regions of the cell colonies, whereas those cultured at a low calcium concentration showed considerable binding (Ponec *et al.*, 1985). Furthermore, only keratinocytes grown at low calcium concentration were capable to bind and internalize LDL–gold conjugates, as revealed by ultrastructural studies (Ponec *et al.*, 1985) (Fig. 3).

Malignant keratinocytes (exhibiting differentiation defects) derived from squamous cell carcinomas, cultured at physiological calcium concentrations, express LDL receptors and reduce cholesterol synthesis in the presence of LDL; LDL receptor binding again being inversely related to the capacity of these cells to differentiate (Ponec *et al.*, 1984). In these cells the LDL receptor expression is regulated at the DNA level. Amplified LDL receptor gene numbers were observed in poorly differentiating SCC-4 and SCC-15 cells, but not in well-differentiating SCC-12F2 cells and in normal keratinocytes (te Pas *et al.*, 1989). The gene amplification is reflected at the mRNA and protein levels (te Pas *et al.*, 1991b). Similarly, as with normal keratinocytes, in malignant keratinocytes both LDL receptor

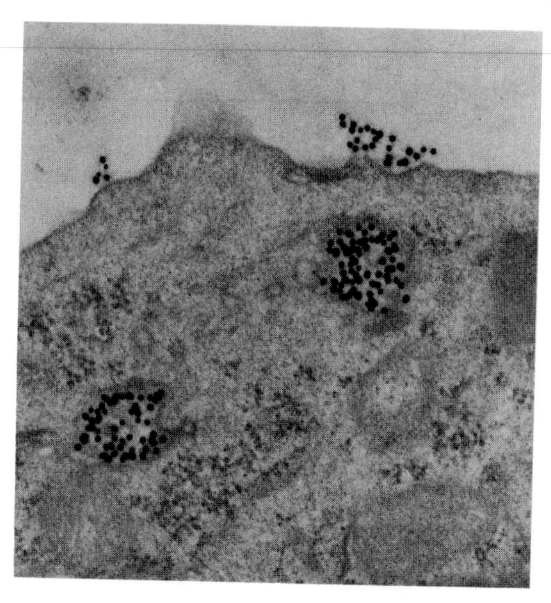

FIG. 3. Visualization of LDL binding and internalization in ultrastructural studies, using LDL–gold particles (20 nm), on keratinocytes cultured at low calcium concentration (x40,000). For details, see Appendix (Section VIII).

expression and LDL-dependent cholesterologenesis vary with cell density (Vermeer *et al.*, 1985, 1986) and can be regulated by extracellular calcium concentration (te Pas *et al.*, 1991b). Finally, induction of differentiation of malignant keratinocytes by raising extracellular calcium concentration causes, as in normal keratinocytes, a decrease of LDL receptor expression at the mRNA level (te Pas *et al.*, 1991b).

Calcium-induced differentiation in normal keratinocytes is associated not only with the loss of functional LDL receptors but also with changes in their cellular distribution. Namely, relatively high levels of intracellular, cytoskeleton-associated LDL receptors have been found in normal human differentiating keratinocytes (te Pas *et al.*, 1990) (Fig. 4). At present it is not clear what the function of these receptor subpopulations may be. Also, the nature of the associated cytoskeletal domain is not known. However, similar observations in the expression (O'Keefe and Payne, 1983; Boonstra *et al.*, 1985; te Pas *et al.*, 1989, 1991a) and cellular distribution (Ponec *et al.*, 1989a; te Pas *et al.*, 1990) of EGF receptors during cell differentiation have been described in both normal keratinocytes and SCC cells, which suggests that a similar mechanism may underlay the regulation of the activity of certain membrane receptors that are internalized via the same endocytotic pathway.

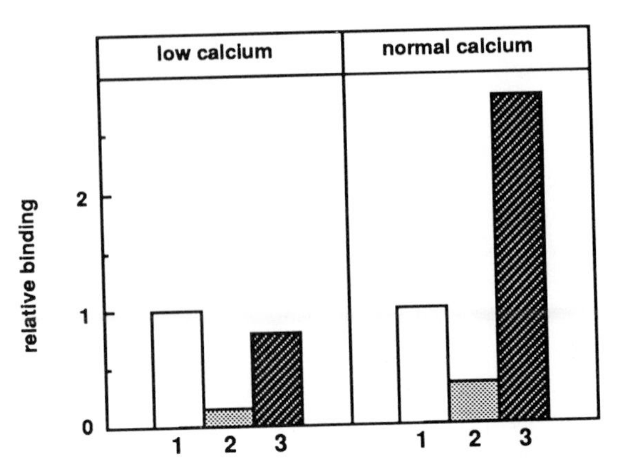

FIG. 4. Differentiation-linked changes in the LDL receptor distribution in normal human keratinocytes. Keratinocytes were cultured to confluence either at low or normal calcium concentration and cytoskeletons were isolated either before or after the binding assay. Results are normalized for each set of data obtained without incubation with Triton X. (1) Binding to intact cells; (2) binding to intact cells followed by Triton X extraction; (3) binding to isolated cytoskeletons. For details, see Appendix (Section VIII). (From te Pas *et al.*, 1991b.)

Differentiation-related modulations of the LDL receptor expression and of the cellular distribution found with cultured keratinocytes parallel observations in epidermis. When the distribution of LDL receptors in the epidermis *in situ* is studied with LDL–gold as the ultrastructural marker, abundant LDL–gold binding and uptake occur only in keratinocytes with the morphologic characteristics of basal cells, and not in the suprabasal, more differentiated keratinocytes (Mommaas-Kienhuis *et al.*, 1987). Though the LDL–gold technique allows visualization of LDL binding to the plasma membrane and subsequent internalization, immunoelectron microscopic studies with LDL receptor antibodies provide information on the localization of LDL receptors on the cell membrane and in the cytoplasm in the absence of a ligand. With this technique it has been found (Mommaas *et al.*, 1990) that in basal cells of normal epidermis, LDL receptors are distributed evenly between the cell surface and the cytoplasm, whereas in suprabasal cells most of the LDL receptors are present inside the cells. Similar observations also have been made recently with keratinocytes cultured at the air–liquid interface (Mommaas *et al.*, 1991), also revealing that cellular LDL receptor distribution is not affected by extracellular lipoproteins. The direct link between the stage of epidermal differentiation and the LDL receptor expression and cellular localization is further supported by observations made in psoriatic epidermis (Mommaas-Kienhuis *et al.*, 1987, 1990). There, LDL receptors are not only present on basal cells but also on suprabasal cells that express markers of hyperproliferation (Lane *et al.*, 1985; Leigh *et al.*, 1985; Dover and Watt, 1987;

Bernard *et al.*, 1988; Stoler *et al.*, 1988; Purkis *et al.*, 1990). Thus it appears that in the basal cells, in which the LDL uptake system is predominantly operative, epidermal cholesterologenesis can be influenced by extracellular LDL cholesterol levels, the extent depending on the amount of LDL reaching this cell layer. In contrast, the sterologenesis in the postmitotic, differentiating layers would not be affected substantially. Monger *et al.* (1988) have demonstrated that sterologenesis in suprabasal epidermis equals or exceeds that found in basal cell layers. The restriction of active LDL receptors predominantly to basal cells and continued active lipid synthesis in upper epidermal layers may explain why epidermal cholesterol synthesis appears to be relatively autonomous with respect to systemic influences: neither dietary nor circulating sterol levels seem to affect cutaneous sterologenesis significantly (Brannan *et al.*, 1975; Andersen and Dietschy, 1977). Despite the autonomy from circulation, epidermal sterologenesis can be regulated by cutaneous barrier requirements (Menon *et al.*, 1985; Grubauer *et al.*, 1987; Feingold *et al.*, 1986), i.e., disruption of the permeability barrier by solvent or detergent treatment or essential fatty deficiency results in increased epidermal sterologenesis. This increase can be, however, abolished if barrier function is restored by application of a water-impermeable membrane to the skin surface.

III. Differentiation-Linked Changes in the Rate of Lipid Synthesis: Effect of Extracellular Lipoproteins

During terminal differentiation, a substantial shift in lipid composition occurs, from a lipid mixture containing high amounts of phospholipids to a mixture lacking phospholipids but containing high amounts of ceramides, free fatty acids, and sterols (for review see Elias, 1983). Although the epidermis is capable of synthesizing a broad spectrum of lipids (Yardley and Summerly, 1981; Lampe *et al.*, 1983a,b), the precise localization of epidermal cell layers where lipid biosynthesis occurs is still not completely clear. Furthermore, the role of extracutaneous sources of lipids in the regulation of epidermal lipogenesis has not yet been fully elucidated. In the epidermis, which is an important site of sterol synthesis (Menon *et al.*, 1985; Feingold *et al.*, 1983, 1986; Monger *et al.*, 1988), the sterologenesis is not influenced by exogenous lipoproteins (Brannan *et al.*, 1975; Andersen and Dietschy, 1977). At present, very little is known about the possible regulatory role of extracutaneous lipids on the synthesis of other lipid classes. Furthermore, information available about the changes in the overall lipid synthesis in relation to keratinocyte differentiation is scarce. To approach the latter, experiments have been performed in which the degree of keratinocyte differentiation was altered again by (1) modulation of extracellular calcium concentration, (2) modulation of cell density, and (3) the use of transformed keratinocytes with defects in their terminal differentiation program. The differentiation-linked changes in the rate of

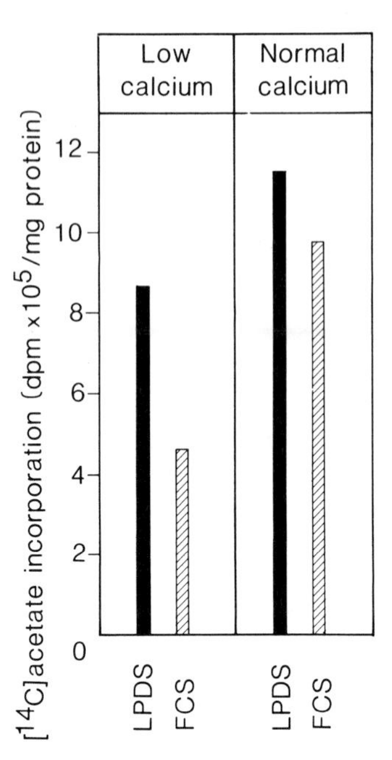

FIG. 5. Calcium-induced effects on the *de novo* lipid synthesis in confluent cultures of normal human keratinocytes grown either at low or normal calcium concentration in the presence (■) or absence (▨) of extracellular lipoproteins. For details, see Appendix (Section VIII). (From Ponec *et al.*, 1987c.)

lipid synthesis were monitored by measuring the rate of incorporation of various radiolabeled (^{14}C or ^{3}H) precursors (acetate, glucose, fatty acids, and water) into various lipid classes.

In normal keratinocytes, the rate of lipid synthesis is related to the capacity of cells to differentiate. However, the profiles of synthesized lipids measured in normal human keratinocytes cultured under submerged conditions, i.e., a high rate of phospholipid synthesis and a very low rate of ceramide synthesis (Ponec *et al.*, 1987c), reflect the incomplete morphological differentiation of these cultures, even when grown in physiological calcium conditions. Namely, under submerged conditions, very few lamellar bodies, the repository of ceramide and glycosphingolipids (for review see Elias, 1983), are formed. In the presence of lipoproteins, induction of differentiation by exposure of low-calcium confluent cultures to medium containing normal calcium concentration is accompanied by a rapid increase in the lipid synthesis; levels are reached that are about two times higher than those in cells grown in low-calcium medium (Fig. 5) (Ponec *et al.*, 1987c). Only in the low-calcium cells, in which the LDL uptake mechanism is fully operative, does the total rate of lipid synthesis increase (about two times) in the absence of extracellular lipoproteins (Fig. 5), an increase predominantly due to the enhanced neutral lipid synthesis (cholesterol and triglycerides) (Ponec *et al.*, 1987c). In contrast, in cells grown under normal calcium conditions, the overall rates of lipid synthesis are only slightly influenced by extracellular lipoproteins. This difference can be ascribed to the absence of any LDL uptake mechanism (Ponec *et al.*, 1983). The only exception is the synthesis of triglycerides, the rate of which decreases in the presence of extracellular lipoproteins. Likewise, when the stage of differentiation of normal human keratinocytes grown at physiological calcium concentration is regulated by the stage of growth, the lipid synthesis parallels the growth activity of the cultures, being highest in the postconfluent differentiated cultures (Williams *et al.*, 1988b). Also in these cultures, the rate of the triglyceride synthesis and the bulk triglyceride content are highest in the differentiated postconfluent cultures. In this system the triglyceride pool does not function primarily as a lipid storage site, because pulse-chase studies reveal a rapid transfer of radioactivity from triglycerides to phospholipids.

It should be noted that among human and murine keratinocytes, differences have been found in their ability to synthesize certain classes of ceramides—namely, linoleic acid-rich acylglucosylceramide (AGC) and acylceramide (AC). These lipids are unique to the epidermis (Gray *et al.*, 1978; Wertz and Downing, 1983; Wertz *et al.*, 1983) and their presence is necessary for the formation of the stratum corneum lipid bilayer structures (Gray *et al.*, 1978; Wertz and Downing, 1982, 1983; Wertz *et al.*, 1989; Swartzendruber *et al.*, 1989) and proper epidermal water barrier function (Prottey *et al.*, 1976; Wertz *et al.*, 1983; Hansen and Jensen, 1985; Nugteren *et al.*, 1985). When human keratinocytes are grown under submerged conditions, the rate of ceramide synthesis (especially that of AGC and AC) is extremely low (Ponec *et al.*, 1987c; Williams *et al.*, 1988a,b), but these lipids are synthesized in sufficient amounts in primary cultures of murine keratinocytes, in which linoleic acid is preferentially incorporated into AGC and AC (Madison *et al.*, 1986). The differences observed among the subcultured human and the primary murine keratinocytes may arise from a decreased activity of enzymes involved in the synthesis of these lipids, caused by the subcultivation of human keratinocytes in the presence of fetal calf serum that contains low amounts of linoleic acid (Isseroff *et al.*, 1985; Stoll and Spector, 1984). Furthermore, primary murine keratinocyte cultures may reflect "organ culture" characteristics, because they show greater propensity to differentiate *in vitro* than do human keratinocytes.

The extent of differentiation of normal human keratinocytes not only parallels the total lipid synthesis (Ponec *et al.*, 1987c; Williams *et al.*, 1988b) but also the production of cholesterol sulfate, a lipid fraction that may influence the normal desquamation process (Long, 1970; Shapiro *et al.*, 1978; Williams and Elias, 1981). Production of cholesterol sulfate in undifferentiated, normal keratinocytes is low, but it dramatically increases (100- to 300-fold) in differentiating cultures,

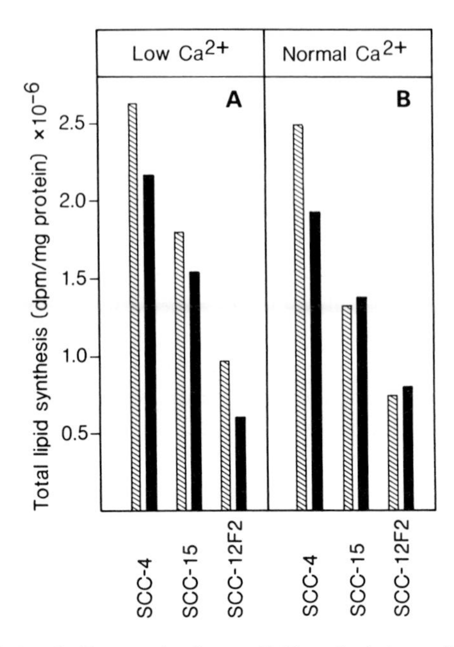

FIG. 6. Calcium-induced effects on the *de novo* lipid synthesis in confluent cultures of malignant keratinocytes (SCC-12F2, SCC-15, and SCC-4 cells) grown either at low or normal calcium concentration in the presence (■) or absence (▨) of extracellular lipoproteins. For details, see Appendix (Section VIII). (From Ponec *et al.*, 1987c.)

due to an increased sulfotransferase activity (Jetten *et al.*, 1989). Cholesterol sulfate synthesized by keratinocytes plays a role not only in the desquamation process but it also may regulate epidermal sterologenesis (Williams *et al.*, 1987b; Williams and Elias, 1987; Ponec and Williams, 1986). This may represent one mechanism whereby cholesterol sulfate regulates the desquamation process, because the cholesterol:cholesterol sulfate ratio is normally about 10:1 (Williams and Elias, 1981).

In malignant keratinocytes (Ponec *et al.*, 1987c) the rate of the total lipid synthesis decreases progressively with an increasing capacity of keratinocytes to differentiate; i.e., it is highest in poorly differentiated (SCC-4) cells, intermediate in the moderately differentiating (SCC-15) cells, and lowest in the well-differentiating (SCC-12F2) cells (Fig. 6). The SCC-12F2 cells reach the levels seen in normal keratinocytes. In malignant keratinocytes the rate of phospholipid synthesis was lower and the rate of triglyceride synthesis was higher, in direct relation to their differentiation capacity. Similarly, as in normal keratinocytes, the rate of sphingolipid synthesis (except of sphingomyelin) was very low. In proliferating cells (cells grown in low-calcium media), in which the LDL receptor is expressed to a great extent (Ponec *et al.*, 1985), the rate of lipogenesis, especially of neutral lipids, responds dramatically to changes in the extracellular lipoprotein concentration. In the presence of lipoproteins a marked decrease in cholesterol and triglyceride synthesis occurs as well as an increase in cholesterol ester synthesis. On the other hand, when SCC cells are cultured at physiological calcium concentrations, only the well-differentiating SCC-12F2 cells (as do normal keratinocytes) synthesized lipids independently of extracellular lipoproteins. As described above, the difference can be ascribed to an absence of the LDL uptake mechanism (Ponec *et al.*, 1983, 1984). The only exception is, again, the synthesis of triglycerides, the rate of which can be modulated to a certain extent by extracellular lipoproteins. These results suggest that a close inverse relationship does exist in both normal and malignant keratinocytes between the rate of lipid synthesis and the regulation of lipogenesis by extracellular lipoproteins, as well as the ability of keratinocytes to differentiate.

IV. Differentiation-Linked Modulations of Keratinocyte Lipid Composition

If the rates of lipogenesis are inversely related to the capacity of keratinocytes to differentiate, is keratinocyte lipid composition modulated by differentiation? Again, such information can be gained in experiments in which the stage of keratinocyte differentiation is altered by (1) modulation of culture conditions, in which the cells are grown either submerged in the culture medium or at the air–medium interface, (2) variations in cell density, and (3) the use of transformed keratinocytes with a defect in the terminal differentiation program.

A. Submerged Cultures

In submerged cultures, normal keratinocytes grown under low or normal calcium conditions display different degrees of differentiation. The degree of differentiation can be estimated from their morphology [Fig. 1a–c] and their capacity to generate various protein differentiation markers, such as cornified envelopes (Watt and Green, 1982). However, the differences in the ability of normal human keratinocytes to differentiate are not accompanied by significant changes in the lipid content (expressed as milligrams of recovered lipid/milligram of protein) or in lipid distribution (Ponec *et al.*, 1988) (Table I). The lipid composition of submerged keratinocyte cultures, however, differs to some extent from that reported for basal and suprabasal human (Lampe *et al.*, 1983a,b) and nonhuman (Yardley and Summerly, 1981; Madison *et al.*, 1986) epidermal cells. The main differences are the virtual absence of AGC and AC, and an increase in the triglyceride content. The results obtained in human keratinocytes conflict with those reported by Madison *et al.* (1986), who described the presence of AGC and AC in primary

Table I
LIPID COMPOSITION OF HUMAN KERATINOCYTES CULTURED UNDER SUBMERGED CONDITIONS[a]

Fraction	Calcium concentration[b]	
	Low (0.06 m*M*)	Normal (1.6 m*M*)
Phospholipids	68.8 ± 5.0	66.5 ± 5.2
Sphingomyelin	5.5 ± 0.2	7.2 ± 0.7
Phosphatidylcholine Phosphatidylserine	32.0 ± 4.7	30.2 ± 5.6
Phosphatidylinositol	8.2 ± 2.2	7.3 ± 0.7
Phosphatidylethanolamine	23.1 ± 2.5	21.8 ± 3.0
Cholesterol sulfate	0.6 ± 0.2	0.7 ± 0.1
Glycosphingolipids	1.2 ± 0.1	0.7 ± 0.1
Acylglucosylceramides	0	0
Acylceramides	0	0
Other ceramides	0.8 ± 0.3	0.8 ± 0.3
Neutral Lipids	28.5 ± 2.9	31.2 ± 3.2
Cholesterol	12.9 ± 4.2	15.6 ± 4.3
Lanosterol	0	0
Free fatty acids	0.5 ± 0.2	0.4 ± 0.2
Triglycerides	10.9 ± 3.0	12.4 ± 3.0
Cholesteryl esters	4.2 ± 1.1	2.8 ± 1.2

[a]From Ponec *et al.* (1988).
[b]The results are given as an average percentage of total lipids ± SEM.

murine cultures. This may reflect either species differences or the effects of subculturing.

Inspection of the lipid composition in more detail reveals even more pronounced differences. For example, the fatty acid content in cultured human keratinocytes markedly differs from that found in noncultured epidermis (Ponec *et al.*, 1988). More than 75% of the fatty acid content of the culture is composed of $C_{16:0}$, $C_{16:1}$, $C_{18:0}$, and $C_{18:1}$ fatty acids, whereas in freshly isolated human epidermis these comprise about 50%. Furthermore, the $C_{18:2}$ (linoleic acid) deficiency observed in cultured cells was striking. The differences in the fatty acid composition might be caused by the use of medium supplemented with fetal calf serum (Isseroff *et al.*, 1985, 1987). These authors demonstrated a gradual but marked decrease in the linoleic acid ($C_{18:2}$) content, a small decrease in the linoleic ($C_{18:3}$) and arachidonic acid ($C_{20:4}$) contents, and an increase in the palmitoleic ($C_{16:1}$) acid content over relatively short periods in primary cultures of mouse keratinocytes when compared to noncultured mouse keratinocytes. Because cultured cells incorporate fatty acids that are present in serum-containing culture medium, the low levels of $C_{18:2}$ and $C_{18:3}$ and the high levels of $C_{16:1}$ and $C_{18:1}$ found in the fetal calf serum might account for the differences in the fatty acid composition in the cultured and noncultured human keratinocytes (Ponec *et al.*, 1988). And the higher amount of $C_{18:2}$ acid found in primary murine keratinocyte cultures may be explained by differences in the length of the cell culture period. Contrary to linoleic acid, the content of arachidonic acid in keratinocytes cultured under submerged conditions does not differ significantly from that seen in noncultured cells (Ponec *et al.*, 1988). This may be due to the fact that contrary to the *in vivo* situation, the conversion of linoleic acid into arachidonic acid takes place in cultured keratinocytes (Isseroff *et al.*, 1987).

Similar differentiation-linked changes in the lipid composition are observed in normal human keratinocyte cultures at different stages of growth (Williams *et al.*, 1988b). In the presence of fetal calf serum, the total lipid content remains unchanged until cultures reach confluence, and the lipid content increases in postconfluent cultures. This increase results largely from striking changes in triglyceride content, reaching 10-fold higher levels in postconfluent versus preconfluent and confluent cultures. The triglyceride accumulation in keratinocyte cultures could be the result of essential fatty deficiency or decreased activity of triacylglycerol hydrolase, an enzyme involved in the triglyceride catabolism (Severson, 1979). This enzyme is probably activated after the extrusion of lamellar bodies (Menon *et al.*, 1986), which are absent in submerged human foreskin keratinocyte cultures. Furthermore, one could speculate that these changes in triglycerides are not related to lipid storage but rather that the triglycerides may serve as an intermediate in phospholipid or sphingolipid synthesis (Williams *et al.*, 1988b; Schürer *et al.*, 1989).

Finally, comparison of total lipid content in normal versus several malignant (SCC) cell lines reveals similar content, regardless whether the cells are cultured under low or normal calcium conditions. However, differences are observed in the content of individual lipid classes (Ponec *et al.*, 1989b): (1) the total phospholipid content of all of the malignant cell lines exceeds that of normal keratinocytes and (2) the total neutral lipid content is lowest in poorly differentiating SCC cell lines and decreases in all cell lines under low-calcium conditions. The differences seen in the neutral lipid class were largely due to the higher triglyceride content in well-differentiating SCC-12F2 cells and also in normal keratinocytes, suggesting once again that the triglyceride accumulation is directly related to the capacity of cells to differentiate.

From these observations it can be concluded that in submerged cultures the differentiation-linked changes in lipid composition are much less pronounced than those found for the expression of various differentiation-specific protein markers, such as keratins (Fuchs and Green, 1980; Woodcock-Mitchell *et al.*, 1982; Bowden *et al.*, 1983; Van Muijen *et al.*, 1987; Watt, 1989), involucrin (Watt and Green, 1982; Magee *et al.*, 1987), rearrangement of keratin (Watt *et al.*, 1984), actin and vinculin filaments (O'Keefe *et al.*, 1987), desmosomes (Watt *et al.*, 1984; Jones and Goldman, 1985), and redistribution of keratinocyte surface glycoproteins (Morrison *et al.*, 1988), suggesting that changes in protein composition and not in lipid composition reflect the earliest events during differentiation.

B. Air-Exposed Cultures

The degree of keratinocyte differentiation can be markedly enhanced by cultivating the cells at the air–medium interface. Under these conditions, both normal keratinocytes and SCC cells form tissues with greater similarities in structure to those of their tissue of origin than occur even in the most mature submerged cultures. Normal keratinocytes form a highly organized multilayered epithelium, i.e., with layers comparable to stratum basale, spinosum, granulosum, and corneum (Fig. 1d) (Régnier *et al.*, 1981). SCC cell lines grown at the air–medium interface form disorganized layers with individual cell keratinization and with a lesser extent of stratification than in normal keratinocytes, inversely related to their capacity to differentiate (Ponec *et al.*, 1989a; Stoler *et al.*, 1988; Régnier *et al.*, 1989). Not only the tissue morphology but also the differences in the lipid composition among normal and malignant keratinocytes become more pronounced when the cells are grown at an air–medium interface. In contrast to submerged conditions, under air-exposed conditions the phospholipid content decreases dramatically in normal keratinocytes and to a certain extent in the well-differentiating SCC-12F2 cells, but it does not change in the least differentiating SCC-4 and SCC-15 cells (Ponec *et al.*, 1989a). These changes are coupled with changes in neutral lipid content. Whereas under submerged conditions no significant differences are observed in the cholesterol sulfate and the ceramide content of normal keratinocytes versus squamous carcinoma cells, in air-exposed cultures the content of these lipid fractions is related to the differentiation state of the cells (Ponec *et al.*, 1989a). However, the content of cholesterol sulfate and ceramides is lower in SCC cells than in normal keratinocytes grown under comparable conditions. In particular, a high content of ceramides and the presence of acylceramides and acylglucosylceramides, the presence of which is of acknowledged importance for the barrier function (Gray *et al.*, 1978; Wertz and Downing, 1983; Elias, 1983), was seen only in normal keratinocytes. In addition, the sterol metabolites, lanosterol and lathosterol, are found only in normal keratinocytes, suggesting that under air-exposed conditions the activity of both the Δ^7 and Δ^4 reductases is lower in comparison to submerged cultures. The decreased activity of these enzymes is postulated (Feingold *et al.*, 1987) to be necessary to provide a sufficient pool of 7-dehydrocholesterol for vitamin D_3 synthesis by the epidermis.

Under the air-exposed conditions, normal keratinocytes reproduce to a high extent the lipids of the native epidermis. Ceramides, the lipid class most indicative of completed differentiation, are present in significant amounts (Table II). However, minor differences also are observed. One of these is a significantly higher triglyceride content in keratinocyte cultures that may be attributed to both an in-

Table II
LIPID COMPOSITION OF RECONSTRUCTED AND FRESHLY ISOLATED HUMAN ADULT EPIDERMIS

Lipid fraction	Epidermis[a]		
	Reconstructed on		Freshly Isolated
	DED[b]	Collagen	
Phospholipids	36.6 ± 4.1	33.6 ± 5.1	48.4 ± 2.8
Cholesterol sulfate	1.2 ± 0.4	1.3 ± 0.1	2.3 ± 2.8
Sphingolipids	11.3 ± 2.7	9.7 ± 1.9	19.6 ± 1.9
Glycosphingolipids	2.3 ± 0.6	0.7 ± 0.1	8.0 ± 1.6
Acylglucosylceramides	0.3 ± 0.1	0.2 ± 0.1	2.0 ± 0.9
Acylceramides	0.8 ± 0.5	0.5 ± 0.2	1.1 ± 0.3
Other ceramides	7.9 ± 0.3	8.3 ± 2.7	8.5 ± 2.1
Neutral Lipids	51.4 ± 4.9	54.4 ± 3.2	29.7 ± 2.6
Cholesterol	12.9 ± 4.2	15.6 ± 4.3	13.1 ± 4.2
Lanosterol	2.7 ± 1.1	3.3 ± 1.1	0.5 ± 0.2
Free fatty acids	2.4 ± 0.8	2.9 ± 0.4	3.0 ± 2.1
Diglycerides	2.2 ± 1.1	3.1 ± 1.2	0.5 ± 0.2
Triglycerides	28.1 ± 2.4	28.2 ± 3.6	9.6 ± 2.6
Cholesteryl esters	3.1 ± 2.4	2.3 ± 1.2	3.0 ± 1.1

[a]The results are given as an average percentage of total lipids ± SEM ($n = 3$).
[b]DED, Deepidermized dermis.

creased rate of proliferation and enhanced differentiation. Namely, under these culture conditions, keratinocytes are seeded at a high density, permitting immediate formation of a continuous keratinocyte layer that covers the "basement membrane surface" of the dermal equivalent, deepidermized dermis (DED) (Régnier *et al.*, 1981). Subsequent lifting of the culture to the air–medium interface results in the development of differentiated epidermis as soon as 1 week after the initiation of the culture (Régnier *et al.*, 1981; Ponec *et al.*, 1988), with the rate of reepithelization approaching that seen during wound healing. Similarly, as occurs in healing epidermis and also in hyperproliferating psoriatic epidermis, keratinocytes grown at the air-medium interface show signs of hyperproliferation, such as expression of "hyperproliferative" keratins K6 and K16 and the presence of involucrin, transglutaminase, and filaggrin in lower suprabasal cell layers (Régnier *et al.*, 1986, 1988; Asselineau *et al.*, 1989; Ponec, 1991a); this has been shown in psoriasis (Weiss *et al.*, 1984; Stoler *et al.*, 1988; Dover and Watt, 1987; Bowden *et al.*, 1983) and wound healing (Lane *et al.*, 1985; Mansbridge and Knapp, 1987; Purkis *et al.*, 1990; M. Ponec, unpublished observations). Moreover, as in psoriatic epidermis (Brody, 1962), intracellular lipid droplets are found in cornified cells of air–medium interface cultures, a feature that is absent in normal stratum corneum.

Another difference in the lipid composition of reconstructed human epidermis

Table III

FATTY ACID COMPOSITION OF MAJOR LIPID FRACTIONS OF HUMAN KERATINOCYTE LIPID EXTRACTS[a]

	Epidermis	
Fatty acid	Reconstructed[b]	Freshly isolated
16:0	22.2	15.8
16:1	13.1	1.8
18:0	10.8	14.8
18:1	45.8	16.6
18:2	1.1	23.3
20:4	0.7	2.1
24.0	3.6	10.2
25:0	0.1	2.7
26:0	1.9	5.3
28:0	0	2.7

[a]From Ponec *et al.* (1988).

[b]Keratinocytes were cultured for 14 days on DED; values are given as percentage of total fatty acids.

and its *in vivo* counterpart is the low content of the essential fatty acids (EFAs), linoleic and arachidonic acids (Table III) (Ponec *et al.*, 1988). The fatty acid profile found in the *in vitro* reconstructed epidermis resembles the fatty acid profile found in epidermal lipids of EFA-deficient (EFAD) animals [enrichment of monoenoic acids $C_{16:1n-7}$ and $C_{18:1n-9}$ and the low content of linoleic acid $C_{18:2n-6}$] (Prottey *et al.*, 1976; Nugteren *et al.*, 1985; Ziboh and Chapkin, 1987; Ziboh, 1989; Bowser *et al.*, 1985; Melton, 1968), suggesting that under the culture conditions used (in the presence of fetal calf serum), supplementation of cells by EFAs is insufficient. Because EFA supplementation is required for normal epidermal morphogenesis and barrier function, normalization of fatty acid composition should occur with EFA supplementation. However, enrichment of culture media with the EFAs did not result in an increase in the EFA content of reconstructed epidermis (Boddé *et al.*, 1990). Furthermore, studies with ^{14}C-labeled fatty acids—linoleic, oleic, palmitic, and stearic acids—reveal that fatty acid incorporation occurs preferentially into triglycerides and phospholipids, with only small amounts being incorporated into sphingolipids (Schürer *et al.*, 1989; Boddé *et al.*, 1990). The incorporation of [^{14}C]linoleic acid was not influenced by the omission of linoleic acid supplementation, suggesting that an enzyme(s) involved in sphingolipid synthesis is not optimally expressed under these culture conditions, or that the air-exposed cultures do not sufficiently duplicate *in vivo* conditions to stimulate the formation of sufficient number of lamellar bodies (Williams *et al.*, 1988a), the principle site of sphingolipid generation. Furthermore, the low content of linoleic acid in the reconstructed epidermis may result in a deficiency or even an absence of linoleic acid metabolites, such as 13-hydroxyoctadecadienoic acid (13-HOD) (Nugteren and Kivits, 1987), which have been shown to regulate epidermal proliferation (Miller and Ziboh, 1990).

It could be speculated that the epidermal lipogenesis in elevated keratinocyte cultures may be affected by the nature of the underlying dermal substrate. In particular, the supply of various nutrients from the culture medium by diffusion may vary when different dermal substrates are used. However, experiments did not confirm this expectation. Namely, with different dermal substrates, such as collagen (Table II) or collagen-coated membranes (Williams *et al.*, 1988a) instead of DED, no substantial changes in lipid composition of human keratinocytes grown under the air-exposed conditions are observed. This suggests that factors other than the dermal substrate may be involved in the regulation of lipid synthesis. These may be, among others, the high relative humidity (100%) in the incubator, because it has been shown that the water gradient across the stratum corneum is one of the factors that can trigger lipogenesis (Menon *et al.*, 1985; Grubauer *et al.*, 1987, 1989).

It also should be noted that the lipid composition of air-exposed, passaged human keratinocyte cultures differs from that of air-exposed, primary murine keratinocyte cultures (Madison *et al.*, 1988, 1989). Murine cultures have less triglycerides than are found in human cultures, in spite of the fact that a slight increase is observed in comparison to freshly isolated, neonatal mouse epidermis. As in the human air-exposed cultures, the linoleic acid content of murine keratinocyte cultures is lower. However, linoleic acid deficiency in murine cultures is much less profound. Hence acylsphingolipids are still substantially enriched in linoleic acid in comparison to other lipid classes, even though their content is decreased to about 50% of the parent tissue content. Furthermore, contrary to human cultures, in murine cultures large amounts of [^{14}C]linoleic acid are incorporated into sphingolipids (Madison *et al.*, 1989). One could speculate that these differences may be due either to prior passaging of human cells or that the activities of the enzymes involved in the synthesis of these lipid classes are higher in neonatal murine skin in comparison to those in juvenile foreskin or adult human epidermis.

Although air-exposed human cultures display the capacity to synthesize all lipid species that are present in the native tissue (Ponec *et al.*, 1988; Williams *et al.*, 1988a; Boddé *et al.*, 1990; Ponec, 1991), some lipids are synthesized and/or metabolized at rates different from those *in vivo*. This leads to the differences observed in the bulk lipid composition of the air-exposed keratinocyte cultures and the native tissue. These differences can be most probably attributed to the culture conditions used, because both the freshly excised skin cultured at the air–medium interface and the air-exposed keratinocyte cultures exhibited similar deviations in lipid composition from noncultured tissue (Ponec, 1991). Therefore, further improvement of culture conditions is needed to approach *in vitro* the lipid composition and distribution of normally proliferating and differentiating epidermis.

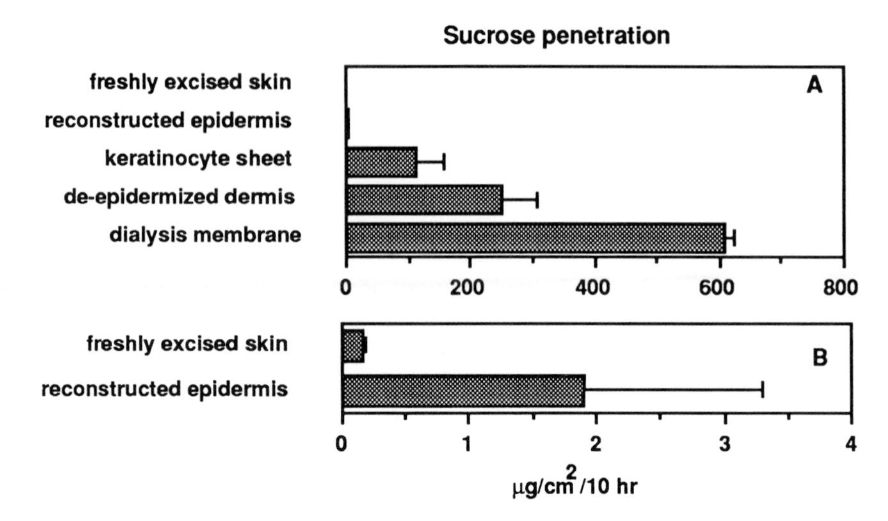

FIG. 7. Penetration of [^{14}C]sucrose through different skin specimens (A). Penetration through freshly excised skin and reconstructed epidermis are illustrated at lower magnification (B). For details, see Appendix (Section VIII). (From Ponec *et al.*, 1990.)

C. BARRIER PROPERTIES OF AIR-EXPOSED CULTURES

Because the main purpose of epidermal differentiation is to generate an efficient permeability barrier, the ultimate criterion to be applied to all attempts at culture optimalization is the presence of a permability barrier (for review see Elias, 1983). Thus, one should not only assess the similarities in lipid composition but also the properties of the stratum corneum. The question arises whether the imperfections in lipid composition in the air-exposed human keratinocyte cultures are reflected in the differences in intercorneocyte lipid deposition and the barrier function. Freeze–fracture electron microscopy of *in vitro* reconstructed cultures are generally similar to that of epidermis, with the exception of localized anomalies in lamellar structures of the intercellular lipids in the cornified layers (Boddé *et al.*, 1990). The abnormalities that may result from the observed differences in the lipid composition (EFA deficiency) may contribute to the enhanced penetration of reconstructed human epidermis by the hydrophilic tracer sucrose and the hydrophobic tracer nitroglycerin (Figs. 7 and 8) (Ponec *et al.*, 1990), or of ^3H_2O (Régnier *et al.*, 1990). The extent of the impairment of barrier function in murine air–medium cultures is similar (Cumpstone *et al.*, 1989) to that in human air–medium cultures, despite the presence of higher amounts of linoleic acid-rich AGC and AC (Madison *et al.*, 1989). Thus the presence of certain critical lipid species alone is not sufficient to prevent imperfections in the barrier function. Furthermore, replacement of linoleic acid by oleic acid in these ceramides (Ponec

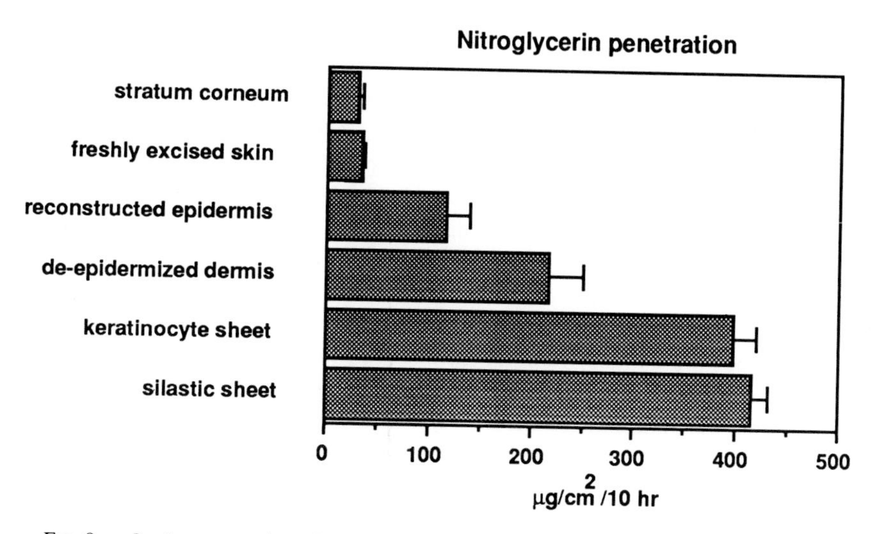

FIG. 8. *In vitro* penetration of nitroglycerin through different skin specimens. For details, see Appendix (Section VIII). (From Ponec *et al.*, 1990.)

et al., 1988; Madison *et al.*, 1989) does not prevent the formation of membrane bilayer structures in the stratum corneum (Madison *et al.*, 1988; Boddé *et al.*, 1990), but it may induce a phase separation of stratum corneum lipids (Kennedy *et al.*, 1990), and cause the observed increase in the water permeability in both human (Régnier *et al.*, 1990) and mouse (Cumpstone *et al.*, 1989) air–medium interface cultures.

As mentioned above, the deviations in lipid composition that likely are a result of the culture microenvironment may be responsible for the observed impairment of the barrier function of air-exposed keratinocyte cultures. Therefore, optimalization of culture conditions is necessary to approach the *in vivo* situation.

V. Changes in the Plasma Membrane Lipid Composition and Fluidity in Relation to Keratinocyte Differentiation

As shown above, LDL receptor-expression, LDL-dependent regulation of the cholesterol, and the bulk lipid synthesis are all highly dependent on the stage of the keratinocyte differentiation. Because the expression and the function of membrane-associated LDL and EGF receptors are inversely related to the degree of differentiation in both the normal and the SCC keratinocyte cell lines (Ponec *et al.*, 1984, 1985; Boonstra *et al.*, 1985), one can assume that the differentiation-associated changes in the plasma membrane lipid composition might mediate the receptor expression/function. In order to obtain further insights into the possible relationship between plasma membrane lipid composition and the differentiation capacity of keratinocytes, studies were performed in which the differentiation-linked changes in plasma membrane lipid composition (Ponec *et al.*, 1987a) were correlated with the physicochemical properties of these membranes. These studies reveal that differentiation-linked differences in the rate of lipid synthesis (Ponec *et al.*, 1987c) are not reflected by differences in plasma membrane composition. For example, in plasma membranes isolated from normal versus malignant cell lines, the cholesterol and phospholipid content and the cholesterol:phospholipid ratios are similar (Table IV). Moreover, the contribution of individual phospholipid species is comparable in the membranes of most of the cell lines. Furthermore, calcium-induced keratinocyte differentiation does not lead to significant changes in plasma membrane composition (Table IV). These findings suggest that no relationship exists between the extent of keratinocyte differentiation and the plasma membrane lipid composition.

The situation is different when the rate of the lipid biosynthesis is modulated by compounds known to affect lipogenesis, e.g., by specific inhibitors of cholesterol synthesis, such as 25-hydroxycholesterol or lovastatin, or by retinoids (retinoic acid or synthetic arotinoid ethyl sulfone). This treatment results in a decrease of both *de novo* cholesterol synthesis and plasma membrane cholesterol content, leading to a decrease in the plasma membrane cholesterol:phospholipid ratio (Ponec *et al.*, 1987a,b; Ponec and Boonstra, 1987). These experimentally induced modulations of plasma composition are accompanied by a decreased ability of keratinocytes to differentiate [assessed by cornified envelope formation (Rice and Green, 1979)] and an increase in epidermal growth factor (EGF) receptor expression (Ponec and Boonstra, 1987). These results suggest that modulation of the plasma membrane composition by various pharmacological agents leads to changes in the differentiation capacity and the expression of plasma membrane-associated receptors, such as the EGF receptor (Ponec *et al.*, 1987a).

Although calcium-induced alterations in keratinocyte differentiation do not lead to significant changes in lipid composition, plasma membrane lipid fluidity, as monitored by lateral lipid mobility using the fluorescence lipid probe C14-diI in fluorescence photobleaching recovery (FPR) measurements, changes upon induction of keratinocyte differentiation (Tertoolen *et al.*, 1988). In normal keratinocytes the lateral diffusion coefficient D is higher in cells cultured in low-calcium media (5.16×10^{-9} cm^2/second) in comparison to that of cells cultured in media containing physiological calcium concentrations (3.27×10^{-9} cm^2/second). Calcium-induced cell differentiation is accompanied by an initial increase of D to 7.07×10^{-9} cm^2/second followed by a gradual, sustained decrease down to values observed in cells cultured continuously under normal calcium conditions. Like-

Table IV

PLASMA MEMBRANE COMPOSITION OF NORMAL AND TRANSFORMED KERATINOCYTES

Component	Cell type							
	SCC-4		SCC-15		SCC-12F2		Normal keratinocytes	
Ca^{2+} (m*M*)	0.06	1.6	0.06	1.6	0.06	1.6	0.06	1.6
Cholesterol/protein (μg/mg)	91 ± 5	85 ± 7	81 ± 6	82 ± 9	72 ± 9	69 ± 7	84 ± 5	72 ± 6
Phospholipids/protein (μg/mg)	214 ± 10	225 ± 15	185 ± 15	190 ± 12	180 ± 12	172 ± 13	200 ± 10	180 ± 15
Cholesterol/phospholipid	0.43	0.38	0.44	0.43	0.40	0.40	0.42	0.40
Relative phospholipid composition (%)								
Sphingomyelin	16.9 ± 1.2	15.4 ± 1.6	18.6 ± 1.2	18.2 ± 1.2	29.1 ± 2.4	32.5 ± 0.8	15.1 ± 0.9	16.8 ± 0.8
Phosphatidylcholine	33.4 ± 2.6	34.4 ± 1.7	30.3 ± 2.3	32.3 ± 0.8	20.9 ± 1.2	18.3 ± 1.6	33.0 ± 1.6	30.4 ± 1.9
Phosphatidylserine	13.8 ± 0.7	14.4 ± 0.7	9.6 ± 1.6	8.1 ± 0.3	7.7 ± 1.6	8.3 ± 0.9	11.9 ± 0.9	10.4 ± 0.9
Phosphatidylinonitol	6.8 ± 1.2	8.9 ± 1.2	8.7 ± 2.0	8.5 ± 0.7	4.5 ± 0.9	4.6 ± 1.5	9.5 ± 1.7	9.0 ± 1.2
Phosphatidylethanolamine	29.1 ± 1.0	26.9 ± 2.8	32.8 ± 2.0	32.9 ± 0.8	37.0 ± 2.1	34.5 ± 0.8	30.5 ± 0.9	33.4 ± 2.0

wise, in malignant keratinocytes (SCC cells), a similar, transient increase in the lateral lipid mobility is observed upon induction of differentiation. However, in contrast to normal keratinocytes, no sustained decrease in lateral diffusion coefficient D is seen upon prolonged culturing under physiological calcium conditions (Tertoolen *et al.*, 1988). The mechanism of the observed differentiation-induced transient changes in the membrane fluidity remains to be unraveled. One can speculate that modulations in lipid turnover may be responsible. For example, an increase in phosphoinositol turnover upon induction of differentiation could lead to an increased rate of hydrolysis of inositol phospholipids, thereby leading to generation of 1,4,5-triphosphate ($IsnP_3$ and 1,2-diacylglycerol (DAG) (Tang *et al.*, 1987, 1988; Ziboh *et al.*, 1984). Such an early release of these intracellular "second messengers" may function as an initial signal, which in turn may act alone or in concert with other modulators [e.g., protein kinase C (Snoek *et al.*, 1988)] to direct proliferating cells to differentiation.

It should be noted that observations made with submerged cultures of normal human keratinocytes parallel the *in vivo* observations. Namely, in guinea pig epidermis, differentiation-linked decrease in membrane fluidity is observed (Tanaka *et al.*, 1989; Hagchisuka *et al.*, 1990). Also in these studies, no significant changes in membrane cholesterol and phospholipid content, cholesterol:phospholipid ratio, and fatty acid composition are observed in differentiating layers of the epidermis. These findings suggest that the differentiation-linked decrease in membrane fluidity is not directly related to the plasma membrane lipid composition. One can speculate that a differentiation-induced decrease in cellular motility (Magee *et al.*, 1987), a differentiation-induced assembly of desmosomes (Hennings and Holbrook, 1983; Watt *et al.*, 1984; Jones and Goldman, 1985), the early rearrangement of keratin (Hennings and Holbrook, 1983) and actin and vinculin (O'Keefe *et al.*, 1987) filaments, and/or redistribution of peanut lectin (PNA)-binding cell membrane glycoproteins (Morrison *et al.*, 1988) may be involved in the changes in plasma membrane physicochemical properties observed upon induction of keratinocyte differentiation.

VI. Retinoid-Induced Modulations of Lipid Composition in Cultured Keratinocytes

A great variety of pharmacological agents used in the treatment of dermatological disorders are known for their modulatory effects on epidermal differentiation. Such drugs include glucocorticoids, retinoids (vitamin A derivatives), and vitamin D_3. The most extensively studied of these are the retinoids, which modulate differentiation of a wide variety of cell types, including epidermal cells (Shapiro,

1986). Exposure of keratinizing cells to retinoids has been demonstrated to inhibit the normal sequence of keratin expression (Fuchs and Green, 1981) and cornified envelope formation (Green and Watt, 1982). As judged from the keratin profile (Fuchs and Green, 1981), the extent of keratinocyte differentiation can be enhanced even under submerged conditions, when the keratinocytes are cultured in the presence of delipidized serum to reduce the concentration of retinoic acid (RA). This enhanced differentiation is also reflected by changes in the lipid composition. As shown above (Table I), in submerged cultures grown in a medium supplemented with fetal calf serum containing about 10^{-9} *M* RA (De Leenheer *et al.*, 1982), no AGC, AC, or lanosterol is synthesized. When keratinocytes are cultured in a medium supplemented with RA-depleted serum, the total ceramide content is increased and small amounts of AC and lanosterol can be detected even under submerged conditions (Brod *et al.*, 1991); in the air-exposed cultures removal of RA is not necessary to induce complete differentiation and stratification and to induce the synthesis of AC and lanosterol (Table II) (Ponec *et al.*, 1988; Ponec, 1989). The differences may result from differences in the manner of supplementation of cells by nutrients. While in submerged cultures, the cells are bathed in culture medium and all nutrients can easily reach all cell layers; in the air-exposed cultures, RA reaches the cells only by diffusion through the dermis, which may act as a filter and limit the amount of available liposoluble substances.

Under both submerged and the air-exposed culture conditions, addition of RA at micromolar concentrations to the cultures grown in RA-depleted medium induces marked changes in lipid synthesis (Ponec *et al.*, 1987b; Ponec and Boonstra, 1987) and in lipid composition (Brod *et al.*, 1991; Ponec, 1989). In submerged cultures, the administration of retinoids leads next to modulations of synthesis of major lipid classes (Ponec *et al.*, 1987b; Ponec and Boonstra, 1987), to decreased cholesterol sulfate production (Jetten *et al.*, 1989), and to a complete suppression of AC and lanosterol synthesis (Brod *et al.*, 1991).

In the air-exposed culture, the addition of RA at micromolar concentrations induces profound changes in tissue architecture, and the epidermal morphology resembles that seen in hypervitaminosis (Shapiro, 1986). In RA-treated cultures, keratin K1, a species normally expressed in suprabasal layers of differentiating epidermis, could hardly be detected (Ponec, 1989; Régnier and Darmon, 1989; Asselineau *et al.*, 1989), and a marked decrease in the cornified envelope formation is also observed (Régnier *et al.*, 1989). Not only stratum corneum morphology is changed but also the number of cells containing keratohyalin granullae and lamellar bodies is markedly decreased, explaining the low content of ceramides, AC, and lanosterol in these cultures (Fig. 9) (Ponec, 1989). These results suggest that liposoluble serum components, like RA, control the synthesis of lipids that are expressed at later stages of epidermal differentiation.

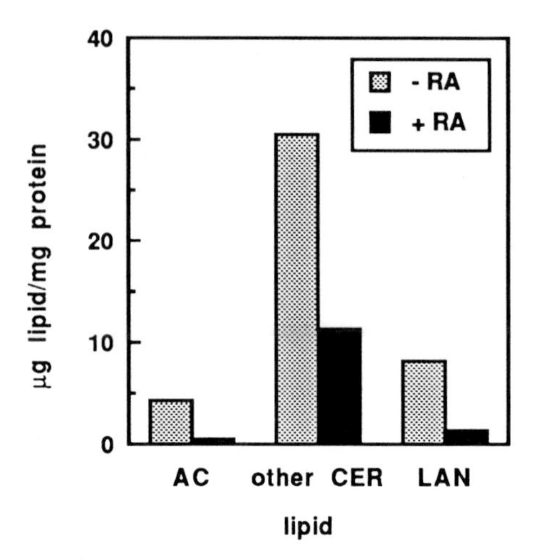

FIG. 9. Retinoic acid-induced modulation of content of acylceramides (AC), ceramides (CER), and lanosterol (LAN) in keratinocytes cultured at the air–liquid interface. For details, see Appendix (Section VIII). (From Ponec, 1989.)

VII. Conclusion

The use of cultured keratinocytes has substantially contributed to the present knowledge of processes involved in the regulation of lipogenesis in the human epidermis. The results obtained with submerged cultures clearly show that only in proliferating basal cells does the uptake mechanism of extracellular lipoproteins take place. These findings offer an explanation why the overall epidermal biosynthesis appears to be relatively autonomous with respect to systemic influences: neither dietary nor circulating sterol levels seem to affect significantly cutaneous sterologenesis and overall lipogenesis. Furthermore, it is clear that the synthesis of lipid classes important for epidermal barrier function (ceramides and especially acylceramides) occurs only when keratinocytes reach a certain stage of differentiation. The synthesis of these lipids is extremely low in submerged cultures, in which the epidermal differentiation is incomplete. The site of synthesis of these lipids is mainly the upper layers of the epidermis—the stratum granulosum—which conclusion is based on observations that before completion of stratification, i.e., (1) at the early growth stages of air-exposed cultures, (2) upon retinoic acid administration, which prevents complete stratification, and (3) in poorly differentiating air-exposed cultures of SCC cells, these lipids are either absent or present at extremely low concentration. The relationship of lipid metabolism and expression of some protein markers to tissue architecture and the overall degree of differentiation is schematically reviewed in Table V.

Table V

EXPRESSION OF EPIDERMAL CHARACTERISTICS IN CULTURES OF NORMAL HUMAN KERATINOCYTES[a]

Epidermal characteristics	Confluent submerged cultures: Low calcium	Confluent submerged cultures: Normal calcium	Air–medium interface culture
Tissue architecture			
Stratification	BC layer	BC layer 2–5 SPB layers	All epidermal strata (SB, SS, SG, SC)
Lamellar bodies	Absent	Absent	Present in lower amounts than *in vivo*
Desmosomes	Absent	Present	Present
Cornified cell envelope	Absent	Present in low numbers	Present in upper cell layers
Expression of differentiation-specific protein markers			
Keratins	BC layer: K5, K14	BC layer: K5, K14; SPB layers: K6, K16, K4, K13, K19, K10, K1[b]	SB K5, K14; SPB layers: K6, K16, K1, K10
Involucrin	In a low number of cells	SPB layers	SPB layers
Transglutaminase	Absent	SPB layers	SPB layers
Lipid markers related to keratinocyte differentiation			
LDL receptor expression	High	Low	In BC layer
LDL regulated sterologenesis	Present	Absent	n.d.
Plasma membrane fluidity	High	Low	n.d.
Lanosterol	Absent	Absent[b]	Present
Sphingolipids	Low amount	Low amount	High amount
AC, AGC	Absent	Absent[b]	Present
Other ceramides	Present	Present	Present
Barrier function	Absent	Absent	Present (lower than *in vivo*)

[a]Abbreviations: BC, basal cell; SPB, suprabasal; SB, stratum basale; SS, stratum spinosum; SG, stratum granulosum; SC, stratum corneum; n.d., not done; K, keratin.

[b]Low amounts detectable in RA-depleted media.

It can be concluded that cultured keratinocytes provide a suitable model to study epidermal lipid metabolism in relation to keratinocyte differentiation. In this respect, air-exposed cultures are most attractive, because their resemblance to intact epidermal tissue is strong. Furthermore, keratinocytes isolated from pathological skin specimens are in many cases able to maintain the morphogenic capabilities of their tissue of origin, and this offers a possibility of comparison with keratinocytes derived from the normal skin. This enables creation of conditions mimicking normal and pathological states *in vitro* and the possibility to evaluate biological responses at the cellular level. In a study on the effects of pharmacological agents, such as retinoids, glucocorticoids, or various additives of cosmetic preparations, two approaches can be used: (1) various modulators of differentiation can be added to the culture medium before elevating the culture to the air–liquid interface, and their effects on the epidermal differentiation program studied thereafter; (2) pharmacological agents can be applied topically after completion of differentiation. Their effects on epidermal differentiation, lipid metabolism, and stratum corneum composition and barrier function can be then studied, thus affording an acceptable alternative to experiments with laboratory animals.

VIII. Appendix

A. Cell Culture

1. *Submerged Culture*

Juvenile foreskin keratinocytes (derived from donors aged 1–2 years) or adult keratinocytes (derived from the skin of healthy human subjects undergoing mammary reduction) and SCC lines (SCC-4, SCC-15, and SCC-12F2) (Rheinwald and Beckett, 1980) were cultured together with irradiated mouse 3T3 fibroblast feeder cells in Dulbecco's modified essential medium (DMEM) and Ham's F12 (3:1) medium supplemented with 5% fetal calf serum (FCS) and 0.4 μg/ml hydrocortisone. Normal keratinocytes were supplemented with 10^{-6} *M* isoproterenol and 10 ng/ml epidermal growth factor (Rheinwald, 1989). For low-calcium conditions, calcium-free DMEM was mixed with standard Ham's F12 (3:1) medium and supplemented with 5% chelex-treated fetal calf serum (Hennings *et al.*, 1980) and other additives as listed above. The final calcium concentration was 0.06 m*M* as determined by flame photometry.

2. *Air-Exposed Culture*

The deepidermized dermis for air-exposed cultures was prepared as described by Régnier *et al.* (1981). Briefly, cadaver skin (stored at 4°C in 85% glycerol) was carefully washed with phosphate-buffered saline (PBS) and incubated for 3–5 days in PBS at 37°C. Subsequently, the epidermis was scraped off and the remaining dermis irradiated (3000 R) and washed several times with culture medium. The dermis was then placed on the stainless-steel grid and 0.5 x 10^6 normal human keratinocytes (second or third passage) were inoculated inside a stainless-steel ring (diameter 1 cm) placed on the top of the dermis. After 24–48 hours, the ring was removed, and the level of culture medium was adjusted to just reach the height of the grid. This method ensures that the cells are exposed to air throughout the remaining period of culture. The medium used for air-exposed cultures was Dulbecco–Vogt and Ham's F12 (3:1) medium supplemented with 5% fetal calf serum (FCS) or 5% delipidized FCS (DLS) (Rothblat *et al.*, 1976),

10^{-6} *M* isoproterenol, and 10 ng/ml epidermal growth factor. In experiments in which the effect of retinoic acid on lipid composition was studied, the cultures were refed on days 3 and 7 with media supplemented with 2 μ*M* retinoic acid, using 0.5 μl/ml medium of freshly prepared stock solution in absolute alcohol. Addition of retinoic acid was performed under yellow light, and the cultures were maintained in the dark. Controls received 0.5 μl ethanol/ml medium only.

In experiments in which collagen lattice was used instead of DED as dermal substrate, the slightly modified method of Bell *et al.* (1983) was used. Collagen lattice was prepared by mixing 1.0 ml of prechilled 10x concentrated DMEM with antibiotics, 1.0 ml fetal calf serum, 8.0 ml rat collagen type I solution (4 mg/ml in 1:1000 acetic acid), 0.2 ml 0.1 *M* NaOH, and 0.2 ml medium with or without 1 x 10^6 human adult skin fibroblasts. After mixing, 0.75 ml collagen solution was pipeted per well (Falcon, 24 well) and the dishes were subsequently placed in a humid incubator at 37°C. The next day the collagen lattice was placed on the stainless-steel grid and the keratinocytes were seeded on the top, as described for DED cultures.

B. Histology

Cultures were fixed in 4% buffered formaldehyde and processed for embedding in paraffin. Vertical sections were stained with hematoxylin and eosin.

C. Cell Studies

1. Cell Association and Degradation of ^{125}I-Labeled LDL

Cells were preincubated for 24 hours with medium containing lipoprotein-deficient newborn calf serum (LPDS) to obtain expression of LDL receptors. ^{125}I-Labeled LDL was then added (10 μg/ml), and the cells were reincubated for 3 hours at 37°C. For assessment of nonspecific binding, 300 μg unlabeled LDL/ml was added to medium containing ^{125}I-labeled LDL. The cellular association and degradation of ^{125}I-labeled LDL were measured according to Goldstein and Brown (1974).

Human LDL (d = 1.03–1.05 g/ml) was isolated from the plasma of a healthy subject by gradient ultracentrifugation (Redgrave *et al.*, 1975). LPDS ($d > 1.215$ g/ml) was prepared by ultracentrifugation at a density of 1.215 g/ml, followed by dialysis against saline. LDL was labeled with ^{125}I according to Bilheimer *et al.* (1972).

In experiments in which the cellular distribution of LDL receptors was studied, cytoskeletons were isolated either before or after the binding assay by exposing cells for 10 minutes at 0°C to 0.5% Triton X-100 in 25 m*M* HEPES (pH 7.4) containing 1 m*M* PMSF (van Bergen en Henegouwen *et al.*, 1989). The binding assay was performed by incubation of cells for 3 hours at 0°C with ^{125}I-labeled LDL (10 μg/ml).

2. Morphological Studies

Prior to incubation with LDL, keratinocytes were preconditioned for 24 hours in medium supplemented with 5% LPDS and subsequently incubated for 30 minutes at 37°C with LDL–gold particles (20 nm) at concentrations between 25 and 50 μg protein/ml. After incubation, the cells were fixed and processed for electron microscopy (Ponec *et al.*, 1985).

3. LDL Receptor mRNA and Protein Expression

a. Isolation of Total Cytoplasmic RNA. Cells were grown to confluency in medium supplemented with FCS. At 48 hours prior to the experiment, the cells were refed with medium containing 5% FCS or 5% LPDS. For total cytoplasmic RNA isolation, a modification of the method described by Brawerman *et al.* (1972) was used (te Pas *et al.*, 1989).

b. Isolation of the LDL-Receptor S1 Nuclease Analysis Probe. Using plasmid pLDLR-2HH1, kindly provided by D. W. Russell (Yamamoto *et al.*, 1984), containing nucleotides 1572–3485 of the LDL receptor cDNA sequence in pSP64 vector, a specific S1 nuclease analysis probe was isolated after digestion with restriction endonuclease HinfI. The probe contained nucleotides 3240–3485 of LDL receptor cDNA and about 200 nucleotides of the vector. After 3′ end labeling (using [^{32}P]ATP and Klenow enzyme), the probe was purified by 5% polyacrylamide gel electrophoresis (PAGE) followed by electroelution (Biotrap; Schleicher & Schuell) from the gel.

c. S1 Nuclease Analysis. The hybridization of the S1 nuclease analysis was performed at 57°C (te Pas *et al.*, 1989).

d. Protein Isolation and Blotting Procedures. For isolation of cellular proteins, a modification of the method described by Daniel *et al.* (1983) was used. Blotting procedures, including gel electrophoresis, electroblotting, and ligand blotting with ^{125}I-labeled β-VLDL, were done, as described elsewhere (te Pas *et al.*, 1991b).

e. Densitometric Analysis. The intensity of the bands obtained after both the S1 nuclease analysis and the ligand blotting experiments were quantified photodensitometrically (DESAGA CD 60).

4. Lipid Extraction and Separation

Cells grown under submerged conditions were harvested after 10 days by trypsinization and washed several times with phosphate-buffered saline. Cells grown on DED were harvested after 10 days of culture, by heating the culture for 1 minute at 60°C, washing in PBS, collection in 2 ml chloroform:methanol (1:2), and storage at -20°C until use. Subsequently, the lipids were extracted according to Bligh and Dyer (1959), with the addition of 0.25 *M* KCl to ensure extraction of polar species.

For lipid separation, CAMAG horizontal development chambers were used. All solvents used for the separation of lipids were of analytical grade (Merck; Darmstadt, Germany). All developments were carried out at 4°C. Following each development step, the high-performance thin-layer chromatography (HPTLC) plate was dried under a stream of air at 40°C on the heat block (Thermoplate, DESAGA; Heidelberg, Germany) for approximately 10 minutes. Lipids were fractionated using two different development systems.

a. Total Lipid Development System. For the analysis of total lipids, 5–50 μg of the total lipid extract was applied to HPTLC plates and developed at 4°C sequentially from the bottom edge of the HPTLC plate: (1) 15 mm, chloroform; (2) 10 mm, chloroform:diethyl ether:acetone:methanol (72:8:10:10); (3) 60 mm, chloroform:acetone:methanol:33% NH_3 (76:16:8:1); (4) 70 mm, chloroform:diethyl ether:acetone:methanol (72:8:10:10); (5) 30 mm, chloroform:ethyl acetate:ethyl methyl ketone:2-propanol:ethanol:methanol:water:acetic acid (34:4:4:6:20:28:4:1); (6) 40 mm, chloroform:ethyl acetate:ethyl methyl ketone:2-propanol:ethanol:methanol:H_2O (46:4:4:6:6:28:6); (7) 80 mm, chloroform:diethyl ether:acetone:methanol (76:4:8:12); (8) 90 mm, hexane:diethyl ether:ethyl acetate (80:16:4).

b. Ceramide Development System. For the analysis of sphingolipids, 5–50 μg of the total lipid extract was applied to HPTLC plates and developed sequentially from the bottom edge of the plate: (1) 15 mm, chloroform; (2) 10 mm, chloroform:acetone:methanol (76:8:16); (3) 70 mm, chloroform:hexyl acetate:acetone:methanol (86:1:10:4); (4) 20 mm, chloroform:acetone:methanol (76:4:20); (5) 75 mm, chloroform:diethyl ether:hexyl acetate:ethyl acetate:acetone: methanol (72:4:1:4:16:4); (6) 90 mm, hexane:diethyl ether:ethyl acetate (80:16:4).

c. Staining and Charring. After accomplishing the lipid separation, the plates were first dried at 40°C under an air stream and subsequently heated for 5 minutes at 130°C. After cooling to 60°C, the lipids were stained with a staining mixture containing acetic acid, H_2SO_4, H_3PO_4, and H_2O (5:1:1:93) and 0.5% $CuSO_4$, and were heated to 160°C.

d. Standards. For identification and quantification of lipids, the following standards were used: lysophosphatidylcholine; sphingomyelin; phosphatidylcholine; phosphatidylserine; phosphatidylinositol; phosphatidylethanolamine;

cerebrosides, type I and type II; lipid standards containing oleic acid, cholesterol, triolein, and cholesterol oleate; lanosterol; lathosterol; cholesterol sulfate; 1,2-diolein and 1,3-diolein; squalene (Sigma); and ceramides (bovine, mixture; Alltech). Although the bovine ceramide standards do not comigrate with all human ceramide bands, they were used for mass calibration purposes. Acylglucosylceramides and acylceramides, kindly provided by Dr. D. Nugteren, were used for qualitative purposes only.

e. Quantification. The HPTLC plates, after HPTLC development and charring, were scanned using the densitometer CD 60 (DESAGA, Heidelberg, Germany), which is equipped with suitable software for automatic measurements and integration of samples and evaluation of standards. Mass determinations of lipids were performed with corresponding standard series of the cochromatographed lipid standards on each HPTLC plate.

5. *Incorporation of [^{14}C]Acetate into Lipids*

Confluent cultures were incubated for 30 hours in culture media supplemented with 5% lipoprotein-deficient serum or 5% fetal calf serum (Ponec *et al.*, 1985). For measurement of *de novo* lipid synthesis, [^{14}C]acetate (2.5 μCi/ml, Amersham International, Amersham) at a final concentration of 0.5 m*M* was added, after which the cells were incubated for 18 hours. The cells were harvested and lipid was extracted and separated, as described above. Lipid fractions, after visualization by iodine vapor, were scraped and radioactivity was determined.

6. *Penetration Studies*

For penetration studies the following skin specimens were used: (1) epidermal sheets obtained by dispase treatment (Green *et al.*, 1979) of keratinocyte cultures grown to confluence under submerged conditions; (2) keratinocyte cultures grown under the air-exposed conditions on DED; (3) freshly excised mamma skin; and (4) stratum corneum isolated from the excised skin (Tiemessen *et al.*, 1989a). For penetration studies, all skin samples were sandwiched between artificial supporting membranes. For nitroglycerin penetration studies, skin samples were placed between two silicone membrane disks (Tiemessen *et al.*, 1989b). For sucrose penetration studies, the skin samples were placed between two dialysis membranes (Ponec *et al.*, 1990). The flux of nitroglycerin or [^{14}C]sucrose through the sandwich (containing a skin specimen) was measured at 32°C using a slightly modified flow-through Franz diffusion cell (Ponec *et al.*, 1990).

ACKNOWLEDGMENTS

The useful discussion and help of P. Elias is gratefully acknowledged.

References

Andersen, R. G. W., and Dietschy, J. M. (1977). *J. Biol. Chem.* **252,** 3652–3657.

Asselineau, D., Bernhard, B., Bailly, C., and Darmon, M. (1985). *Exp. Cell Res.* **159,** 536–539.

Asselineau, D., Bernhard, B., Bailly, C., and Darmon, M. (1989). *Dev. Biol.* **133,** 322–335.

Bell, E., Sher, S., Hull, B., Merrill, C., Rosen, S., Chamson, A., Asselineau, D., Dubertret, L., Coulomb, B., Lapiere, C., Nusgens, B., and Neveux, Y. (1983). *J. Invest. Dermatol.* **82,** 2s–10s.

Bernard, B. A., Asselineau, D., Schaffar-Deshayes, L., and Darmon, M. Y. (1988). *J. Invest. Dermatol.* **90,** 801—805.

Bernstam, L. I., Vaughan, F. L., and Bernstein, I. A. (1986). *In Vitro* **22,** 695–705.

Bilheimer, D. W. S., Eisenberg, S., and Levi, K. I. (1972). *Biochim. Biophys. Acta* **260,** 212–221.

Bligh, E. G., and Dyer, W. J. (1959). *Can. J. Biochem. Physiol.* **37,** 911–917.

Boddé, H. E., Holman, B., Spies, F., Weerheim, A., Kempenaar, J., Mommaas, A. M., and Ponec, M. (1990). *J. Invest. Dermatol.* **95,** 108–116.

Bohnert, A., Hornung, J., Mackenzie, I. C., and Fusenig, N. E. (1986). *Cell Tissue Res.* **244,** 413–429.

Boonstra, J., de Laat, S. W., and Ponec, M. (1985). *Exp. Cell Res.* **161,** 421–433.

Bowden, P. E., Wood, E. J., and Cunliffe, W. J. (1983). *Biochim. Biophys. Acta* **743,** 172–176.

Bowser, P. A., Nugeteren, D. H., White, R. J., Houtsmuller, U. M. T., and Protteey, C. (1985). *Biochim. Biophys. Acta* **834,** 419–428.

Brannan, P. G., Goldstein, J. L., and Brown, M. S. (1975). *J. Lipid Res.* **16,** 7–11.

Brawerman, G., Mendecki, J., and Lee, S. Y. (1972). *Biochemistry* **11,** 637–641.

Brod, J., Bavelier, E., Justine, P., Weerheim, A., and Ponec, M. (1991). *In Vitro Cell. Dev. Biol.* **27A,** 163–168.

Brody, I. (1962). *J. Ultrastruct. Res.* **6,** 314–353.

Brysk, M. M., Miller, J., and Walker, G. K. (1984). *Exp. Cell Res.* **150,** 329–337.

Cumpstone, M. B., Kennedy, A. H., Harmon, C. S., and Potts, R. O. (1989). *J. Invest. Dermatol.* **82,** 598–600.

Daniel, T. O., Schneider, W. J., Goldstein, J. L., and Brown, M. S. (1983). *J. Biol. Chem.* **258,** 4606–4611.

De Leenheer, A. P., Lambert, W. E., and Clayes, I. (1982). *J. Lipid Res.* **23,** 1362–1367.

Dover, R., and Watt, F. M. (1987). *J. Invest. Dermatol.* **89,** 349–352.

Elias, P. M. (1983). *J. Invest. Dermatol.* **80s,** 44–49.

Elias, P. M., and Friend, D. S. (1975). *J. Cell Biol.* **65,** 180–191.

Elias, P. M., Brown, B. E., Fritsch, P., Goerke, J., Gray, G. M., and White, R. J. (1979). *J. Invest. Dermatol.* **73,** 338–348.

Feingold, K. R., Brown, B. E., Lear, S. R., Moser, A. H., and Elias, P. M. (1983). *J. Invest. Dermatol.* **81,** 365–369.

Feingold, K. R., Brown, B. E., Lear, S. R., Moser, A. H., and Elias, P. M. (1986). *J. Invest. Dermatol.* **87,** 588–591.

Feingold, K. R., Williams, M. L., Pillai, S., Menon, G. K., Halloran, B. P., Bikle, D. D., and Elias, P. M. (1987). *Biochim. Biophys. Acta* **930,** 193–200.

Fuchs, E., and Green, H. (1980). *Cell* **19,** 1033–1042.

Fuchs, E., and Green, H. (1981). *Cell* **25,** 617–625.

Goldstein, J. L., and Brown, M. S. (1974). *J. Biol. Chem.* **149,** 5153–5162.

Goldstein, J. L., and Brown, M. S. (1984). *J. Lipid Res.* **25,** 1450–1461.

Gray, G. M., and Yardley, H. J. (1975). *J. Lipid Res.* **16,** 441–447.

Gray, G. M., White, E. J., and Majer, J. R. (1978). *Biochim. Biophys. Acta* **528,** 127–137.

Green, H., and Watt, F. (1982). *Mol. Cell. Biol.* **2,** 1115–1117.

Green, H., Kehinde, O., and Thomas, J. (1979). *Proc. Natl. Acad. Sci. U.S.A.* **76,** 5665–5668.

Grubauer, G., Feingold, K. R., and Elias, P. M. (1987). *J. Lipid Res.* **28,** 746–752.
Grubauer, G., Elias, P. M., and Feingold, K. R. (1989). *J. Lipid Res.* **30,** 323–330.
Hagchisuka, H., Nomura, H., Mori, O., Nakano, S., Okubo, K., Kusuhara, M., Karashima, M., Tanikawa, E., Higuchi, M., and Sasai, Y. (1990). *Cell Tissue Res.* **260,** 207–210.
Hansen, H. S., and Jensen, B. (1985). *Biochim. Biophys. Acta* **834,** 357–363.
Hennings, H., and Holbrook K. (1983). *Exp. Cell Res.* **143,** 127–142.
Hennings, H., Michael, D., Cheng, C., Steinert P., Holbrook, K., and Yuspa, S. H. (1980). *Cell* **19,** 245–254.
Holbrook, K. A., and Hennings, H. (1983). *J. Invest. Dermatol.* **81,** 11s–24s.
Isseroff, R., Martinez, D. T., and Ziboh, V. A. (1985). *J. Invest. Dermatol.* **85,** 131–134.
Isseroff, R., Ziboh, V. A., Chapkin, R. S., and Martinez, D. T. (1987). *J. Lipid Res.* **28,** 1342–1349.
Jetten, A. M., George, M. A., Nervi, C., Boone, L. R., and Rearick, J. I. (1989). *J. Invest. Dermatol.* **92,** 203–209.
Jones, J. C. R., and Goldman, R. D. (1985). *J. Cell Biol.* **101,** 506–517.
Kennedy, A., Francoeur, M., Jakowski, A., Potts, R., Ongpipattanakul, B., and Burnette, R. (1990). *J. Invest. Dermatol.* **94,** 541.
Kim, K. H., Schwartz, F., and Fuchs, E. (1985). *Proc. Natl. Acad. Sci. U.S.A.* **81,** 4280–4284.
Kopan, R., Traska, G., and Fuchs, E. (1987). *J. Cell Biol.* **105,** 427–440.
Lampe, M. E., Williams, M. L., and Elias, P. M. (1983a). *J. Lipid Res.* **24,** 131–140.
Lampe, M. E., Burlingame, A. L., Whitney, J., Williams, M. L., Brown B. E., Roitman, E., and Elias, P. M. (1983b). *J. Lipid Res.* **24,** 120–130.
Lane, B. E., Bartek, J., Purkis, P. E., and Leigh, I. M. (1985). *Ann. N.Y. Acad. Sci.* **455,** 241–258.
Leigh, I. M., Pulfors, K. A., Ramaerkers, F. C. S., and Lane, E. B. (1985). *Br. J. Dermatol.* **113,** 53–64.
Lillie, J. H., MacCallum, D. K., and Jepson, A. (1980). *Exp. Cell Res.* **125,** 153–165.
Long, V. J. W. (1970). *Br. J. Dermatol.* **55,** 269–273.
Mackenzie, I. C., and Fusenig, N. E. (1983). *J. Invest. Dermatol.* **81,** 189s–194s.
Madison, K. C., Wertz P. W., Strauss J. S., and Downing, D. T. (1986). *J. Invest. Dermatol.* **87,** 253–259.
Madison, K. C., Swartzendruber, D. C., Wertz, P. W., and Downing, D. T. (1988). *J. Invest. Dermatol.* **90,** 110–116.
Madison, K. C., Swartzendruber, D. C., Wertz, P. W., and Downing, D. T. (1989). *J. Invest. Dermatol.* **93,** 10–17.
Magee, A. I., Lytton, N. A., and Watt, F. M. (1987). *Exp. Cell Res.* **172,** 42–53.
Malmqvist, K. G., Carlsson, I. E., Forslind, B., Roomans, G. M., and Akselsson, K. R. (1984). *Nucl. Instrum. Methods Phys. Res.* **B3,** 611–617.
Mansbridge, J. N., and Knapp, A. M. (1987). *Br. J. Dermatol.* **111,** Suppl. 27, 253–262.
Melton, D. (1968). *Am. J. Anat.* **122,** 337–341.
Menon, G. K., Feingold, K. R., Moser, A. H., Brown, B. E., and Elias, P. M. (1985). *J. Lipid Res.* **26,** 418–427.
Menon, G. K., Grayson, S., and Elias, P. M. (1986). *J. Invest. Dermatol.* **86,** 591–597.
Miller, C. C., and Ziboh, V. A. (1990). *J. Invest. Dermatol.* **94,** 353–358.
Mommaas-Kienhuis, A. M., Grayson, S., Wijsman, M. C., Vermeer, B. J., and Elias, P. M. (1987). *J. Invest. Dermatol* **89,** 513–517.
Mommaas, A. M., Tada, J., Wijsman, M. C., Onderwater, J. J. M., and Vermeer, B. J. (1990). *J. Dermatol. Sci.* **1,** 15–22.
Mommaas, A. M., Tada, J., and Ponec, M. (1991). *J. Dermatol. Sci.* **2,** 97–105.
Monger, D. J., Williams, M. L., Feingold, K. R., Brown, B. E., and Elias, P. M. (1988). *J. Lipid Res.* **29,** 603–612.
Morrison, A. I., Keeble, S., and Watt, F. M. (1988). *Exp. Cell Res.* **177,** 247–256.
Nugteren, D. H., and Kivits, G. A. A. (1987). *Biochim. Biophys. Acta* **921,** 135–141.
Nugteren, D. H., Christ-Hazelhof, E., van der Beek, A., and Houtsmuller, U. M. T. (1985). *Biochim. Biophys. Acta* **843,** 429–436.

O'Keefe, E. J., and Payne, R. E. (1983). *J. Invest. Dermatol.* **81,** 213–235.
O'Keefe, E. J., Briggaman, R. A., and Herman, B. (1987). *J. Cell Biol.* **105,** 807–817.
Ponec, M. (1989). *In* "Pharmacology and the Skin," (B. Shroot, and H. Schaefer, eds.), pp. 45–51. Karger, Basel.
Ponec, M. (1991). *Toxicol. In Vitro* (in press).
Ponec, M., and Boonstra, J. (1987). *Dermatologica* **175,** Suppl. 1, 66–72.
Ponec, M., and Williams, M. L. (1986). *Arch. Dermatol. Res.* **279,** 32–36.
Ponec, M., Havekes, L., Kempenaar, J., and Vermeer, B. J. (1983). *J. Invest. Dermatol.* **81,** 125–130.
Ponec, M., Havekes, L., Kempenaar, J., Lavrijsen, S., and Vermeer, B. J. (1984). *J. Invest. Dermatol.* **83,** 436–440.
Ponec, M., Havekes, L., Kempenaar, J., Lavrijsen, S., Wijsman, M., Boonstra, J., and Vermeer, B. J. (1985). *J. Cell. Physiol.* **125,** 98–106.
Ponec, M., Weerheim., A., Kempenaar, J., and Boonstra, J. (1987a). *J. Cell. Physiol.* **133,** 358–364.
Ponec, M., Weerheim, A., Havekes, L., and Boonstra, J. (1987b). *Exp. Cell Res.* **171,** 426–435.
Ponec, M., Kempenaar, J., and Boonstra, J. (1987c). *Biochim. Biophys. Acta* **921,** 512–521.
Ponec, M., Weerheim, A., Kempenaar, J., Mommaas, A. M., and Nugteren, D. H. (1988). *J. Lipid Res.* **29,** 949–962.
Ponec, M., te Pas, M. F. W., and Boonstra, J. (1989a). *In* "The Dermis. From Biology to Diseases," (G. Piérard and C. Piérard-Franchimont, eds.), pp. 60–71. Monogr. Dermatopathol., Liege.
Ponec, M., Weerheim, A., Kempenaar, J., Elias, P. M., and Williams, M. (1989b). *In Vitro Cell. Dev. Biol.* **25,** 689–696.
Ponec, M., Wauben-Penris, P. J. J., Burger, A., Kempenaar, J., and Boddé, H. E. (1990). *Skin Pharmacol.* **3,** 70–85.
Prottey, C., Hartop, P. J., Black, J. G., and McCormack, J. I. (1976). *Br. J. Dermatol.* **94,** 579–587.
Purkis, P. E., Leigh, I. M., and Lane, E. B. (1990). *J. Invest. Dermatol.* **95,** 484.
Redgrave, T. G., Roberts, D. C. K., and West, C. (1975). *Anal. Biochem.* **65,** 42–49.
Régnier, M., and Darmon, M. (1989). *In Vitro* **25,** 1000–1008.
Régnier, M., Prunieras, M., and Woodley, D. (1981). *Front. Matrix Biol.* **9,** 4–35.
Régnier, M., Schweizer, J., Michel, S., Bailly, C., and Prunieras, M. (1986). *Exp. Cell Res.* **165,** 63–72.
Régnier, M., Desbas, C., Bailly, C., and Darmon, M. (1988). *In Vitro* **24,** 625–632.
Régnier, M., Eustache, J., Shroot, B., and Darmon, M. (1989). *Skin Pharmacol.* **2,** 1–9.
Régnier, M., Asselineau, D., and Lenoir, M. C. (1990). *Skin Pharmacol.* **3,** 70–85.
Rheinwald, J. G. (1989). *In* "Cell growth and Division. A Practical Approach" (R. Baserga, ed.), pp. 81–94. IRL Press, Oxford.
Rheinwald, J. G., and Beckett, M. A. (1980). *Cell* **22,** 629–632.
Rheinwald, J. G., and Beckett, M. A. (1981). *Cancer Res.* **41,** 1657–1663.
Rice, R. H., and Green, H. (1979). *Cell* **18,** 681–694.
Rothblat, G. H. Arbogast, L. Y., Ovellett, L., and Howard, B. V. (1976). *In Vitro* **12,** 554–557.
Schürer, N. Y., Monger, D. J., Hincenbergs, M., and Williams, M. L. (1989). *J. Invest. Dermatol.* **92,** 196–202.
Severson, D. L. (1979). *J. Mol. Cell. Cardiol.* **11,** 569–583.
Shapiro, L. J., Weiss, R., Webster, D., and France, J. T. (1978). *Lancet* **1,** 70–72.
Shapiro, S. (1986). *In* "Retinoids and Cell Differentiation" (M. I. Sherman, ed.), pp. 29–59. CRC Press, Boca Raton, Florida.
Snoek, G., Boonstra, J., Ponec, M., and de Laat, S. W. (1987). *Exp. Cell Res.* **172,** 146–157.
Stoler, A., Kopan, R., Duvic, M., and Fuchs, E. (1988). *J. Cell Biol.* **107,** 427–446.
Stoll, L. L., and Spector, A. A. (1984). *In Vitro* **20,** 732–738.
Swartzendruber, D. D., Wertz, P. W., Kitko, D. J., Madison, K. C., and Downing, D. T. (1989). *J. Invest. Dermatol.* **92,** 251–257.
Tanaka, T., Ogura, R., Hidaka, T., and Sugiyama, M. (1989). *J. Invest. Dermatol.* **93,** 682–686.
Tang, W., Ziboh, V. A., Isseroff, R. R., and Martinez, D. (1987). *J. Cell. Physiol.* **132,** 131–136.

Tang, W., Ziboh, V. A., Isseroff, R. R., and Martinez, D. (1988). *J. Invest. Dermatol.* **90,** 37–43.
te Pas, M. F. W., Boonstra, J., Havekes, L., Hesseling, S. C., and Ponec, M. (1989). *Cell Biol. Int. Rep.* **13,** 237–249.
te Pas, M. F. W., Ponec, M., van Bergen en Henegouwen, P. M. P., Lombardi, P., Havekes, L., and Boonstra, J. (1990). *Cell Biol. Int. Rep.* **14,** 989–999.
te Pas, M. F. W., van Bergen en Henegouwen, P. M. P., Boonstra, J., and Ponec, M. (1991a). *Arch. Dermatol. Res.* **283,** 125–130.
te Pas, M. F. W., Lombardi, P., Havekes, L., Boonstra, J., and Ponec, M. (1991b). *J. Invest. Dermatol.* **97,** 334–339.
Tertoolen, L. G. J., Kempenaar, J., Boonstra, J., de Laat, S. W., and Ponec, M. (1988). *Biochem. Biophys. Res. Commun.* **152,** 491–496.
Tiemessen, H. L. G. M., Boddé, H. E., Mollee, H., and Junginger, H. E. (1989a). *Int. J. Pharmacol.* **56,** 87–94.
Tiemessen, L. G. M., Boddé, H. E., and Junginger, H. E. (1989b). *Int. J. Pharmacol* **53,** 119–127.
van Bergen en Henegouwen, P. M. P., Defize, L. H. K., De Kroon, J., Van Damme, H., Verkleij, A. J., and Boonstra, J. (1989). *J. Cell. Biochem.* **39,** 455–465.
Van Muijen, G. N. P., Warnaar, S. O., and Ponec, M. (1987). *Exp. Cell Res.* **171,** 331–345.
Vermeer, B. J., Reman, F. C., and van Gent, C. M. (1979). *J. Invest. Dermatol.* **73,** 303–305.
Vermeer, B. J., Wijsman, M. C., Mommaas-Kienhuis, A. M., and Ponec, M. (1985). *Eur. J. Cell Biol.* **38,** 353–360.
Vermeer, B. J., Wijsman, M. C., Mommaas-Kienhuis, A. M., and Ponec, M. (1986). *J. Invest. Dermatol.* **86,** 195–200.
Vessby, B., Gustafson, S., Chapman, M. J., Hellsing, K., and Lithell, H. (1987). *J. Lipid Res.* **28,** 629–641.
Watt, F. M. (1989). *Curr. Opinion Cell Biol.* **1,** 1107–1111.
Watt, F. M., and Green, H. (1982). *Nature (London)* **295,** 434–436.
Watt, F. M., Mattery, D. L., and Garrod, D. R. (1984). *J. Cell Biol.* **99,** 2211–2215.
Weiss, R. A., Eichner, R., and Sun, T.-T. (1984). *J. Cell Biol.* **98,** 1397–1406.
Wertz, P. W., and Downing, D. T. (1982). *Science* **217,** 1261–1262.
Wertz, P. W., and Downing, D. T. (1983). *J. Lipid Res.* **24,** 759–765.
Wertz, P. W., Cho, E. S., and Downing, D. T. (1983). *Biochim. Biophys. Acta* **753,** 350–355.
Wertz, P. W., Swartzendruber, D. C., Madison, K. C., and Downing, D. T. (1987). *J. Invest. Dermatol.* **89,** 419–425.
Wertz, P. W., Swartzendruber, D. C., Kitko, D. J., Madison, K. C., and Downing, D. T. (1989). *J. Invest. Dermatol.* **93,** 169–172.
Williams, M. L., and Elias, P. M. (1981). *J. Clin. Invest.* **68,** 1404–1410.
Williams, M. L., and Elias, P. M. (1987). *CRC Crit. Rev. Ther. Drug Carrier Syst.* **3,** 95–122.
Williams, M. L., Mommaas-Kienhuis, A. M., Rutherford, S. L., Grayson, S., Vermeer, B. J., and Elias, P. M. (1987a). *J. Cell. Physiol.* **132,** 428–440.
Williams, M. L., Rutherford, S. L., and Feingold, K. R. (1987b). *J. Lipid Res.* **28,** 955–967.
Williams, M. L., Brown, B. E., Monger, D. J., Grayson, S., and Elias, P. M. (1988a). *J. Cell. Physiol.* **136,** 103–110.
Williams, M. L., Rutherford, S. L., Ponec, M., Hincenbergs, M., Placzek, D. R., and Elias, P. M. (1988b). *J. Invest. Dermatol.* **91,** 86–91.
Woodcock-Mitchell, J., Eichner, R., Nelson, W. G., and Sun, T. T. (1982). *J. Cell Biol.* **95,** 580–588.
Yamamoto, T., Davis, C. G., Brown, M. S., Schneider, W. J., Casey, M. L., Goldstein, J. L., and Russell, D. W. (1984). *Cell* **39,** 29–38.
Yardley, H. J., and Summerly, R. (1981). *Pharmacol. Ther.* **13,** 357–383.
Ziboh, V. A. (1989). *In* "Pharmacology of the Skin" (M. W. Greaves and S. Shuster, eds.), Vol. 1, pp. 59–68. Springer-Verlag, Berlin.
Ziboh, V. A., and Chapkin, R. S. (1987). *Arch. Dermatol.* **123,** 1686a–1690a.
Ziboh, V. A., Isseroff, R. R., and Pandey, R. (1984). *Biochem. Biophys. Res. Commun.* **122,** 1234–1240.

ADVANCES IN LIPID RESEARCH, VOL. 24

Lipid Modulators of Epidermal Proliferation and Differentiation

WALTER M. HOLLERAN

Department of Dermatology
University of California School of Medicine
San Francisco, California 94143

I. Introduction

The importance of lipids in the maintenance and integrity of cellular membrane structures is well established. In the epidermis, investigation has focused most diligently on lipids in relation to production and maintenance of the hydrophobic interstitial domains, which represent the major barrier to water loss in mammals. Early studies demonstrated that significant changes in cellular lipid type and distribution accompany epidermal differentiation *in vivo* (see Elias and Menon, this volume). These dramatic changes in cellular lipid content reflect their function as bulk constituents that regulate the epidermal permeability barrier. However, more recent studies suggest that lipids are not only important as membrane components, but also that small quantities of lipids, generated at specific sites within the epidermis, may be involved in inflammation, as well as in directing cellular proliferation and differentiation.

This review describes a diverse group of lipids that have been implicated as effector molecules in various tissues, often including the epidermis. Lipids with more established roles will be presented first, including platelet-activating factor in cellular inflammation. Next, those lipids involved with signal transduction processes, such as phosphatidylinositol, phosphatidylcholine, and diacylglycerol, will be examined. Then, recent data concerning a range of sphingolipid compounds as possible biologic effectors, including gangliosides and sphingosine base derivatives, will be discussed, including possible sphingolipid involvement with growth factor-mediated signal transduction. Finally, a number of other lipids that are reported to affect a diverse array of cellular functions will be presented. Specifically excluded herein is coverage of eicosanoid lipids in inflammatory and regulatory processes, a topic far beyond the scope of this review. However, the relationship between phosphoinositide turnover and eicosanoids will be discussed, particularly in relation to the pathophysiology of psoriasis. The reader is referred to extensive recent reviews on eicosanoid metabolism and functions in the skin and other tissues (Ruzicka, 1990; Cunningham, 1990; Kupper, 1990; Ziboh and Tang, 1991b).

This review is intended as an introduction to some of the diverse cellular activities recently ascribed to lipids, often including preliminary findings. Although few data are available, the evidence for lipids as active modulators in epidermal tissues will be described. Finally, current speculations are included as to the possible importance of the various lipid effectors for epidermal function.

II. Ether Phospholipids: Platelet-Activating Factor

Platelet-activating factor (PAF) is an ether phospholipid with potent physiologic effects. Although this compound initially was described to stimulate platelet activation (Benveniste *et al.*, 1972), later it also was shown to activate macrophages, polymorphonuclear leukocytes, and monocytes, as well as to mediate numerous other processes, including inflammatory responses (for reviews see Snyder, 1987; Braquet *et al.*, 1987; Cunningham, 1990). During tissue inflammatory responses, PAF is released from mast cells and triggers the release of histamine from platelets (for review see Braquet *et al.*, 1987). Structural characterization of PAF has shown it to be a phosphorylcholine derivative of acetylglycerol (1-*O*-alkyl-2-acetyl-*sn*-glycero-3-phosphocholine) (Benveniste *et al.*, 1979). The ether linkage at the *sn*-1 position is critical for optimal activity (Demopoulos *et al.*, 1979; for reviews see Hanahan, 1986; Snyder, 1987). In addition, either increasing the chain length at the *sn*-2 position beyond three carbons or modifying the polar head groups diminishes the relative potency of PAF homologues (Snyder, 1987).

The activity of PAF is receptor mediated, with at least two receptors involved (Hwang, 1988; for review see Prescott *et al.*, 1990). Binding of PAF results in G-protein activation (Houslay *et al.*, 1986; Hwang *et al.*, 1986), as well as activation of phosphoinositide-specific phospholipase C (Snyder, 1987, Chap. 9), leading to subsequent activation of protein kinase C (PKC). Numerous PAF receptor antagonists have been described (Snyder, 1987; Saunders and Handley, 1987), and endogenous inhibitors of PAF also may be lipid molecules (Mina *et al.*, 1987; Nakayama *et al.*, 1987).

PAF is produced by human dermal fibroblasts (Michel *et al.*, 1988) as well as by human and murine epidermal cells (Csato *et al.*, 1987; Cunningham *et al.*, 1987; Michel *et al.*, 1988, 1990). When injected intradermally, PAF is a potent proinflammatory agonist (Archer *et al.*, 1984, 1985), supporting the view that the pathogenesis of inflammatory skin disorders (Czarnetzki, 1984), as well as late allergic response (Hemocq and Vargaftig, 1986; Michel *et al.*, 1988) may involve PAF generation and release. In fact, PAF has been found in psoriatic scale (Mallet and Cunningham, 1985) and in fluid collected in chambers placed over lesional tissues (Mallet *et al.*, 1987). However, abrasion of normal skin also leads to similar PAF levels (Cunningham, 1990). In addition, since other lipid and nonlipid mediators of cellular proliferation also have been identified in psoriatic lesions, including arachidonic acid metabolites (Hammarstrom *et al.*, 1975; Brain *et al.*, 1985), interleukin-1 (Camp *et al.*, 1986), and the complement fragment C5a (Takematsu *et al.*, 1986), the relative importance of PAF in psoriasis remains unknown.

Finally, because PAF is known to induce the release and metabolism of arachidonic acid, eicosanoids may mediate some of the activities of PAF (Voelkel *et al.*, 1982; Ezra *et al.*, 1987). PAF also has been shown to stimulate phosphoinositide turnover and arachidonic acid mobilization in keratinocytes, but it did not exhibit similar effects in cultured dermal fibroblasts (Fisher *et al.*, 1989). In addition, PAF may be released from the same ether–phospholipid precursor for arachidonic acid and its metabolites (Swedsen *et al.*, 1983; see Fig. 1). Thus, both the site of PAF action and the specific role of this class of lipid mediator in normal and pathologic skin conditions await further elucidation.

III. Phospholipase-Initiated Signaling Mechanisms

Cellular signaling processes require the transduction of messages received at the cell surface to numerous intracellular targets, including the cell nucleus. Many extracellular signals act by inducing the turnover of phospholipids at or within the plasma membrane of the cell. The discovery that hormones can induce phosphoinositide metabolism was made by Hokin and Hokin (1953, 1955). Since then, a large number of agonists have been shown to stimulate phosphoinositide turnover (for review see Berridge, 1984). In the current general model for receptor-mediated signal transduction, binding of the effector molecule or agonist to the cell surface receptor, through receptor-specific GTP-binding proteins, initially triggers the activation of a specific family of phospholipase C (PLC) isoenzymes (Fig. 1). This initiates the breakdown of phosphatidylinositol-4,5-bisphosphate (PIP_2) to form inositol-1,4,5-triphosphate (IP_3) and 1,2-diacylglycerol (DAG). IP_3 and DAG are now accepted to function as second messengers, which

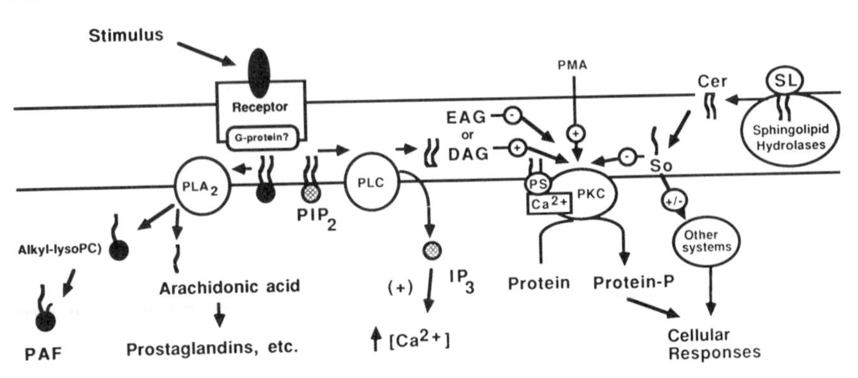

FIG. 1. Signal transduction pathways involving lipid biomodulators. In response to cell stimuli, phospholipase C (PLC) is activated to yield diacylglycerol (DAG) or 1-alkyl-2-acylglycerols (EAG) that activate or inhibit protein kinase C (PKC), respectively. Phorbol esters (PMA) added exogenously also activate PKC. If PLC acts on phosphatidylinositol diphosphate (PIP_2), the resulting phosphoinositols (IP_3) increase cytosolic calcium. PKC is inhibited by exogenous sphingosine (So), which is also present from turnover of exogenous sphingolipids (SL) via ceramides (Cer); free sphingosine may affect additional systems. Phospholipase A_2 (PLA_2) releases arachidonic acid as the precursor of prostaglandins, leukotrienes, and so on, and lysophospholipids, which can affect protein kinase C. If the substrate is 1-alkyl-2-acylphosphatidylcholine, reacylation using acetyl-A produces platelet-activating factor (PAF). [Reprinted with permission from Merrill, A. H., Jr. (1989). *Nutr. Rev.* **47,** 161–169. Copyright © 1989, ILSI–Nutrition Foundation.]

in turn mediate cellular responses through effects on intracellular calcium concentration and via activation of protein kinases, respectively (Kishimoto *et al.*, 1980; Nishizuka, 1983, 1984, 1986; Berridge, 1987).

The inositol phosphates, liberated by PLC activation, are nonlipid compounds that regulate calcium homeostasis in numerous cell types through both influx channels and release from intracellular stores (for review see Berridge, 1987). Inositol phosphate levels rapidly increase following receptor-mediated ligand binding (Berridge, 1984, 1987). The P_3 generated by PI–PLC is rapidly metabolized to form numerous inositol phosphate derivatives, a pathway that includes the regeneration of PIP. An in-depth review of the role of inositol phosphates in cell proliferation and transmembrane signaling is beyond the scope of this review, thus the reader is referred to recent reviews by Whitman and Cantley (1988), Rana and Hokin (1990), and Ziboh and Tang (1991b).

Activation of PKC appears to be the primary regulatory effect of phospholipase C-generated DAG. Structure–activity studies indicate that PKC is specifically activated by the *sn*-1,2-diacylglycerol stereoisomer and requires a 3-hydroxyl group and ester-linked fatty acids in the 1- and 2-positions, with acyl chain lengths of at least six carbons, or a long-chain acyl group at the 1-position with an acetyl group at the 2-position (Cabot and Jaken, 1984; Rando and Young, 1984; Davis *et al.*, 1985; Ganong *et al.*, 1986). *In vitro*, DAG acts as an allosteric activator of PKC, in association with membrane translocation of the enzyme. It should be noted, however, that at least one isoform of PKC may not require DAG for activation (Nishizuka, 1989). At least four distinct groups of lipid activators of protein kinase C isozymes have been identified to date (for reviews see Chauhan *et al.*, 1990; Nelsestuen and Bazzi, 1991). The first group includes the 1,2-diacyl-*sn*-glycerols mentioned above, which are the putative *in vivo* modulators, because their K_m values are in the physiologic range and they exhibit stereospecific binding to PKC (Rando and Young, 1984; Ganong *et al.*, 1986). The second group includes phorbol esters, including tetradecanoylphorbol acetate (TPA), and other tumor promoters (e.g., bryostatin). Indeed, PKC appears to be the putative cellular receptor for TPA (Kishimoto *et al.*, 1980). Moreover, although TPA competitively binds to DAG sites on PKC, TPA appears to exert other distinctive effects on cellular activities as well. The third group of lipid PKC activators consists solely of PIP_2, which exhibits a 10- to 50-fold higher affinity for PKC compared to DAG (Chauhan and Brockerhoff, 1988), and may be involved in signal transduction processes. The fourth and final group includes a wide range of low-affinity compounds, including fatty acids (for review see Chauhan *et al.*, 1990), where relevance for the *in vivo* regulation of PKC remains unknown.

Keratinocyte differentiation has been correlated with increased hydrolysis of phosphatidylinositol to generate IP_3 and DAG (Ziboh *et al.*, 1984; Chapkin and Ziboh, 1988). When cultured murine keratinocytes were induced to differentiate by transfer from a medium of low (<0.1 m*M*) Ca^{2+} concentration to one of high (1.8 m*M*) Ca^{2+} concentration, DAG levels increased fourfold in comparison to the cells maintained in low-calcium medium (Ziboh *et al.*, 1984). Jaken and Yuspa (1988) also showed that increasing the extracellular Ca^{2+} concentration from 0.1 to 1.0 m*M* resulted in increased inositol phosphate levels (IP, IP_2, and IP_3) with a concomitant decrease in the levels of PIP_2, PIP, and PI. In addition, the calcium ionophores, ionomycin and A23187, mimicked the effect of increasing Ca^{2+} concentration, suggesting that intracellular free calcium levels could be responsible for activation of PI-dependent phospholipase C. Thus, it seems likely that the hydrolysis of the lipid phosphatidylinositol 4,5-bisphosphate, with the subsequent generation of inositol triphosphates and DAG, also may modulate cellular events in the epidermis that are involved in the regulation of keratinocyte differentiation.

The view that agonist-induced increases in DAG result solely from increased breakdown of PIP_2 has recently been challenged. Exton's group has shown that both the magnitudes and kinetics of agonist-induced increases in DAG and IP_3 frequently differ (Exton, 1987, 1988). They also showed that the fatty acid compositions of inositol phosphates often are distinct from that of DAG (Exton, 1988). The conclusion drawn from these and other studies suggests that a greater quantity of DAG may be formed *in vivo* from phosphoryl choline (PC) hydrolysis than that generated from PIP_2 (Exton, 1987, 1988). Distinct phospholipase C isoenzymes, which display different binding specificities and calcium sensitivities, hydrolyze PC and PIP_2 (for review see Exton, 1990). Thus, turnover of PC

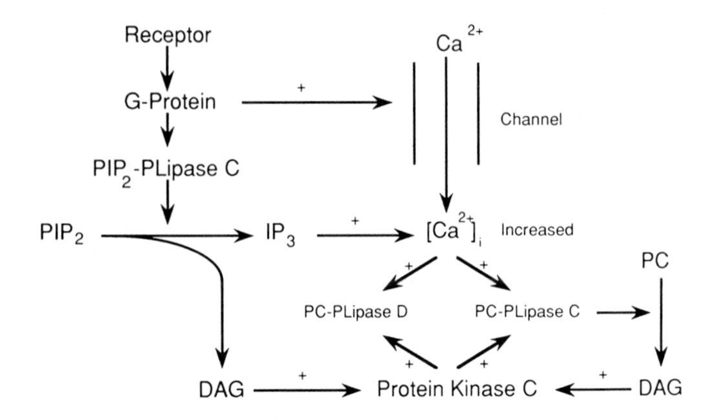

FIG. 2. Proposed mechanism for the activation of phosphoryl choline (PC) phospholipase (PC-PLipase) involving protein kinase C and Ca^{2+}. This scheme indicates that the activation of the PC phospholipases is initiated by the production of DAG from PIP_2 due to the activation of PIP_2 phospholipase C (PIP_2-PLipase C) through a receptor/G-protein mechanism. The DAG activates protein kinase C, which acts on the PC phospholipases either directly or indirectly. Maintenance of the activation of the PC phospholipases could be mainly due to the production of DAG from PC either directly via PC phospholipase C or indirectly via PC phospholipase D plus phosphatidic acid phosphohydrolase. This would represent a positive-feedback loop. For purposes of brevity and clarity, the hydrolysis of PC by phospholipases A_2 and its possible mechanisms of control by Ca^{2+} and protein kinase C have been omitted. Regulation of both PC phospholipases C and D is shown, but it is likely that the phospholipases are affected differentially by protein kinase C and Ca^{2+}, depending on the cell type. The scheme also indicates that the activation of the PC phospholipases is the consequence of the elevation of cytosolic Ca^{2+} that results from the mobilization of internal Ca^{2+} stores due to IP_3 production and the opening of plasma membrane Ca^{2+} channels. The latter events are both postulated to be mediated by receptor/G-protein mechanisms. The Ca^{2+} ions might activate the PC phospholipases directly or act via a Ca^{2+}-binding protein or a Ca^{2+}-responsive protein kinase. The mechanisms involving Ca^{2+} might act in concert with that involving protein kinase C. An important point to note is that the cellular sites at which Ca^{2+} acts are unknown and may not include the plasma membrane. [Modified from Exton, J. H. (1990). *J. Biol. Chem.* **265,** 1–4.]

may be important in the regulation of intracellular calcium as well as protein kinase activities (Fig. 2).

In addition, Ziboh and Tang (1991a,b) have recently proposed that an alternative proliferative signaling pathway may exist in keratinocytes, as may be the case in other tissues (Whitman and Cantley, 1988). This assertion resulted from the finding that leukotriene B_4 (LTB_4) induced a novel phosphatidylinositol 3,4,5-triphosphate, generated by a phosphatidylinositol 3-kinase activity, in adult human keratinocytes (Ziboh and Tang, 1991a). Although the activity of phosphatidylinositol 3-kinase has been described previously in other tissues, detection of 3-kinase products has been limited by their presence in minor quantities relative to other inositol phosphates. Because an association between phosphatidylinositol 3-kinase, growth factor receptors, and several protooncogene products has been described (Whitman and Cantley, 1988), regulation of this kinase may indeed prove to be important in mitogenic signaling in keratinocytes as well as in other cellular systems.

The link between phosphoinositide metabolism, eicosanoids, and psoriasis is currently under investigation (for review see Kragballe *et al.*, 1990). An early study by Hammarstrom *et al.* (1975) reported psoriatic lesions to have high levels of free arachidonic acid (AA). Later, Chapkin, and associates (1985; Chapkin and Ziboh, 1984) demonstrated that normal and psoriatic epidermis is deficient in both Δ^6- and Δ^5-desaturase activities, indicating that arachidonic acid synthesis is not likely to occur in the epidermis. Because phosphatidylinositides contain a predominance of AA esterified at the 2-position, increased phospholipid metabolism might account for the increased free AA observed in psoriatic tissues. Consistent with this, elevated PI-specific phospholipase C activity (Bartel *et al.*, 1986), increased PIP_2 and DAG hydrolysis (Fisher *et al.*, 1990), as well as increased phospholipase A_2 (Forster *et al.*, 1983; Verhagen *et al.*, 1984; Bartel *et al.*, 1986) have been reported. Taken together, these data suggest a strong link between arachidonic acid levels and phospholipid metabolism. Better understanding of the relationship between eicosanoids, phospholipids, and hyperproliferative disorders is clearly required.

IV. Sphingolipids as Effector Molecules

Membrane glycosphingolipids, especially gangliosides, have been implicated in a wide variety of biological functions. These include mediators of cell-to-cell interactions; membrane receptors for toxins, hormones, and growth factors; cell surface antigens (i.e., transplantation and blood group antigens); and, finally, regulators of cell growth and differentiation (Fishman and Brady, 1976; for review see Hakomori, 1981; Hakomori and Young, 1983). However, more recent studies have shown that the core structure of sphingolipids, namely the sphingoid base and ceramide, may possess biological activities beyond the cell recognition functions noted above. Hannun and Bell (1989) and Merrill (1989, 1991; Merrill and Stevens, 1989) have recently reviewed the effects of a diverse array of sphingolipids on cellular proliferation and differentiation. The activities of a number of these compounds for structures are discussed below (refer to Fig. 3).

A. Gangliosides

Gangliosides, or sialic acid-containing glycosphingolipids (Fig. 3), have long been implicated in the regulation of cell growth and differentiation (Hakomori, 1981; Hakomori and Young, 1983). The expression of specific cell surface gangliosides also has been correlated with tumor progression, thus gangliosides are often employed as differentiation or tumor cell markers. Early studies showed

	SPHINGOLIPID	LYSOSPHINGOLIPID
Structure	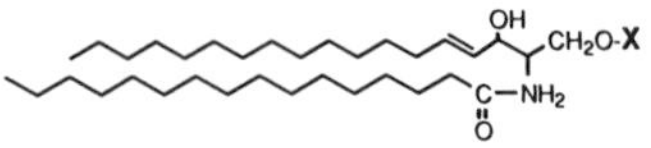 OH, CH_2O-**X**, C—NH_2, O	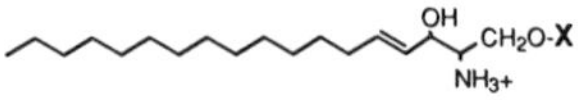 OH, CH_2O-**X**, NH_3+
X = H –	Ceramide	Sphingosine
X = Glucose –	Glucosylceramide	Glucosylsphingosine
X = Gal – Glc – (Gal–SA)	GM_3	Lyso GM_3
X = GalNAc – Gal – Glc – (Gal–SA)	GM_2	Lyso GM_2
X = Gal – GalNAc – Gal – Glc – (Gal–SA)	GM_1	Lyso GM_1
X = Phosphoryl choline –	Sphingomyelin	Lysosphingomyelin

FIG. 3. Structure of sphingolipids and lysosphingolipids. Sphingolipids are derived from ceramide by different substitutions at the 1-position (labeled X). Ceramide is composed of a sphingoid base with an amide-linked fatty acyl chain. With the exception of sphingomyelin, which has a phosphoryl choline head group at the 1-position linked through a phosphodiesteric bond, sphingolipids have a glycosidic bond at C-1. These sugar head groups can vary in complexity from a single glucose (glucosylceramide) or galactose, as occurs in cerebrosides, to more complex structures, such as lactosylceramide with two sugars, trihexosides with three, and higher order carbohydrate moieties. Certain subclasses are recognized by additional components, such as gangliosides (GM_1–GM_3), which contain sialic acid (SA) residues, and sulfatides, which contain sulfate (not shown). Lysophingolipids are based on sphingosine in a manner analogous to the way sphingolipids are based on ceramide. For each parental sphingolipid there is a corresponding lysosphingolipid that has an identical head group at the 1-position but that lacks the amide-linked fatty acyl group at the 2-position. All lysosphingolipids share two important structural features with sphingosine: a charged amine at position 2 and a hydrophobic hydrocarbon tail. [Modified with permission from Hannun, Y. A., and Bell, R. M. (1989). *Science* **243,** 500–507.]

that supplementation of culture medium with exogenous gangliosides reduced cellular proliferation (Keenan *et al.*, 1975; Langenbach *et al.*, 1977) by increasing the G phase of the cell cycle (Icard-Liepkalns *et al.*, 1982). Gangliosides are now believed to participate in a variety of other cellular regulatory processes, including contact inhibition and cell-to-cell communication (for review see Hannun and Bell, 1987). However, the mechanisms by which gangliosides exert their effects remain poorly understood.

A major role of gangliosides may be to modulate epidermal growth factor (EGF)-induced cellular responses. The presence of both endogenous EGF and EGF receptors has been well-described in skin (for review see Fisher and Lakshmanan, 1990) and extensively studied in dermal fibroblast cell lines. EGF is a mitogen, but appears to be involved in a variety of cell functions (Fisher and Lakshmanan, 1990). Thus, the precise role of EGF and EGF receptors in normal function and pathogenesis in the skin is not established. In culture, basal keratinocytes elaborate high EGF receptor levels (King *et al.*, 1988), whereas the number of EGF receptors is related to epithelial cell growth (Green *et al.*, 1983) and decreases with the degree of epidermal differentiation (Nanney *et al.*, 1984; Green and Couchman, 1985; Ponec *et al.*, 1987). Increased EGF receptor levels also have been described in various skin disorders, including psoriasis (Nanney *et al.*, 1987; Ellis *et al.*, 1987). The recent findings that other growth factors, such as transforming growth factor α (TGFα), share a common receptor with EGF expands the complexity of growth factor interactions within the skin and other tissues (for review see King *et al.*, 1990).

Although it seems clear that EGF and the expression of the EGF receptor are important for mammalian development and function, the interaction of gangliosides and other sphingolipids with the EGF receptor complex has only recently been examined (refer to Fig. 4). The best characterized effect of exogenously supplied gangliosides is the modulation of EGF receptor autophosphorylation and EGF-dependent mitogenesis (Bremer and Hakomori, 1982; Bremer *et al.*, 1984, 1986; Hanai *et al.*, 1987, 1988a,b). The growth of mouse fibroblast (3T3) cells was inhibited by the monosialo-gangliosides GM_3 or GM_1, an effect that correlated well with inhibition of phosphorylation of the platelet-derived growth factor receptor (Bremer and Hakomori, 1982; Bremer *et al.*, 1984). Exogenous GM_3 also inhibited epidermal growth factor-dependent phosphorylation of the EGF receptor in the A431 epidermoid carcinoma cell line (Bremer *et al.*, 1986). Moreover GM_3, and to a lesser extent, GM_1, inhibited A431 cell growth by approximately 90%. Finally, the addition of GM_3 or GM_1, but not other gangliosides or nonsialic acid-containing glycosphingolipids, influenced fibroblast growth factor-, EGF-, and/or platelet-derived growth factor-dependent stimulation of cellular growth (Bremer *et al.*, 1986). These results suggest that specific membrane lipids, particularly GM_3, could modulate the phosphorylation in keratinocytes of the EGF receptor, thereby leading to modulations of cell growth.

In contrast, modified GM_3 can exert quite opposite effects on tyrosine phosphorylation of the EGF receptor (Hanai *et al.*, 1987, 1988b). Whereas GM_3 and lyso-GM_3 were again shown to inhibit EGF-dependent tyrosine phosphorylation of the EGF receptor, de-*N*-acetyl-GM_3 strongly promoted tyrosine phosphorylation of the EGF receptor in both A431 cells and Swiss 3T3 fibroblasts (Hanai *et al.*, 1987, 1988b). The authors postulated that N-deacylation of either the acetyl group from sialic acid or the *N*-acyl group from the ceramide of gangliosides is an important event in EGF-dependent cell growth, and possibly in subsequent events in transmembrane signaling. More recently, Song *et al.* (1991) confirmed GM_3 inhibition of EGF receptor autophosphorylation, but found that de-*N*-acetyl-GM_3 stimulated EGF receptor autophosphorylation only in the presence of triton X-100. They concluded that only GM_3 ganglioside, and not its deacytylated analog, can effect EGF receptor kinase activity in intact membranes.

Rosner *et al.* (1990) looked at the expression of GM_3 ganglioside in a human skin fibroblast cell line (HH-8) and found a cell density-dependent increase in

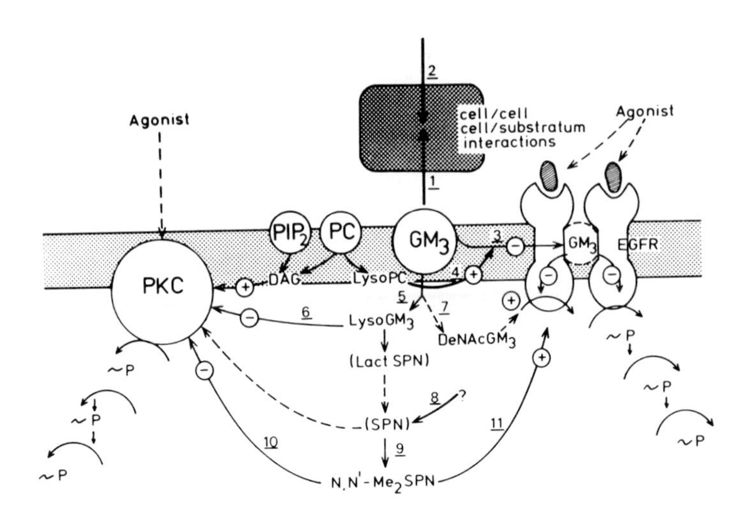

FIG. 4. Bimodal function of glycosphingolipids. Participation of gangliosides in cellular interaction and transmembrane signal transduction is illustrated in this scheme, using GM_3 as an example; other gangliosides may have similar bimodal functions. This scheme also illustrates the relationships between the effects of GM_3 and its catabolities, based mainly on a series of studies with A431 cells as described in the text. Parentheses indicate compounds that are absent or chemically undetectable in A431 cells. Broken lines indicate a weak effect or ambiguous synthetic route. Details in similar schemes for other types of cells including keratinocytes (i.e., placement of parentheses and broken lines) would presumably differ from those shown here, although the overall system may be the same. GM_3 is recognized by complementary carbohydrate structures or by its binding protein (arrows 1 and 2). On the other hand, GM_3 inhibits EGF-dependent receptor kinase (arrow 3), possibly by inhibiting dimerization of the receptor (EGFR). The inhibition of GM_3-dependent EGF receptor autophosphorylation is promoted by lyso-PC (reaction 4), a degradation product of phosphoryl choline (PC) (Igarashi *et al.*, 1990). Diacylglycerol (DAG), which is derived mainly from phosphatidylinositol diphosphate (PIP_2) and partially from PC (see Figs. 1 and 2), is known to promote protein kinase C (PKC) activity. GM_3 is degraded through reactions 5 and 7 to give lyso-GM_3 and de-*N*-acetyl-GM_3, respectively. Lyso-GM_3 strongly inhibits PKC (through reaction 6), whereas de-*N*-acetyl-GM_3 enhances receptor kinase activity (the degree of enhancement is variable and is therefore shown by a broken line). Lactosylsphingosine (Lact SPN), the intermediate degradation product of lyso-GM_3, has no apparent reactivity with either PKC or receptor kinase, but the further degradation product, sphingosine (SPN), is known to inhibit PKC activity. Sphingosine could be derived from ceramide through a "sphingomyelin cycle" (Okazaki *et al.*, 1989) or from other glycosphingolipids (Felding-Habermann *et al.*, 1990); its real degradative or synthetic route (reaction 8) is unknown. However, sphingosine is present only in trace quantities (indicated by parentheses), because it is immediately converted to dimethylsphingosine (the major sphingoid base) by a specific transmethylase (reaction 9) (Igarashi *et al.*, 1989). Dimethylsphingosine strongly inhibits PKC activity (reaction 10) and promotes receptor kinase activity (reaction 11) in a stereospecific way (Igarashi *et al.*, 1989), whereas the effect of sphingosine on PKC is nonspecific for stereoisomers (Merrill *et al.*, 1989). [Reprinted with permission from Hakomori, S. I. (1990). *J. Biol. Chem.* **265,** 18713–18716.]

GM_3 expression, which at confluence was sixfold greater than in preconfluent cultures. Immunostaining with a monoclonal antibody, which recognizes the GM_3 ganglioside epitope (M2590) (Hirabayashi *et al.*, 1985), revealed that GM_3-expressing cells increased in parallel with cell density. In addition, a dose-dependent inhibition of cell growth was observed after a 5-day incubation of cells with a mixture of bovine brain gangliosides, which curiously did not include GM_3 (21% GM_1, 40% GD_{1a}, 16% GD_{1b}, and 19% GT_{1b}) (Rosner *et al.*, 1990). Maximal growth inhibition (evidenced by cell counting only) was observed at the highest ganglioside concentration tested (1.0 m*M*). These studies again suggest that gangliosides are involved with the regulation of cellular growth in human skin fibroblasts.

Direct studies with epidermal tissues or keratinocytes have begun only recently. Preliminary studies showed that the major ganglioside of epidermal keratinocytes is GM_3 (Skrivanek *et al.*, 1985, 1986). Recently, Paller *et al.* (1990) confirmed these findings by quantitating gangliosides extracted from whole cultured keratinocytes. Paller *et al.* (1989) also showed that GM_3 expression in keratinocytes may be related to proliferative potential. First, using a monoclonal antibody (8G9D8) to the α2,6-*N*-acetylsialic acid of GM_3, increased immunostaining was observed in skin biopsies from patients with hyperproliferative disorders, such as psoriasis (Paller *et al.*, 1989). In addition, when keratinocytes in culture were induced to differentiate by increasing the Ca^{2+} concentration in the medium from low to physiologic levels, the expression of GM_3 was diminished (Paller *et al.*, 1990). Interestingly, the highest GM_3 levels appear to be associated with proliferating keratinocytes, whereas a dose-dependent inhibition of keratinocyte growth by (NeuAc)-GM_3 (in the range of 25–100 μg/ml in the medium) was reported (Paller *et al.*, 1990).

Tatsumura *et al.* (1988) also have reported GM_3 to be the major ganglioside in a human squamous cell carcinoma line (Sq CC/Y1). Differentiation of these cells was associated with a marked accumulation of GD_3, a disialoganglioside frequently associated with cells committed to terminal differentiation or undergoing neoplastic transformation (Siddiqui *et al.*, 1984; Seyfried and Yu, 1985). Because GD_3 is derived from GM_3, it has been postulated that the sialyltransferase responsible for this transformation may be regulated by cellular differentiation (Tatsumura *et al.*, 1988).

In contrast to the above studies, Hersey *et al.* (1988) were unable to detect GM_3 immunostaining in the epidermis of normal skin sections using a different monoclonal antibody (M2590) to GM_3 (Hirabayashi *et al.*, 1985). However, studies with melanoma tissues revealed that nearly 60% of primary and 75% of metastatic melanoma expressed GM_3 to a varying extent (Hersey *et al.*, 1988).

The significance of these findings relative to epidermal differentiation and function is not known. It is clear that tissue expression of gangliosides in epidermis is related to cellular growth and differentiation. Because the majority of cellular gangliosides are expressed in the plasma membrane, their potential for influences on cell–cell communication and regulation of cell growth is evident. Moreover, gangliosides and their metabolites also appear to regulate phosphorylation events that regulate growth factor receptor-mediated processes, which holds obvious implications for the pathogenesis of hyperproliferative disorders. However, current understanding of the mechanisms whereby gangliosides effect cell growth and differentiation, and their possible role *in vivo*, is far from complete.

B. Sphingosine and Lysosphingolipids

The first link between sphingolipids and signal transduction was made when Hannun and associates (1986, 1987) reported that sphingosine base and lysosphingolipids potently and reversibly inhibited protein kinase C. They postulated that sphingolipid breakdown products could act as endogenous negative regulators of PKC (Fig. 1). Merrill *et al.* (1986) also demonstrated that sphingosine base inhibited phorbol ester-induced differentiation in human promyelocytic leukemic (HL-60) cells. Subsequent studies showed that both sphingosine and lysosphingolipids influenced a variety of cellular processes, including inhibition of epidermal growth factor activity, neutrophil and platelet activation, phorbol ester-induced responses, as well as modulation of receptor function (for reviews see Hannun and Bell, 1989; Merrill, 1989, 1991; Merrill and Stevens 1989; Hakomori, 1990). Figure 4 provides another current, but still controversial model for potential effects of sphingolipids on the proteins involved in agonist induced signal transduction.

Structural characterization studies have shown that the key elements required for sphingosine effects are the presence of a positively charged amino group (Hannun *et al.*, 1986, 1987) and the length of the sphingosine chain (Merrill *et al.*, 1989). However, there is currently a debate concerning the optimum structure of the bioactive sphingosine molecule(s). Though Hannun, Bell, and associates (Khan *et al.*, 1990), as well as Merrill and associates (1989), support the view that the free sphingosine base (see Fig. 3) represents the active lipid species *in vivo*, Hakomori and associates (Igarashi *et al.*, 1989, 1990; Igarashi and Hakomori, 1989) have proposed that N-methylation of sphingosine to *N,N*-dimethylsphingosine is required to achieve the observed sphingosine bioactivity.

The potential significance of sphingosine *in vivo*, and its relationship to PKC in the epidermis, was suggested recently by Gupta *et al.* (1988), who showed that the effects of topical application of the tumor-promoting phorbol ester TPA were inhibited by coapplication of topical D-sphingosine. Topical application of TPA to skin induces an increase in both PKC activity and ornithine decarboxylase (ODC) activity. Topical coapplication of excess sphingosine base (20 μmol) blocked the TPA-induced increased in PKC and ODC activities. The implication that sphingosine could directly block the activity of PKC *in vivo* is consistent with the model for locally generated sphingosine acting to modulate signaling events within the epidermis (Figs. 1 and 4).

Addition of sphingosine to intact A431 human epidermoid carcinoma cells caused an inhibition of phorbol ester-stimulated phosphorylation of two protein kinase C substrates (Faucher *et al.*, 1988). Phosphorylation of both the threonine residue 654 on the epidermal growth factor receptor, and the serine residue 24 of the transferrin receptor, was blocked when cells were pretreated with 10 μ*M* sphingosine before treatment with the phorbol ester, phorbol myristate acetate (PMA) (10 n*M*). Although these data suggest that exogenous sphingosine inhibits PKC in A431 cells, further experimentation also demonstrated that sphingosine regulated EGF receptor by a mechanism independent of PKC. Exposure of A431 cells to 10 μ*M* sphingosine alone caused an increase in threonine phosphorylation of the EGF receptor, increased the affinity of EGF receptor, and increased the number of EGF-binding sites expressed at the cell surface (Faucher *et al.*, 1988). *N*-Acetylsphingosine, which does not inhibit PKC *in vitro*, did not produce the effects seen with sphingosine, indicating that these effects were specific. Thus, the authors concluded that sphingosine is a potent, bioactive molecule that not only inhibits protein kinase C activities, but also stimulates tyrosine kinase activity and protein phosphorylation independent of PKC, as well as increasing the affinity and number of cell surface EGF receptors. Therefore, although numerous reports have linked sphingosine to an inhibition of PKC-related activities, the action of this molecule appears far more complex.

Numerous studies also have shown that sphingosine may exert effects independent of PKC (Wincovic and Gershengorn, 1988; Krishnamurthi *et al.*, 1989; Zhang *et al.*, 1990a,b). Cellular targets for sphingosine other than PKC have included the EGF receptor (Faucher *et al.*, 1988; Davis *et al.*, 1988; Igarashi *et al.*, 1990), thryrotropin releasing hormone (TRA) receptor (Kolesnick, 1989), phospholipase D (Lavie and Liscovitch, 1990), calmodulin kinases (Jefferson and Schulman, 1988), and possibly cyclic AMP-dependent kinases. Civan *et al.* (1988) demonstrated that D-sphingosine inhibited the insulin-stimulated transport of Na^+ across frog skin, presumably through inhibition of PKC. However, sphingosine also reduced the cellular response to vasopressin, a cAMP-dependent agonist, suggesting that cAMP-dependent kinase activity also was inhibited. Spiegel and coworkers have shown that low concentrations of sphingosine (<10 μ*M*) stimulate proliferation (mitogenesis) in quiescent and confluent Swiss 3T3 fibroblasts (Zhang *et al.*, 1990a). These studies suggest that sphingosine may act as a positive regulator of cell growth, again via a protein kinase C-independent pathway. More recently, Speigel and associates (Zhang *et al.*, 1990b) have shown that mitogenic concentrations of sphingosine (10 μ*M*) increase cytosolic phosphatidic acid, which is known to be a potent mitogen for Swiss 3T3 fibroblasts (Murayama and Ui, 1987; Yu *et al.*, 1988; see Section V,A). In addition, the cAMP response of these cells was diminished along with increased inositol phosphate turnover (Zhang *et al.*, 1990b).

In order to support a role for sphingosine in regulating events in the skin, its presence *in vivo* needed to be demonstrated. An earlier study implied that free sphingosine is generated in skin, because free hydroxy fatty acids were found in the stratum corneum (Bowser *et al.*, 1985), which presumably derived from hydrolysis of ceramide compounds. However, the presence of free sphingosine was recently described in both pig and human stratum corneum (Wertz and Downing, 1989, 1990a). Moreover, these authors described a neutral ceramidase in porcine epidermis capable of generating free sphingosine, with a pH optimum of >7.0 (Wertz and Downing, 1990b). In addition, the levels of free sphingosine were found to be higher in stratum corneum than in the more metabolically active cell layers of the epidermis (Wertz and Downing, 1989). The presence of this gradient is consistent with a possible role for sphingosine in mediating differentiation-related events in the epidermis *in vivo*.

In summary, sphingosine and lysosphingolipids have been implicated as effectors of signal transduction and cellular differentiation. However, despite the demonstration of free sphingosine in the epidermis, along with the enzymatic mechanism for its production, the exact role of sphingosine and its metabolic derivatives remains to be determined. Whereas a significant body of data exists to support sphingosine as a second messenger in various cell types, the majority of the work in the epidermis is still circumstantial. Elucidation of *in vivo* activities of sphingosine in a tissue that is known to synthesize large quantities of this compound for bulk lipid functions, such as the permeability barrier (see Elias and Menon, this volume), will be difficult. Distinguishing between the pools of lipid generated at the site of signal transduction events, from the much larger pool of lipid generated for other epidermal functions, presents a formidable challenge. At present, the most promising approach for clarification of these events within the epidermis involves the use of cultured keratinocytes, in which effects on cell growth and differentiation may be more easily pinpointed.

C. Other Ceramide Derivatives

Ceramide, the N-acylated derivative of sphingosine, has recently been suggested to be a lipid modulator of cellular differentiation. Okazaki *et al.* (1990) have shown that exogenous ceramide regulates the 1α,25-dihydroxy Vitamin D_3 [1,25-$(OH)_2D_3$]-induced differentiation of HL-60 cells. Treatment of HL-60 cells with 1,25-$(OH)_2D_3$ is known to induce myeloid differentiation, producing macrophage/monocyte-like cells (Rovera *et al.*, 1979). A "sphingomyelin cycle" also has been described in HL-60 cells in response to 1,25-$(OH)_2D_3$ treatment (Okazaki *et al.*, 1989). Following treatment, an acute increase in sphingomyelinase activity occurred, with a concurrent decrease in the levels of sphingomyelin. In addition, total cellular ceramide levels increased approximately 50% after addition of 1,25-$(OH)_2D_3$, which mirrored the decrease in sphingomyelin levels. The increased ceramide levels were not merely secondary to the induced differentiation, because subsequent studies with exogenous, cell-permeant ceramides also resulted in myeloid differentiation to an extent comparable to that induced by 1,25-$(OH)_2D_3$ alone (Okazaki *et al.*, 1990). Thus, sphingomyelin and/or ceramide, or a metabolic derivative of ceramide, could regulate HL-60 differentiation. Because previous work has shown that sphingosine is a potential regulator of PKC (see above), the ceramide effect may be linked to increased levels of sphingosine through the subsequent action of ceramidase. Whether the increase in ceramide levels is part of a common or universal mechanism for differentiation in other cell types has yet to be determined.

In the epidermis, there is one reference to the effects of ceramides on cellular activities. Uchida *et al.* (1990) showed that exogenous ceramide had no effects of the activity of transglutaminase, a key marker of keratinocyte differentiation. In addition, no induction of cornified envelope formation was observed after ceramide treatment. However, further experimentation with different ceramide structures and other differentiation markers is necesssary before firm conclusions can be drawn.

The epidermis is also known to contain a unique class of ceramides, containing very long-chain *N*-acyl groups (C_{24}–C_{30}), that are further N-acylated at the ω-terminus (Bowser *et al.*, 1985). These unique sphingolipids are thought to be critical for the structure and integrity of the epidermal barrier to water loss (see Elias and Menon, this volume). However, recent preliminary studies suggest that acylceramides may subserve functions in the epidermis independent of the permeability barrier, i.e., in the regulation of differentiation. Because human keratinocytes produce these acylceramides in culture in association with the most fully differentiating culture conditions, Ponec and co-workers (Brod *et al.*, 1991; see Ponec, this volume) have proposed that the acylceramides are markers of keratinocyte differentiation. In addition, when fetal rat keratinocytes (FRSKs) were exposed in culture to exogenous acylceramide (1–10 μg/ml), moderate induction of certain differentiation-related markers was observed (Uchida *et al.*, 1990, 1991). The total cellular transglutaminase activity was increased by nearly 100% over untreated controls (an increase similar to that induced by TPA), and cornified envelope formation also was increased (56–73%) (Uchida, 1990). Keratin synthesis also was increased in FRSKs in response to exogenous acylceramide (Uchida, 1991). Although these preliminary results may still prove anomalous, they provide further hints about the potential roles of a wide variety of lipids in mediating cellular functions.

V. Miscellaneous Lipid Effectors

A. Phosphatidic Acid

Phosphatidic acid (PA) has mitogenic effects when supplied exogenously to cells in culture (see Section IV,B). PA also can be formed from diacylglycerol through the action of kinases, or *de novo* in response to growth factor action (Farese *et al.*, 1987). The mitogenic activity of PA was thought to be due primarily to a calcium ionophore effect, because exogenously supplied PA stimulates Ca^{2+} uptake and can induce calcium-mediated responses (Salmon and Honeyman, 1980; Putney *et al.*, 1980; Harris *et al.*, 1981; Ohsako and Deguchi, 1981, 1983; Barrit *et al.*, 1981). Using human A431 carcinoma cells, Moolenaar *et al.* (1986) reported that PA (50 μg/ml) triggers the hydrolysis of phosphoinositides, resulting in the subsequent release of calcium from intracellular stores. PA also displayed effects similar to growth factors in raising cytosolic pH, inducing c-*fos* and c-*myc* protooncogenes, and stimulating DNA synthesis in A431 cells (Moolenaar *et al.*, 1986). Stacey and co-workers had previously shown that cellular *ras* (c-*ras*) protooncogene protein is required for the mitogenic action of serum growth factors (Mulcahy *et al.*, 1985). Phospholipids, including phosphatidic acid, have been postulated to be an important biochemical link between growth factor receptors and cellular *ras* activity (Yu *et al.*, 1988). In addition, PA levels are increased in quiescent 3T3 cells treated with low concentrations ($\leq 10\ \mu M$) of sphingosine (Zhang *et al.*, 1990b), which Zhang and co-workers contend may explain the mitogenic effect of sphingosine treatment. Taken together, these studies suggest a role for PA in growth factor-induced mitogenesis, including protooncogene-regulated mechanisms.

B. Palmitoyl Carnitine

Palmitoyl carnitine is a long-chain fatty acid ester of carnitine that has been shown to accumulate in cardiac tissues during ischemia (Patmore *et al.*, 1989). Because its activities resemble certain Ca^{2+} channel activators, palmitoyl carnitine may be an endogenous modulator of intracellular calcium, at least during ischemia. Although the mechanism of palmitoyl carnitine activity has not been established, it is known to competitively inhibit purified protein kinase C (Wise *et al.*, 1982). Combined with other evidence, it is thought that palmitoyl carnitine may function in a PKC-dependent fashion in cardiac tissues. However, in demonstrating that palmityol carnitine decreased EGF binding to isolated rat pancreatic acini, Brockenbrough and Korc (1987) showed this effect likely to be a PKC-independent phenomenon. The role of this compound in modulating tissues of dermal–epidermal origin has not been investigated. However, a recent report by Williams *et al.* (1991) suggests that the growth of keratinocytes in culture may be carnitine-dependent, an effect that has initially been ascribed to a substrate-limiting phenomenon.

VI. Summary

The importance of lipids within the skin as components of the permeability barrier has been appreciated for quite some time. However, the more recent work reviewed here suggests numerous alternative bioactive functions for lipid molecules within the skin and other tissues. The precise roles of lipids in epidermal proliferation and differentiation have only begun to be studied and are far from being defined.

References

Archer, C. B., Page, C. P., Paul, W., Morley, J., and MacDonald, D. M. (1984). *Br. J. Dermatol.* **110,** 45–79.
Archer, C. B., Page, C. P., Moorley, J., and MacDonald, D. M. (1985). *Br. J. Dermatol.* **112,** 285–290.
Barrit, G. J., Dalton, K. A., and Whiting, J. A. (1981). *FEBS Lett.* **125,** 137–140.
Bartel, R., Marcelo, C. L., Gorsulowsky, D., and Voorhees, J. J. (1986). *J. Invest. Dermatol.* **86,** 462.
Benveniste, J., Henson, P. M., and Cochrane, C. G. (1972). *J. Exp. Med.* **136,** 1356–1377.
Benveniste, J., Tence, M., Varenne, P., Bidault, J., Boullet, C., and Polonsky, J. (1979). *C. R. Hebd. Seances Acad. Sci., Ser. D* **289,** 1037–1040.
Berridge, M. J. (1984). *Biochem. J.* **220,** 345–360.
Berridge, M. J. (1987). *Annu. Rev. Biochem.* **56,** 159–193.
Bowser, P. A., Nugteren, D. H., White, R. J., Houtsmuller, U. M., and Prottey, C. (1985). *Biochim. Biophys. Acta* **834,** 419–428.
Brain, S. D., Camp, R. D. R., Kobza-Black, A., Dowd, P. M., Greaves, M. N., Ford-Hutchinson, A. W., and Charleson, S. (1985). *Prostaglandins* **29,** 611–619.
Braquet, P., Touqui, L., Shen, T. Y., and Vargaftig, B. B. (1987). *Pharmacol. Rev.* **39,** 97–145.
Bremer, E. G., and Hakomori, S. (1982). *Biochem. Biophys. Res. Commun.* **106,** 711–718.
Bremer, E. G., Hakomori, S., Bowen-Pope, D. F., Raines, E., and Ross, R. (1984). *J. Biol. Chem.* **259,** 6818–6825.
Bremer, E. G., Schlessinger, J. S., and Hakomori, S. (1986). *J. Biol. Chem.* **261,** 2434–2440.
Brockenbrough, J. S., and Korc, M. (1987). *Cancer Res.* **47,** 1805–1810.
Brod, J., Bavelier, E., Justine, P., Weerheim, A., and Ponec, M. (1991). *In Vitro Cell. Dev. Biol.* **27A,** 163–168.
Cabot, M., and Jaken, S. (1984). *Biochem. Biophys. Res. Commun.* **125,** 163–169.
Camp, R. D. R., Chu, A., Cunningham, F. M., Fincham, N. J., Greaves, M. W., and Morris, J. (1986). *Br. J. Pharmacol.* **88,** 244B.
Chapkin, R. S., and Ziboh, V. A. (1984). *Biochem. Biophys. Res. Commun.* **124,** 784–792.
Chapkin, R. S., and Ziboh, V. A. (1988). *J. Nutr.* **117,** 1360–1370.
Chapkin, R. S., Ziboh, V. A., Marcelo, C. L., and Voorhees, J. J. (1986). *J. Lipid Res.* **27,** 945–954.
Chauhan, V. P. S., and Brockerhoff, H. (1988). *Biochem. Biophys. Res. Commun.* **155,** 18–23.
Chauhan, V. P. S., Chauhan, A., Deshmukh, D. S., and Brockerhoff, H. (1990). *Life Sci.* **47,** 981–986.
Civan, M. M., Peterson-Yantorno, K., and O'Brien, T. G. (1988). *Proc. Natl. Acad. Sci. U.S.A.* **85,** 963–967.
Csato, M., Rosenbach, T., Moller, A., and Czarnetzki, B. M. (1987). *J. Invest. Dermatol.* **89,** 440.
Cunningham, F. (1990). *J. Lipid Mediators* **2,** 61–74.
Cunningham, F. M., Leigh, I., and Mallet, A. I. (1987). *Br. J. Pharmacol.* **90,** 117P.
Czarnetzki, B. (1984). *Clin. Exp. Immunol.* **54,** 486–492.
Davis, R., Ganong, B., Bell, R., and Czech, M. (1985). *J. Biol. Chem.* **260,** 5315–5322.
Davis, R. J., Girones, N., and Faucher, M. (1988). *J. Biol. Chem.* **263,** 5373–5379.
Demopoulos, C. A., Pinckard, R. N., and Hanahan, D. J. (1979). *J. Biol. Chem.* **254,** 9355–9358.
Ellis, D. L., Kafka, S. P., Chow, J. C., Nanney, L. B., Inman, W. H., McCadden, M. E., and King, L. E., Jr. (1987). *N. Engl. J. Med.* **317,** 1582–1587.
Exton, J. H. (1987). *In* "Cell Membranes: Methods and Reviews" (E. L. Elson, W. A. Frazier, and L. Glaser, eds.), Vol. 3, pp. 113–182. Plenum, New York.

Exton, J. H. (1988). *FASEB J.* **2,** 2670–2676.

Exton, J. H. (1990). *J. Biol. Chem.* **265,** 1–4.

Ezra, D., Laurindo, F. R. M., Czaja, J. F., Snyder, F., Goldstein, R. E., and Feuerstein, G. (1987). *Prostaglandins* **34,** 41–57.

Farese, R. V., Konda, T. S., Davis, J. S., Standaert, M. L., Pollet, R. J., and Cooper, D. R. (1987). *Science* **236,** 586–589.

Faucher, M. F., Girones, N., Hannun, Y. A., Bell, R. M., and Davis, R. J. (1988). *J. Biol. Chem.* **263,** 5319–5327.

Felding-Habermann, B., Igarashi, Y., Fenderson, B. A., Parks, L. S., Radin, N. S., Inokuchi, J., Strassmann, G., Handa, K., and Hakomori, S. (1990). *Biochemistry* **29,** 6314–6322.

Fisher, D. A., and Lakshamanan, J. (1990). *Endocr. Rev.* **11,** 418–442.

Fisher, G. J., Talwar, H. S., Ryder, N. S., and Voorhees, J. J. (1989). *Biochem. Biophys. Res. Commun.* **163,** 1344–1350.

Fisher, G. J., Talwar, H. S., Baldassare, J. J., Henderson, P. A., and Voorhees, J. J. (1990). *J. Invest. Dermatol.* **95,** 428–435.

Fishman, P. H., and Brady, R. O. (1976). *Science* **194,** 906–915.

Forster, S., Ilderson, E., Summerly, R., and Yardley, H. J. (1983). *Br. J. Dermatol.* **108,** 103–105.

Ganong, B., Loomis, C., Hannun, Y., and Bell, R. (1986). *Proc. Natl. Acad. Sci. U.S.A.* **83,** 1184–1188.

Green, M. R., and Couchman, J. R. (1985). *J. Invest. Dermatol.* **85,** 239–245.

Green, M. R., Basketter, D. A., Couchman, J. R., and Rees, D. A. (1983). *Dev. Biol.* **100,** 506–512.

Gupta, A. K., Fisher, G. J., Elder, J. T., Nickoloff, B. J., and Voorhees, J. J. (1988). *J. Invest. Dermatol.* **91,** 486–491.

Hakomori, S. (1981). *Annu. Rev. Biochem.* **50,** 733–764.

Hakomori, S. (1990). *J. Biol. Chem.* **265,** 18713–18716.

Hakomori, S., and Young, W. W., Jr. (1983). *In* "Sphingolipid Biochemistry" (J. N. Kanfer and S. Hakomori, eds.), pp. 381–436. Plenum, New York.

Hammarstrom, S., Hamburg, M., Samuelsson, B., Duell, E. A., Stawiski, M., and Voorhees, J. J. (1975). *Proc. Natl. Acad. Sci. U.S.A.* **72,** 5130–5134.

Hanahan, D. J. (1986). *Annu. Rev. Biochem.* **55,** 483–509.

Hanai, N., Nores, G., Torres-Mendez, C. R., and Hakomori, S. I. (1987). *Biochem. Biophys. Res. Commun.* **147,** 127–134.

Hanai, N., Dohi, T., Nores, G. A., and Hakomori, S. I. (1988a). *J. Biol. Chem.* **263,** 6296–6301.

Hanai, N., Nores, G. A., MacLeod, C., Torres-Mendez, C. R., and Hakomori, S. I. (1988b). *J. Biol. Chem.* **263,** 10915–10921.

Hannun, Y. A., and Bell, R. M. (1987). *Science* **234,** 670–674.

Hannun, Y. A., and Bell, R. M. (1989). *Science* **243,** 500–507.

Hannun, Y. A., Loomis, C. R., Merrill, A. H., Jr., and Bell, R. M. (1986). *J. Biol. Chem.* **261,** 12604–12609.

Hannun, Y. A., Greenberg, C. S., and Bell, R. M. (1987). *J. Biol. Chem.* **262,** 13620–13626.

Harris, R. A., Schmidt, J., Hitzemann, B. A., and Hitzemann, R. J. (1981). *Science* **212,** 1290–1291.

Hemocq, E., and Vargaftig, B. B. (1986). *Lancet* **1,** 1378–1379.

Hersey, P., Jamal, O., Henderson, C., Zardawi, I., and D'Alessandro, G. (1988). *Int. J. Cancer* **41,** 336–343.

Hirabayashi, Y., Hamaoka, A., Matsumoto, M., Matsubara, T., Tagawa, M., Wakabayashi, S., and Tanaguchi, M. (1985). *J. Biol. Chem.* **260,** 13328–13333.

Hokin, L. E., and Hokin, M. R. (1955). *Biochim. Biophys. Acta* **18,** 102–110.

Hokin, M. R., and Hokin, L. E. (1953). *J. Biol. Chem.* **203,** 967–977.

Hong, Z., Buckley, N. E., Gibson, K., and Spiegel, S. (1990). *J. Biol. Chem.* **265,** 76–81.

Houslay, M. D., Bojanic, D., and Wilson, A. (1986). *Biochem. J.* **234,** 737–740.
Hwang, S.-B. (1988). *J. Biol. Chem.* **263,** 3225–3233.
Hwang, S.-B., Lam, M.-H., and Pong, S.-S. (1986). *J. Biol. Chem.* **261,** 532–537.
Icard-Liepkalns, C., Liepkalns, V. A., Yates, A. J., and Stephens, R. E. (1982). *Biochem. Biophys. Res. Commun.* **105,** 225–230.
Igarashi, Y., and Hakomori, S. (1989). *Biochem. Biophys. Res. Commun.* **164,** 1141–1146.
Igarashi, Y., Hakomori, S., Toyokuni, T., Dean, B., Fujita, S., Sugimoto, M., Ogawa, T., El-Ghendy, K., and Racker, E. (1989). *Biochemistry* **28,** 6796–6800.
Igarashi, Y., Kitamura, K., Toyokuni, T., Dean, B., Fenderson, B., Ogawa, T., and Hakomori, S. (1990). *J. Biol. Chem.* **265,** 5385–5389.
Jaken, S., and Yuspa, S. H. (1988). *Carcinogenesis* **9,** 1033–1038.
Jefferson, A. B., and Schulman, H. (1988). *J. Biol. Chem.* **263,** 15241–15244.
Keenan, T. W., Schmid, E., Franke, W. W., and Weigandt, H. (1975). *Exp. Cell Res.* **92,** 259–270.
Khan, W. A., Dobrowsky, R., El Touny, S., and Hannun, Y. A. (1990). *Biochem. Biophys. Res. Commun.* **172,** 683–691.
King, L. E., Gates, R. E., Stoscheck, C. M., and Nanney, L. B. (1990). *J. Invest. Dermatol.* **95,** 10S–12S.
King, L. E., Ellis, D. L., Gates, R. E., Inman, W. J., Stoscheck, M., Fava, R. A., and Nanney, L. B. (1988). *Am. J. Med. Sci.* **296,** 154–158.
Kishimoto, A., Takai, Y., Mori, T., Kikkawa, U., and Nishizuka, Y. (1980). *J. Biol. Chem.* **255,** 2273–2276.
Kolesnick, R. N. (1989). *J. Biol. Chem.* **264,** 11688–11692.
Kragballe, K., Fisher, G. J., and Voorhees, J. J. (1990). *In* "Eicosanoids and the Skin" (T. Ruzicka, ed.), pp. 67–78. CRC Press, Boca Raton, Florida.
Krishnamurthi, S., Patel, Y., and Kakkar, V. V. (1989). *Biochim. Biophys. Acta* **1010,** 258–264.
Kupper, T. S. (1990). *J. Clin. Invest.* **86,** 1783–1789.
Langenbach, R., Malick, L., and Kennedy, S. (1977). *Cancer Lett.* **4,** 13–19.
Lavie, Y., and Liscovitch, M. (1990). *J. Biol. Chem.* **265,** 3868–3872.
Mallet, A. I., and Cunningham, F. M. (1985). *Biochem. Biophys. Res. Commun.* **126,** 192–198.
Mallet, A. I., Cunningham, F. M., Wong, E., and Greaves, M. W. (1987). *Adv. Prostaglandin, Thromboxane, Leukotrine Res.* **17B,** 640–642.
Merrill, A. H., Jr. (1989). *Nutr. Rev.* **47,** 161–169.
Merrill, A. H., Jr. (1991). *J. Bioenergetics and Biomembranes* **23,** 83–104.
Merrill, A. H., and Stevens, V. L. (1989). *Biochim. Biophys. Acta* **1010,** 131–139.
Merrill, A. H., Jr. Sereni, A. M., Stevens, V. L., Hannun, Y. A., and Bell, R. M. (1986). *J. Biol. Chem.* **261,** 12610–12615.
Merrill, A. H., Jr., Nimkar, S., Menaldino, D., Hannun, Y. A., Loomis, C. R., Bell, R. M., Tyagi, S. R., Lambeth, J. D., Stevens, V. L., Hunter, R., and Liotta, D. C. (1989). *Biochemistry* **28,** 3138–3145.
Michel, L., Denizot, Y., Thomas, Y., Benveniste, J., and Dubertret, L. (1988). *Lancet* **2,** 404.
Michel, L., Denizot, Y., Thomas, Y., Jean-Louis, F., Heslan, M., Benveniste, J., and Dubertret, L. (1990). *J. Invest. Dermatol.* **95,** 576–581.
Mina, M., Hill, C., Kumar, R., Sugatani, J., Olson, M.-S., and Hanahan, D. J. (1987). *J. Biol. Chem.* **262,** 527–530.
Moolenaar, W. H., Kruijer, W., Tilly, B. C., Verlaan, I., Bierman, A. J., and de Laat, S. W. (1986). *Nature (London)* **323,** 171–173.
Mulcahy, L. S., Smith, M. R., and Stacey, D. W. (1985). *Nature (London)* **313,** 241–243.
Murayama, T., and Ui, M. (1987). *J. Biol. Chem.* **262,** 12463–12467.
Nakayama, R., Yasuda, K., and Saito, K. (1987). *J. Biol. Chem.* **262,** 13,174–13,179.

Nanney, L. B., Stoscheck, C. M., and King, L. E., Jr. (1984). *J. Invest. Dermatol.* **83,** 385–393.
Nanney, L. B., Stoscheck, C. M., Magid, M., and King, L. E., Jr. (1987). *J. Invest. Dermatol.* **86,** 260.
Nelsestuen, G. L., and Bazzi, M. D. (1991). *J. Bioenerg. Biomembr.* **23,** 43–61.
Nishizuka, Y. (1983). *Trends Biochem. Sci.* **8,** 13–16.
Nishizuka, Y. (1984). *Science* **225,** 1365–1370.
Nishizuka, Y. (1986). *Science* **233,** 305–312.
Nishizuka, Y. (1989). *Cancer (Philadelphia)* **63,** 1892–1903.
Ohsako, S., and Deguchi, T. (1981). *J. Biol. Chem.* **256,** 10945–10948.
Ohsako, S., and Deguchi, T. (1983). *FEBS Lett.* **152,** 62–66.
Okazaki, T., Bell, R. M., and Hannun, Y. A. (1989). *J. Biol. Chem.* **264,** 19076–19080.
Okazaki, T., Bielawska, A., Bell, R. M., and Hannun, Y. A. (1990). *J. Biol. Chem.* **265,** 15823–15831.
Paller, A. S., Siegel, J. N., Spalding, D. E., and Bremer, E. G. (1989). *J. Invest. Dermatol.* **92,** 240–246.
Paller, A. S., Ladner, A. J., and Bremer, E. G. (1990). *Clin. Res.* **38,** 637A.
Patmore, L., Duncan, G. P., and Spedding, M. (1989). *Br. J. Pharmacol.* **97,** 443–450.
Ponec, M., Kempenaar, J., Weerheim, A., and Boonstra, J. (1987). *J. Cell Phys.* **133,** 358–364.
Prescott, S. M., Zimmerman, G. A., and McIntyre, T. M. (1990). *J. Biol. Chem.* **265,** 17381–17384.
Putney, J. W., Weiss, S. J., Van De Walle, C. M., and Haddas, R. A. (1980). *Nature (London)* **284,** 345–347.
Rana, R. S., and Hokin, L. E. (1990). *Psychol. Rev.* **70,** 115–164.
Rando, R., and Young, N. (1984). *Biochem. Biophys. Res. Commun.* **122,** 818–823.
Rosner, H., Greis, C., and Rodemann, H. P. (1990). *Exp. Cell Res.* **190,** 161–169.
Rovera, G., Santoli, D., and Damsky, C. (1979). *Proc Natl. Acad. Sci. U.S.A.* **76,** 2779–2783.
Ruzicka, T., ed. (1990). "Eicosanoids in the Skin." CRC Press, Boca Raton, Florida.
Salmon, D. M., and Honeyman, T. W. (1980). *Nature (London)* **284,** 344–345.
Saunders, R. N., and Handley, D. A. (1987). *Annu. Rev. Pharmacol. Toxicol.* **27,** 237–255.
Seyfried, T. N., and Yu, R. K. (1985). *Mol. Cell. Biochem.* **68,** 3–10.
Siddiqui, B., Buehler, J., DeGregorio, M. W., and Macher, B. (1984). *Cancer Res.* **44,** 5262–5265.
Skrivanek, J. A., King, D., Phelps, R., Schwartz, E., and Fleischmajer, R. (1985). *J. Invest. Dermatol.* **84,** 318A.
Skrivanek, J. A., King, D., DaSilva, D., Phelps, R., Schwartz, E., and Fleischmajer, R. (1986). *J. Invest. Dermatol.* **86,** 507A.
Snyder, F. (1987). "Platelet-Activating Factor and Related Lipid Mediators." Plenum, New York.
Song, W., Vacca, M. F., Welti, R., and Rintoul, D. A. (1991). *J. Biol. Chem.* **266,** 10174–10181.
Takematsu, H., Ohkohchi, K., and Tagami, H. (1986). *Br. J. Dermatol.* **14,** 1–6.
Tatsumura, T., Ariga, T., Yu, R. K., and Sartorelli, A. C. (1988). *Cancer Res.* **48,** 2121–2124.
Uchida, Y., Iwamori, M., and Nagai, Y. (1990). *Biochem. Biophys. Res. Commun.* **170,** 162–168.
Uchida, Y., Ogawa T., Iwamori, M., and Nagai, Y. (1991). *J. Biochem.* **109,** 462–465.
Verhagen, A., Bergers, M., Van Erp, P. E., Gommans, J. M., van de Kerkhoff, P. C., and Mier, P. D. (1984). *Br. J. Dermatol* **110,** 731–732.
Voelkel, N. F., Worthen, S., Reeves, J. T., and Henson, P. M. (1982). *Science* **218,** 286–288.
Wertz, P. W., and Downing, D. T. (1989). *Biochem. Biophys. Acta* **1002,** 213–217.
Wertz, P. W., and Downing, D. T. (1990a). *J. Invest. Dermatol.* **94,** 159–161.
Wertz, P. W., and Downing, D. T. (1990b). *FEBS Lett.* **268,** 110–112.
Whitman, M., and Cantley, L. (1988). *Biochim. Biophys. Acta* **948,** 327–344.
Williams, M. L., Hanley, K. P., and Hincenbergs, M. (1991). *Clin. Res.* **39,** 531A.
Wincovic, I., and Gershengorn, M. C. (1988). *J. Biol. Chem.* **263,** 12179–12182.
Wise, B. C., Glass, D. B., JenChou, C.-H., Raynor, R. L., Katoh, N., Schatzman, R. C., Turner, R. S., Kibler, R. F., and Kuo, J. F. (1982). *J. Biol. Chem.* **257,** 8489–8495.

Yu, C.-L., Tsai, M.-H., and Stacey, D. W. (1988). *Cell* **52,** 63–71.

Zhang, H., Buckley, N. E., Gibson, K., and Spiegel, S. (1990a). *J. Biol. Chem.* **265,** 76–81.

Zhang, H., Desai, N. N., Murphey, J. M., and Spiegel, S. (1990b). *J. Biol. Chem.* **265,** 21,309–21,316.

Ziboh, V. A., and Tang, W. (1991a). *Adv. Prostaglandin, Thromboxanae, Leukotriene Res.* **21,** 863–866.

Ziboh, V. A., and Tang., W. (1991b). *In* "Seminal Monographs in Pharmacology and Toxicology." CRC Press, Boca Raton, Florida, in press.

Ziboh, V. A., Isseroff, R. R., and Pandley, R. (1984). *Biochem. Biophys. Res. Commun.* **122,** 1234–1240.

ADVANCES IN LIPID RESEARCH, VOL. 24

X-Ray Diffraction and Electron Paramagnetic Resonance Spectroscopy of Mammalian Stratum Corneum Lipid Domains

SUI YUEN E. HOU,* SELWYN J. REHFELD,*
AND WILLIAM Z. PLACHY†

*Department of Dermatology
University of California School of Medicine
San Francisco, California 94143
†Department of Chemistry and Biochemistry
San Francisco State University
San Francisco, California 94132

I. Introduction

Knowledge about the function of mammalian stratum corneum (SC) has expanded tremendously in the last three decades. However, much of the information available still comprises observations. The structural basis for the properties of the SC, especially those properties afforded by the intercellular lipid-enriched domain, has only begun to emerge in the last decade or so (see Elias and Menon, this volume). The ultimate determinants of the characteristics of the SC and its functions are the cellular and biochemical processes that give rise to this compartment. However, it is also important to learn about the properties of this compartment, because this constituent directly determines tissue function. Also, because of the interfacial location of the SC between the body proper and the environment, the properties of the SC are influenced by the environment. The use of physical and especially spectroscopic methods to study SC lipid domains is simply a natural extension of these methods, as widely employed for the study of other model or biological membranes (for review see Andersen, 1978), into study of a tissue, the stratum corneum, which has not yet received much attention. In this review, we discuss the application of X-ray diffraction and electron paramagnetic resonance (EPR) spectroscopy to elucidate the structure and organization of the lipid-enriched cellular peripheral domains of the SC. The two techniques provide confirmatory and complementary information about structure and physical properties on a molecular level. Temperature is commonly used as a parameter in these studies, and most of the detailed information available on SC lipids using these two techniques to date has included their thermotropic behavior. Although thermal phenomena are emphasized in this review, we also describe the lyotropic behavior of SC lipids and perturbations induced by exogenous chemicals and diseases. Traditionally, differential scanning calorimetry (DSC) is widely used to obtain information about the heat-induced phase changes in biological membranes (for review see McElhaney, 1982), but it must be emphasized that calorimetry does not probe membrane structure. Nevertheless, we correlate information derived from DSC with those from the other two techniques wherever appropriate.

II. X-Ray Diffraction

A. Introduction

The stratum corneum can be described by a simple operational model consisting of two predominant phases: one phase is keratin enriched (bricks), surrounded by the other phase, a lipid matrix (mortar) (Michaels *et al.*, 1975). This simple model has received wide recognition. Freeze–fracture (Elias and Friend, 1975) and thin-section transmission electron microscopy (Swartzendruber *et al.*, 1987) techniques demonstrated that the lipid matrix is organized into stacks of lamellae in the intercellular spaces. This organization is particularly well-suited for structural studies using X-ray diffraction due to enhancement in the observed diffraction intensity. The advantage is that diffraction at rather high angles could be observed for periodic structures of large repeats, revealing finer details about membrane structure.

In this article, we will only briefly discuss the type of information that can be obtained on lipids and membranes in general using X-ray diffraction. Instead, we will review in more detail current information about SC lipids as studied by X-ray diffraction, the unresolved problems that could be addressed with this technique, and the implications of this work for SC function. For reviews of a general scope on the use of X-ray diffraction for the study of biological membrane structure, the reader is referred to Worthington (1973), Levine (1973), Shipley (1973), Mitsui

(1978), Franks and Levine (1981), Blaurock (1982), Makowski and Li (1983), Pape (1985), and Laggner (1988). Although X-ray diffraction is an equally fruitful technique for the study of protein structure in the SC, this topic is excluded in this review except where such a discussion is relevant to the lipid-enriched domains. The review by Potts (1989) includes a consideration of this topic.

B. The Diffraction Technique

1. Theory as Applied to Biological Membranes

The theory of X-ray diffraction is discussed in detail in several texts: Hosemann and Bagchi (1962), Guinier (1963), James (1965), Cantor and Schimmel (1980), and Cowley (1981). A general theory on the X-ray diffraction of biological membranes and other lamellar systems is given by Welte and Kreutz (1979). To understand how X-ray diffraction provides information on membrane and lipid structure, we begin with a geometric description using Bragg's law. X-Rays are scattered by electrons. In essence, Bragg's law states that constructive interference of X-rays scattered from crystals would occur in certain angles with respect to the incident radiation. The scattered X-rays in these directions can be thought of as arising from reflection of the incident radiation from planes of constant electron density, but such reflections only occur at the aforementioned angles such that the path difference between the incident X-rays and the diffracted X-rays are integral multiples of the wavelength of the radiation:

$$2d \sin \theta = n\lambda \qquad n = 1,2,3,... \tag{1}$$

where d is the spacing between these planes, θ is the glancing angle of reflection, and λ is the wavelength of the radiation. The integer n is called the order of the reflection. In stacked biological membranes or lipid-enriched bilayers, as in the case of the SC intercellular domains, these planes of constant electron density run parallel to the bilayers or at some angle to the molecular axes of the lipid molecules. This direction is imposed by the polar head groups of the lipid molecules, which have a higher electron density than the nonpolar regions, and hence there is a variation in electron density in the direction normal to the plane of the bilayers. Thus, parallel planes of equal electron density extend in this direction. Within the plane of the bilayer(s), the lipid molecules may be found in various arrangements, and if these are ordered, the same argument applies. Franks and Lieb (1981) and Blaurock (1982) provide more graphic illustrations than what is described here. We see from Eq. (1) that the spacing is inversely proportional to the sine of the Bragg angle, hence low-angle (or small-angle) diffraction gives us information on structures of relatively larger spacing whereas high-angle (or wide-angle) diffraction provides data about relatively smaller spacing. The cutoff in this terminology is arbitrary and is usually set at θ corresponding to a spacing of 20–30 Å for Cu K_α radiation ($\lambda = 1.54$ Å). Because the magnitude of $sin\ \theta$ in Eq. (1) cannot be larger than unity, d cannot be smaller than $\lambda/2$, hence the theoretical resolution is dictated by the wavelength of the radiation used. As should be apparent from this discussion, the most straightforward information provided by X-ray diffraction patterns is the spatial distribution of diffracted intensities, and hence the Bragg spacings that can be calculated from them. The Bragg angle determines where the diffracted intensity is observed, and because of its inverse relationship with real distance in the specimen, the term *reciprocal space* is used to describe distance within the diffraction pattern. The actual intensities of the pattern provide information on the distribution of electron density in the specimen. This result and Bragg's law can be derived by considering Fraunhofer diffraction of X-rays, which involves the mathematics of Fourier transforms (see, e.g., Hosemann and Bagchi, 1962). In short, the diffracted intensity is proportional to the square of the structure factor in reciprocal space, which is the Fourier transform of the electron density distribution in real space. As indicated above, due to the relationship between the structure of molecular groups and their electron density, the electron density profile can be interpreted in terms of molecular structure. In general, electron densities of molecules are proportional to their mass densities or specific gravities (Blaurock, 1982). However, definitive assignment is possible only if exact chemical composition is known and in most cases this only applies to model systems.

Isolated, single-bilayer fragments (e.g., when suspended in solution) provide a continuous intensity distribution with radial symmetry. If the membranes are stacked and oriented such that their planes are parallel to any axis that is perpendicular to the direction of the beam, discrete reflections can be seen along an axis (the meridian), perpendicular to the former axis in reciprocal space (on the recording plane or film) at various angles of orientation (about the first axis) where the Bragg condition is satisfied. If no ordered structure exists within the plane of the membrane, the observed intensity arises from the electron density distribution of the membrane projected onto the normal to the membrane plane. For membranes with in-plane ordered structure, reflections would be seen along an axis (the equator) normal to the meridian. If the stacks are randomly oriented, multiple Bragg reflections are seen which spread into concentric circles on the recording plane.

The diffraction pattern is analyzed in terms of intensity or relative intensity distribution (magnitude) as a function of reciprocal space coordinate (position in the recording plane). Because the intensity distribution is proportional to the Fourier transform of the electron density distribution in the specimen, it is important for calculating the latter. A correction, often called the Lorentz factor, has to be applied to the observed intensities for the particular geometry of the experiment (Welte and Kreutz, 1979; Laggner, 1988). One of the major objectives in low-angle X-ray diffraction analysis of membranes is to determine the electron density profile of the specimen. However, phase information is lost in intensity collection and its determination is the major problem in X-ray structure analysis. Various methods have been developed, though they do not provide unique solutions per se (Franks and Levine, 1981). Information from other sources, such as electron microscopy, may be needed to provide unambiguous answers. X-Ray diffraction is capable of revealing structure at the molecular level and, as will be seen, it can be applied to tissues in their native state without extensive chemical processing. These attributes make X-ray diffraction a powerful and almost ubiquitous technique in studies on biological structure.

2. *Experimental Considerations*

a. Instrumentation. Mitsui (1978) provided a brief summary of X-ray diffraction instrumentation suitable for use on membranes and lipids, including X-ray sources, camera optics, and detection devices. We have further summarized this subject in Table I. Briefly, cameras using pinhole collimation can give any desired angular resolution, but the transmitted intensities are low and hence are

Table I

LITERATURE SOURCES FOR X-RAY DIFFRACTION INSTRUMENTATION APPLICABLE TO STUDYING STRATUM CORNEUM

Aspect of instrumentation	Reference	Comments
X-Ray sources	Stout and Jensen (1989)	Short introductory account
	Philips (1985)	Practical aspects of X-ray generators
	Yoshimatsu and Kozaki (1977)	Advanced design considerations
X-Ray cameras	Fraser and MacRae (1973)	Brief but informative summary
	Witz (1969)	More detailed comparison
Collimating camera	Fraser and MacRae (1973)	References therein have more detail
Cylindrical double-mirror camera	Franks (1955, 1958)	Design and practical considerations
	Harrison *et al.* (1985)	Resolution and intensity for a camera for protein crystals with large unit cell compared to collimating case
	Philips and Rayment (1985)	Method for aligning Franks mirrors
Toroidal double-mirror camera	Elliot (1965)	—
Synchrotron radiation	Rosenbaum and Holmes (1980); Sweet and Woodhead (1989)	Theory and instrumentation

not suitable for spacings greater than 50 Å. Focusing cameras overcome this obstacle, with the torroidal camera being more suitable for high-angle studies. Fraser and MacRae (1973) provided a good discussion of these types of cameras. Although photographic detection of intensity provides higher resolution than do electronic detectors, film inherently possess a smaller dynamic range and also it requires longer exposure time. Synchrotron radiation, due to its high intensity, requires very short exposure times (order of milliseconds to seconds for lipid systems), but there is the increased possibility of specimen damage, which, however, can be monitored and prevented (Gruner, 1987; Caffrey, 1989).

b. Specimen Preparation. Mammalian stratum corneum can be isolated as intact sheets using trypsin, but the isolation conditions and duration vary for tissues derived from different species. Thickened SC, such as callus or scales in various skin diseases, need only be cut or scraped from the surface of the skin using a blade. However, callus may not be representative of normal SC. The intercellular region of the SC can be isolated in the form of membrane complexes, composed of sandwiched intercellular lipids between fragments of opposing cornified envelopes, and devoid of cytoplasmic components (Grayson and Elias, 1982). Whole SC sheets then can be cut into small fragments and packed into thin-walled X-ray capillary tubes. Whole sheets can also be rolled around capillary tubes or cut into narrow strips, stacked, and clamped down. The latter method has been observed to provide slightly oriented patterns (intensified arcs on circles) in the stacking direction of the SC sheets, with the beam parallel to the planes of the sheets. Orientation better than this cannot be achieved due to the rippled topography of the corneocyte surface. If dry SC fragments are cut into sizes smaller than the diameter of the capillary tube, the same orientation effect as in the clamped case can be achieved by repeatedly packing small amounts of SC fragments and centrifuging. White *et al.* (1988) did not observe any orientation effects when the sheets were rolled.

In theory, it should be possible to obtain information on ordering in the membrane planes by deploying the beam normal to the planes of the stacked SC sheets, but in actual experiments this does not provide additional information, partly for the same reason stated above. A compromise has to be struck between absorption and scattering of the incident radiation by the sample so that an acceptable pattern can be obtained in a reasonable amount of time. For most biological membranes, the ideal sample thickness is 0.5–1 mm. This applies also to highly hydrated samples of SC. Because the size of the beam striking the sample provided by most focusing cameras is 100–200 μm wide, a 1-mm sample size usually suffices. At ambient humidities, for SC sheets stacked and clamped under moderate pressure, the ideal sample thickness is about 3 mm. Finally, the samples can be kept under differing conditions of vapor pressures and degrees of hydration or temperature with suitably designed holders. Sample deterioration has not been found to be a problem at room temperature with X-ray irradiation, using common laboratory X-ray sources for most experiments. But when the sample is heated, or with the use of synchrotron radiation, the sample must be monitored for deterioration (see above).

C. Diffraction Studies on Stratum Corneum

1. Historical Overview

Despite the enormous interest in stratum corneum as a diffusional and mechanical barrier, very limited X-ray diffraction work has focused on the organization and properties of the lipid-enriched domains in this tissue. The only review to date of work in this area is that by Potts (1989), in which information on both lipids and proteins in the SC was reviewed and correlated with thermal analysis and infrared spectroscopy data. Low-angle X-ray diffraction was first applied to the study of SC supramolecular and molecular structure by Swanbeck (1959). [For earlier reports, which were all high-angle studies and mainly concerned keratin, the reader is referred to the references in Swanbeck (1959).] A detailed study on the keratins was conducted and a quantitative model for keratin organization was proposed. Within this model, Swanbeck also proposed that lipid molecules are associated with the keratin filaments in the form of a bilayer sheath that coats the filaments. Goldsmith and Baden (1970) studied the high-angle diffraction of isolated keratin and also suggested that there is lipid closely associated with the keratin. Wilkes *et al.* (1973) studied the thermal behavior of the crystalline lipid structure in human and neonatal rat SC using wide-angle X-ray diffraction. They found that the thermal behavior of the lipids varied with the source of the SC. When Elias *et al.* (1983) obtained the same high-angle diffraction patterns from neonatal mouse SC membrane complexes that they obtained with whole SC, they concluded that the lipid patterns, previously attributed to a coating around keratin filaments, actually emanated from the intercellular bilayers. Friberg *et al.* (1985) observed a very broad peak attributable to lipids in the low-angle diffraction of human SC. The most comprehensive and informative study on lipid-enriched structures in the SC interstices to date has been the work of White *et al.* (1988) on hairless mouse SC and isolated membrane complexes. They found that the SC intercellular domains gave lamellar diffraction with a repeat of 131 ± 2 Å. Because extracted lipids did not produce a similar diffraction pattern, they suggested that another component of the intercellular domain, which could be a protein, determined the organization of this domain. Also, patterns were observed that could be attributed to the cornified envelope, but none to keratin. Bouwstra *et al.* (1990a,b) used X-rays from a synchrotron source to study mouse, human, and pig SC, and the influence of temperature, hydration, and topically applied skin penetration enhancers.

Table II
LAMELLAR SPACING OF STRATUM CORNEUM OF VARIOUS ORIGIN OBTAINED BY LOW-ANGLE X-RAY DIFFRACTION[a]

Tissue origin	Lamellar spacing (Å)
Hairless mouse	131[b]
Pig	52[c]
Human	65[c], 130[d], 50–80[e]

[a]Stratum corneum isolated with trypsin. All human tissues were from postmastectomy skin.
[b]White *et al.* (1988). Hydrated.
[c]Bouwstra *et al.* (1990a). No change in spacing upon hydration or treatment with propylene glycol.
[d]Bouwstra *et al.* (1990b).
[e]Friberg and Osborne (1985). Broad Peak. Hydration state not reported, but probably fully hydrated.

2. *Low-Angle Studies*

As seen in electron micrographs, the intercellular regions of the SC are filled with lipids organized into lamellar structures. Cross-section images of these lamellae preserved in thin sections, using ruthenium tetroxide as a secondary fixative, with neonatal mouse SC (Swartzendruber *et al.*, 1987), reveal a repeating pattern with a period of 128 Å. In most cases, the low-angle X-ray diffraction pattern of the SC is dominated by discrete reflections that can be attributed to these lamellar structures (White *et al.*, 1988; Bouwstra *et al.*, 1990b). Moreover, the reflections occur in multiple orders of a given spacing due to the stacking of repetitive lamellar units. The lamellar spacing observed in the SC by X-ray diffraction from various species is summarized in Table II. However, Friberg *et al.* (1985) only observed a broad diffraction peak at 50–80 Å in trypsin-isolated SC from a postmastectomy specimen of human skin. Bouwstra *et al.* (1990a) also initially reported lamellar spacings of 65 Å for trypsin-isolated SC from human abdominal and breast skin. However, in a subsequent, more detailed report (Bouwstra *et al.*, 1990b), the authors indicated that they were unsure whether the 65-Å peak represented a first-order peak or the second order of a 130-Å repeat. The latter interpretation is very likely and can be explained by the very high background scattering of the low-angle region, which could have obscured the first-order peak. Moreover, electron micrographs show a similar pattern or spacing between murine and human SC intercellular lamellae (Swartzendruber *et al.*, 1989). This is also true for porcine SC in terms of the overall lamellar pattern (Swartzendruber *et al.*, 1989). The reason that X-ray diffraction of porcine SC gave a lamellar spacing (52 Å) (Bouwstra *et al.*, 1990a: Table II) that is less than half of those of human or murine tissue is unclear. Although there is similarity in the lamellar repeats of human and murine SC, more data are needed to determine whether the

130-Å (or dimensions close to this) lamellar repeat is characteristic of the intercellular domains among different species.

The lamellar reflections are mostly spread into circles or at best arcs (see Section II,B,1,b). These patterns cannot be made to orient further by changing the orientation of the SC sample due to the shape of the corneocyte surface. It seems that there is no evidence of large two-dimensional lattices in the plane of the lamellae, because the only spacing observed at low angles is attributable to stacking.

White *et al.* (1988) observed five orders of Bragg reflections with a repeat period of 131 ± 2 Å in the low-angle diffraction pattern of hairless mouse SC (isolated using trypsin) in excess water. It was stated that the observed lines were very sharp, indicating that the diffraction originated from an ordered lattice with many unit cells. At 45°C, the pattern became very diffuse, and at 70°C the pattern is similar to that of amorphous material. These changes with temperature are completely reversible even for durations of several days. They suggested that this might be due to the presence of a "second component," presumably protein, which has a tight structural relationship with the lipids and contributes to the formation of the lamellar structure. Because the high-angle reflections due to the lateral packing of the alkyl chains in intact SC and in its lipid extract are identical (see below), this second component might form an aggregate containing a large number of lipid molecules so that the lateral packing of the lipids is unchanged in the presence of this second component. The dimension of the lamellar repeat distance indicates that the unit cell probably consists of two bilayers. Models were suggested wherein each bilayer could be asymmetric or symmetric but symmetrically disposed in the unit cell in both cases. Relative intensities were calculated for these two models by assuming the electron density profiles to be step functions (strip models). Although not determined quantitatively, the relative magnitudes of the intensities observed in murine SC are consistent with the bilayers being asymmetric. The asymmetry could be due to asymmetric distribution of lipids or proteins. Isolated membrane complexes consisting of the intercellular material sandwiched by adjacent cornified envelopes gave a diffraction pattern that is largely continuous with a small admixture of discrete diffraction. Peaks occur at 126, 63, 32, and 23 Å. The low-angle diffraction patterns of the extracted lipids are totally different from those of intact SC and SC membrane complexes. They seem to fit two-dimensional lattices rather than multiple lamellar phases. The various types of lattice at different temperatures are indicated in Table III. Again these data point to the possible existence of a second component as discussed above.

The thermal behavior of the lamellar structure in trypsin-isolated human SC (from masectomy) has recently been investigated by Bouwstra *et al.* (1990b). The prominent diffraction peak at 65 Å disappeared at temperatures higher than 90°C.

Table III

LOW-ANGLE PATTERN OBSERVED IN EXTRACTED STRATUM CORNEUM LIPIDS (HYDRATED) AS A FUNCTION OF TEMPERATURE[a]

Temperature	Lattice structure (two-dimensional)	Interpretation
25°C	Oblique: $a = 59$ Å, $b = 90.5$ Å, $\gamma = 98°$	Assuming a rippled bilayer structure at 25 and 45°C, a corresponds to the lamellar repeat distance and b corresponds to the in-plane repeat distance of the ripples
45°C	Rectangular: $a = 47$ Å, $b = 150$ Å	—
75°C	Hexagonal: $A = 67$ Å	Hexagonal II phase of polar head group-lined tubes filled with water

[a]Summarized from White *et al.* (1988).

This corresponded to the highest gel–liquid transition observed using differential thermal analysis (DTA) with a transition temperature of 87°C. This is in contrast to murine SC where the lamellar pattern is already very diffuse at 45°C (White *et al.*, 1988; see above). Upon cooling to 45°C, three diffraction peaks appeared that were much sharper than those originally observed. In murine SC, the lamellar pattern was sharp only at 25°C. When the continuous scattering curve at 120°C was subtracted from the curve of the sample cooled to 25°C, a fourth peak showed up at 130 Å and this turned out to be the first-order peak of this spacing, whereas the other three are the second- to fourth-order peaks. The first order peak was not observed on the unsubtracted pattern probably due to high scattering at the very low-angle region. The authors' interpretation of these observations was that the lipids became more ordered after the proteins were denatured on heating to 120°C (denaturation temperature, 107°C). Heating and cooling the sample between 25 and 90°C repeatedly did not result in the above observations. They also stated that after protein denaturation, only one lamellar phase was observed and cited DTA observations of two gel–liquid transitions at 68 and 87°C before protein denaturation (Bouwstra *et al.*, 1989). But it was not obvious from the X-ray data that there were two lamellar phases to begin with.

Friberg *et al.* (1985) observed a broad peak between 50 to 80 Å in the low-angle diffraction pattern of human SC (from mastectomy) isolated with trypsin. Friberg and Osborne (1985) found that a mixture of lipids of the composition found in human SC did not give a lamellar structure. However, if 41% of the free fatty acids in the mixture were neutralized, a lamellar liquid-crystal structure was formed with 32 to 45 wt% water and this gave an X-ray pattern similar to that of human SC. They suggested that it is this fatty acid/soap/water combination and not the sphingolipids (ceramides) that provides the lamellar structure in the SC. Sphingolipids alone did not give a lamellar liquid crystal with water. However, this is understandable, because the sphingolipids hydrate very slowly at room temperature (for cerebrosides, see Ruocco *et al.*, 1981). The evidence for this fatty acid/soap/water system to be representative of human SC intercellular lamellar structure is still premature, because the X-ray pattern observed here disagrees with that obtained by Bouwstra *et al.* (1990a,b). Also, the distribution of lipid species may not be uniform within the intercellular region or at different levels of the tissue. The distribution of lipids in extracts from the whole tissue does not take this into account. Moreover, there is a contribution from lipids located elsewhere in the tissue. However, the requirement for the SC intercellular lipids to have a lamellar structure in order to provide water barrier properties seems apparent and has recently been supported experimentally by Friberg *et al.* (1990). Again, a lipid mixture of human SC lipid composition was used as a model lipid system in addition to fatty acids alone. The constituent fatty acids in these lipid systems were partially neutralized, providing the lamellar structure. They found that human corneocytes dispersed by solvent extraction and homongenization, when reaggregated with these model lipid systems, gave similar water permeabilities. These values were five to nine times higher than the literature value for human epidermis, but they demonstrated that these lamellar lipid systems provide a barrier to water permeation in the tissue.

3. High-Angle Studies

In biological membrane or model lipid systems, lateral packing of hydrocarbon chains in lipid molecules gives rise to high-angle reflections. Two spacings are commonly found, but not necessarily together: a sharp reflection at 4–4.2 Å, indicative of crystalline packing, and a broad diffuse reflection at 4.5–4.8 Å, indicative of fluidlike packing. In *Acholeplasma.* (*Mycoplasma*) *laidlawii* and *Escherichia coli* membranes, the former (4.2 Å) is present below the transition temperature, whereas the latter (4.7 Å) is present above this temperature (Worthington, 1973). Worthington (1973) also stated that all biological membranes in the natural state show the broad 4.7-Å reflection and that the 4.2-Å sharp reflection is observed in a variety of air-dried membranes. In these two types of membranes, the crystalline packing is of the hexagonal type. In lipid systems, other subcells can be assigned, depending on the number and spacing of the reflections present (Small, 1986; Abrahamsson *et al.*, 1978). Various long-chain lipids in the solid state exhibit spacings of 4.2 and 3.8 Å in their powder pattern, indicative of an orthorhombic subcell (Chapman, 1965). As discused below, these observations also apply to the SC lipid domains.

In human callous SC, Swanbeck (1959) found a broad reflection from 5 to 4 Å, with a preferred orientation on the meridian. In addition, a sharp reflection was observed at 4.2 Å, again oriented along the meridian, and a weaker one at 3.7 Å. These reflections were attributed to fatty acids and some derivatives thereof. Goldsmith and Baden (1970) reported a 4.15-Å reflection, accentuated on the meridian, in oriented, stretched, and dried SC of unspecified mammalian source.

Table IV
THERMAL BEHAVIOR OF LIPID AND PROTEIN DOMAINS IN SC USING WIDE-ANGLE X-RAY DIFFRACTION[a]

		Spacings (Å) and behavior as a function of temperature			
Origin of SC	Assignment	25°C	40°C	69–70°C	80–90°C
Neonatal rat	Lipid	3.7	Absent but reversible up to at least 157°C		
		4.2	4.2	Absent but reversible to 157°C	
	Protein	4.6		No change up to 77°C	
		9.8		No change up to 77°C	
Human	Lipid	3.7	Absent	Absent	Absent and irreversible
		4.2	4.2	4.2	Absent
		4.6	4.6	4.6	Absent
	Protein	4.6		No change up to 77°C	
		9.8		No change up to 77°C	

[a]As found by Wilkes *et al.* (1973).

In isolated keratin fibers, they also observed arcs at this spacing, which were localized in a direction parallel to the axis of the fibers. Extraction of whole SC caused this 4.15-Å spacing to disappear from the diffraction pattern. They attributed these patterns to lipids oriented with their molecular axes perpendicular to the axes of keratin fibers. However, because whole SC was used for the initial isolation, the actual origin of these lipids is uncertain and the bulk of it might have come from the SC interstices. The results of subsequent, more detailed high-angle studies of SC are summarized in Tables IV (Wilkes *et al.*, 1973) and V (White *et al.*, 1988). Wilkes *et al.* (1973) observed two lipid spacings (4.2 and 3.7 Å) in both neonatal rat and human SC. There is an additional lipid band at 4.6 Å in human SC. The latter (presumably a relatively sharp band, characteristic of triclinic parallel chain packing) melted out at 40°C, allowing it to be distinguished from the aforementioned fluidlike hydrocarbon chain-packing spacing (4.6 Å). In addition, halos present at 4.6 and 9.8 Å were assigned to protein structures, presumably because they are stable even at 77°C. However, this assignment to the 4.6-Å halo is equivocal, because one could interpret the data as indicating the presence of two different types of chain packing. This possibility could be tested by checking whether the halo turns into a sharp reflection at ~4.2 Å when the temperature is lowered. As with Goldsmith and Baden (1970), Wilkes *et al.* (1973) also observed the same orientation effect of the lipid reflections with respect to the keratin filaments in the SC, which lie more or less parallel to the plane of the SC, and came to the same conclusion that the lipids are oriented perpendicular to the keratin filaments. Elias *et al.* (1983) found the same prominent reflections in murine whole SC (4.159 and 3.749 Å) and in SC membrane complexes (4.153 and 3.746 Å). Moreover, sphingolipids, isolated from lipid extracts of the membrane complexes, gave reflections at 4.195 and 3.888 Å, which were

Table V

HIGH-ANGLE SPACING (Å) OBSERVED IN STRATUM CORNEUM AS A FUNCTION OF TEMPERATURE[a]

Temperature[b]			
25°C	45°C	75°C	Interpretation
9.4s	9.4s	9.4s	Both of the sharp lines, 9.4 and 4.6 Å, originate from protein in the corneocyte envelopes
4.6s	4.6s	4.6s	
4.6b	4.6b	4.6b	Liquid alkyl chains
4.16s	Absent	Absent	Both 4.16- and 3.75-Å spacings are due to the crystalline alkyl chains organized as an orthorhombic perpendicular subcell. There may be a distribution of alkyl chains in the gel state because the 4.16-Å line is relatively wide
Absent	4.12s	Absent	The 4.12-Å spacing is due to gel-state alkyl chains organized as hexagonal subcell—transition from crystalline state at 25°C (4.16 and 3.75 Å)
3.75s	Absent	Absent	See above

[a]Summarized from White *et al.* (1988).

[b]Letters after the spacings refer to the width of the reflection qualitatively: s, sharp; b, broad.

stated to be statistically different from the corresponding ones in both whole SC and SC membrane complexes. It is clear now that the reflections at 4.15–4.2 Å observed in all three of the studies cited above arose primarily from oriented lipid bilayers in the SC interstices, because most of the SC lipids are found in the interstices (Grayson and Elias, 1982). Because the lipids in the interstices are arranged in bilayers parallel to the plane of the SC, their alkyl chains are more or less perpendicular to the plane of the SC and the keratin filaments, and hence this reflection, which is due to the lateral packing of the alkyl chains, should be oriented as indicated above. However, it is still possible that there may be additional intracellular lipids associated with the keratins in an ordered fashion, because lipids are not totally localized to the interstices. Finally, in hairless mouse SC and isolated membrane complexes, White *et al.* (1988) also found two sharp lines at 3.75 and 4.16 Å and a broad diffuse line at 4.6 Å, suggesting that crystalline and liquid alkyl chains coexisted at 25°C (Table V). On raising the temperature to 45°C, the 3.75- and 4.16-Å lines disappeared and a sharp line at 4.12 Å appeared, which also disappeared at 75°C. The 4.16-Å line was stated to be wide, such that the existence of a distribution of gel-state alkyl chains could not be ruled out. These observations were interpreted to be crystalline-to-gel (45°C) and gel-to-liquid crystalline transitions (75°C) of the alkyl chain packing of the intercellular lipids.

The high-angle diffraction pattern of murine SC is not solely contributed by lipids. In isolated murine SC membrane complexes, two lines, at 9.4 and 4.6 Å, which are present in the pattern for whole SC, persist after solvent extraction. In addition, a faint 3.9-Å line appears. White *et al.* (1988) suggested that these originate from the cornified envelope, which is composed of highly cross-linked [ϵ-(γ-glutamyl)lysine bonds] proteins. They also noted the similarity of the envelope diffraction pattern to that of crystals of β-glutamate salts. Electron micrographs of solvent-extracted whole SC or SC membrane complexes show the presence of an electron-lucent layer external to the cornified envelope (Elias *et al.*, 1977; Swartzendruber *et al.*, 1987). Swartzendruber *et al.* (1987) suggested that this layer represents covalently bound ceramides (ester linkage between ω-hydroxyceramides and glutamate residues of the cornified envelope), which could be removed by saponification (Wertz and Downing, 1986). Whether this lipid structure contributes to features in the diffraction pattern of extracted SC or SC membrane complexes is still not clear.

4. Applications

Much of the interest in using X-ray diffraction for the study of SC has originated from the desire to investigate the structure or properties of human SC in diseased states (e.g., psoriasis and ichthyosis), or how chemicals affect the structure and function of the SC (e.g., skin softeners or percutaneous drug penetration enhancers). These studies have attempted to correlate structure with function, but interpretations are limited due to a lack of correlative subcellular localization and biochemical data.

a. Effects of Hydration. Bouwstra *et al.* (1990b) found the same lamellar spacing in trypsin-isolated human SC at various levels of hydration (0–40%). They stated that the intensity of the peak at 65 Å decreased with decreasing water content, but it is not clear whether the intensities had been properly scaled to make the comparison among samples meaningful and valid. The lamellar spacing of the SC of essential fatty acid-deficient mice, that is, mice with their diet depleted of linoleic acid, was found to be constant whether the tissue was vacuum dried or with excess water (Hou *et al.*, 1991). Although the mice were in an altered physiological state (excessive transepidermal water loss) and their SC showed scaling, the lamellar spacing was not different from that of normal mice.

b. Human Stratum Corneum in Disease States. Swanbeck (1959) studied scales from psoriasis and one type of recessively inherited ichthyosis, congenital ichthyosiform erythroderma. The low-angle patterns showed reflections due to crystalline cholesterol (33.6 and 16.4 Å) and also possibly triglyceride (14.8 Å). In the high-angle patterns, sharp but very weak reflections were present at 4.9, 5.2, and 5.7 Å, in addition to the spacings found in normal SC. These were interpreted to be due to cholesterol and lecithin. Swanbeck (1959) suggested that these lipids may function as a "glue," causing the persistent cohesion of the SC in these diseases. This implies that the crystallized lipids reside in the intercellular domain but it was not demonstrated in the study. Later, Swanbeck and Thyresson (1961) examined scales from various dermatoses with low-angle X-ray diffraction and found that except for scarlatina scales, all scales showed a deficient aggregation of keratin filaments and the presence of crystalline lipids. They proposed that the deficient aggregation of tonofilaments was caused by the intracellular crystallization of lipids, which in turn was caused by a decreased water content in the scales due to either increased insensible perspiration and/or excessive content or abnormal lipid composition. To further investigate the causes of lipid crystallization, Swanbeck and Thyresson (1962) studied lipids extracted from scales in these diseases in their crystalline form. Their rationale was to eliminate factors contributed by the environment of the lipids in the tissue and to examine variables introduced by the lipids alone. They showed that the lipids extracted from callous SC crystallized from solution (ether) into two phases: cholesterol (34-Å repeat) and an unidentified phase (123-Å repeat). In contrast, in lipids extracted from psoriasis vulgaris scales, an additional phase (47 Å) was present, which was interpreted to originate from a saturated triglyceride that they thought represented lipid droplets visible in electron micrographs. Again, there was no direct evidence for this assignment.

From these studies, the need to localize lipid abnormalities to specific cellular domains of the tissue is apparent. Moreover, lipid composition has to be delineated. Underlying these considerations is the need to determine whether consistent observations can be made within specific diseases, and such an attempt would require large patient populations. As with all studies involving human subjects, a large number of factors that are not disease specific come into play. These factors need to be controlled or considered in order to distinguish whether the abnormality observed is primary or secondary to some factors that may or may not be disease specific. Despite all these caveats, whole-tissue studies have their place, particularly under controlled-for sample conditions (history) and supported by studies on subcellular fractions and/or model lipid systems.

c. Effects of Applied Chemicals. In serving as a permeability barrier to the penetration of exogenous chemicals into the body, the SC also provides a barrier for the topical administration of therapeutic agents across the skin. In the last two decades, various chemicals (e.g., dimethyl sulfoxide, *n*-decyl methyl sulfoxide, 1-dodecylazacylcoheptan-2-one (Azone), pyrrolidones, and oleic acid) have been investigated for their skin permeability-enhancing action. The mechanism of action of these agents is still unclear and researchers have turned to physical techniques to investigate how these chemicals enhance SC permeability. Bouwstra *et al.* (1990b) examined the low-angle diffraction pattern of hydrated human SC previously immersed in a propylene glycol solution with 10% w/w *n*-alkyl azacycloheptan-2-ones [*n*-alkyl chain length of C_6, C_{12} (Azone), and C_{16}]. No changes were observed in the lamellar spacing for SC treated with propylene glycol. Moreover, no significant differences were found in the scattering profile for SC treated with C_6-heptan-2-one in propylene glycol alone. However, treatment with either Azone (C_{12}-) or C_{16}-heptan-2-one in propylene glycol obliterated the first-order diffraction peak. The spacing for this peak was not stated in the original report (Bouwstra *et al.*, 1990a). It is probably the 65-Å peak that is being described, because the 130-Å peak was not observed in the untreated tissue. Yet, freeze–fracture electron microscopy still showed lamellar structures to be present in treated SC. Thus, the authors suggested that the effect of the treatment was to disorder rather than obliterate the intercellular lamellae. The fact that enhancers with longer alkyl chains appeared to have stronger effects on the lamellae than the ones with shorter chains was supported by DTA observations that showed a decrease in the enthalpy of gel–liquid transition with increasing alkyl chain length of the enhancers (Bouwstra *et al.*, 1989). The above interpretations are not definitive, however, because the apparent disorder of the lamellae produced by treatment with the long-chain enhancers could be due to changes in the contrast (electron density difference) with excessive application of these enhancers. The lamellar reflections (relative to air or water, depending on the hydration state of the tissue) can be reduced from either high concentration of these enhancers on the SC surface or within the SC interior. If true, partial removal of the excess enhancer just by blotting the surface with tissue paper or rinsing with ethanol should partially restore the lamellar reflections, and this has been shown to be true for Azone (S. Y. E. Hou, unpublished observations). Moreover, structural disorder should theoretically affect the high-angle reflections more than low-angle ones, whether it be their intensity or shape. Thus, care must be exercised in interpreting X-ray diffraction patterns in these types of studies.

D. Perspectives

X-Ray diffraction studies have provided various types of information about the organization of the intercellular lipid domains in the SC. However the picture is still very limited due to the paucity of studies performed to date, and the still-sketchy information about the composition of the SC intercellular domains. Certain fundamental questions have to be addressed before more fruitful results can be obtained from applications of the technique to experimentally perturbed or to diseased SC. The major problem in the low-angle X-ray diffraction of the SC is the elucidation of the electron density profile of the SC intercellular lamellar structure. As indicated above, both amplitude and phase data are needed for determining the electron density profile (Fourier synthesis). Amplitude is obtained from diffracted intensity, but phase information is lost in the diffraction process. If the lamellar lattice swells without affecting the membrane structure, this could provide a means for ascertaining the phases (Franks and Levine, 1981). Although there are preliminary indications that lamellar dimensions do not change with increasing hydration, more detailed studies are needed. There are various indirect methods for obtaining phase information, based on X-ray diffraction or with the aid of electron microscopy (Franks and Levine, 1981). Swartzendruber *et al.*

(1989) proposed molecular models for the lamellar structure of the SC intercellular region based on electron microscopy images of fixed and stained tissue samples. However, as indicated, these models would have to be supported by X-ray diffraction data, which have a major role in the determination of such molecular architecture. In the high-angle region, the effects of hydration on the packing of the lipid hydrocarbon chains have to be studied. Proton magnetic resonance (^{1}H NMR) (Rehfeld and Plachy, 1991) and EPR data (Rehfeld *et al.*, 1990a) (see below) indicate that the motion or disorder of the chains increases with hydration, but infrared studies (see Potts *et al.*, this volume) have not supported this finding.[1]

III. Electron Paramagnetic Resonance

A. Introduction

Electron paramagnetic resonance, also referred to as electron-spin resonance (ESR), is a spectral technique that reveals the presence of unpaired electrons in a gas, liquid, or solid by their resonant absorption of microwave radiation when the sample is placed in a strong magnetic field (3000 G; ~0.3 T). In-depth descriptions of EPR can be found in textbooks by Wertz and Bolton (1986) and in the two series by Berliner (1976) and Berliner and Reuben (1989). For those unfamiliar with the EPR technique, an introductory chapter in Campbell and Dwek (1984) also is helpful. EPR spectroscopy detects only those molecules containing one or more unpaired electrons, which are referred to as paramagnetic. Due to an alignment of the electron magnetic moments with an applied magnetic field, these species are weakly attracted into the magnetic field. On the other hand, molecules in which all the electrons are paired are referred to as diamagnetic and are weakly repelled from the magnetic field. Molecules possessing unpaired electrons, called free radicals, are normally short-lived intermediates unless captured at extremely low temperatures (less than ~100K).

B. The Spin Probe Technique

Organic free radicals containing the nitroxide functional group are called spin probes or spin labels. The synthesis and chemistry of these stable spin probes are described by Gaffney (1976) and Keana (1984), and that of the more spherical hydrophobic probes, by Plachy *et al.* (1989). Here we show examples of two types of spin probes: type I, nitroxide radicals covalently bonded to biomimetic lipid molecules (Fig. 1), and type II, smaller lipophilic probes approaching spherical geometry (Fig. 2). In biological membranes, characteristic EPR spectra

[1]Note Added in Proof. An X-ray diffraction study of human SC using synchroton radiation was reported after this review had been completed (Garson *et al.*, 1991).

$CH_3-(CH_2)_m-C(-O-, -N(-O)-)-(CH_2)_n-COOH$

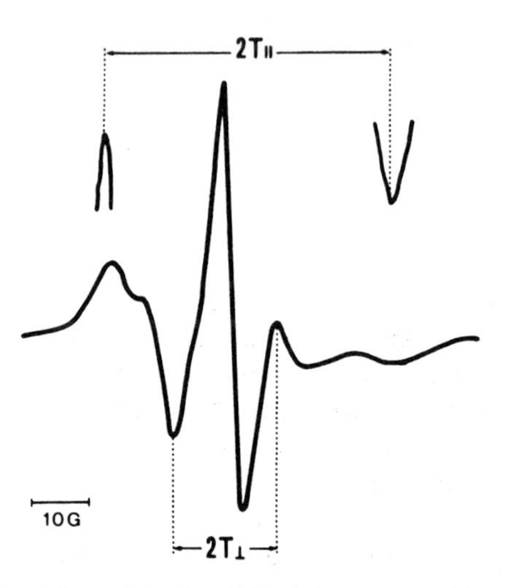

FIG. 1. EPR spectra of 5-doxyl stearic acid-labeled cow snout epidermal cells ($m = 12$, $n = 3$ in structure shown) (Tanaka *et al.*, 1986). The order parameter, S, is defined as the ratio of the observed anisotropy to the maximum anisotropy (Hubbell and McConnell, 1971; Gaffney and McConnell, 1974; Gaffney, 1976). $S = 1.732(T_{\parallel} - T_{\perp} - C)/(T_{\parallel} + 2T_{\perp} + 2C)$, where $C = 1.4 - 0.053(T_{\parallel} - T_{\perp})$ (Tanaka *et al.*, 1986, reproduced with permission of Elsevier Science Publishing Co., Inc. from *J. Invest. Dermatol.* Copyright 1986 by the Society for Investigative Dermatology, Inc.).

(Fig. 1) of partially immobilized probes are observed for type I spin probes, whereas the more highly symmetrical type II spin probes usually give rise to isotropic spectra (Fig. 2). Further, the spin-labeled natural lipid molecules perturb natural lipids as demonstrated in air–water interface monolayer studies (Cadenhead and Muller-Landau, 1973; Tinoco *et al.*, 1972). These types of spin probes are not perfect analogs of pure lipids. However, the thermal transitions determined using the order parameter for these types of anisotropic probes, e.g., nitroxide radical covalently bonded to diphosphatidylcholine (DPPC) in pure DPPC, compare extremely well with the phase transitions for the pure lipid examined by DSC alone (Small, 1986).

Perdeuterated di-*tert*-butyl nitroxide (pdDTBN) and two other small perdeuterated hydrophobic probes synthesized by Plachy *et al.* (1989) (Fig. 2), some de-

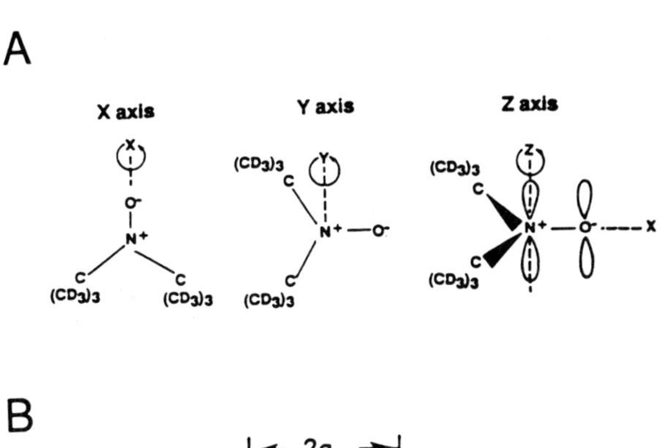

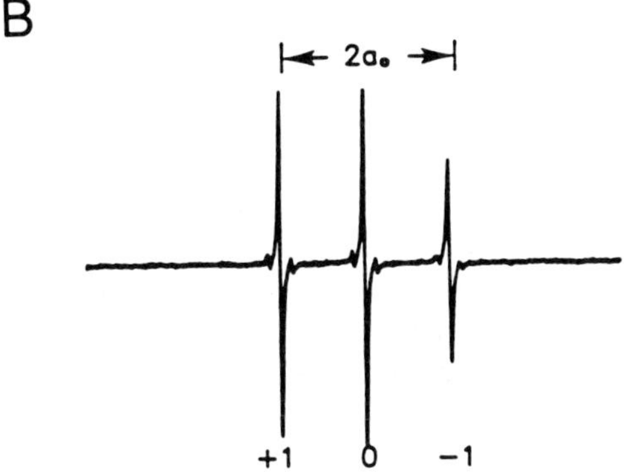

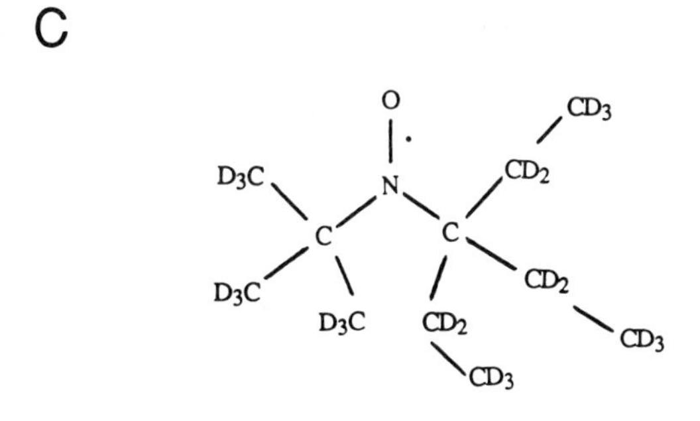

FIG. 2. (A) Perdeuterated di-*tert*-butyl nitroxide (pdDTBN) structure showing molecular orientation axes. (B) EPR spectrum (first derivative) observed for pdDTBN dissolved in vacuum-dried murine SC (Rehfeld *et al.*, 1990b). The narrow line widths show rapid motion averaging over all spin probe axes. The small peaks on either side of the lines are due to the presence of ^{13}C in pdDTBN. (C) A perdeuterated spin probe similar to pdDTBN synthesized by Plachy *et al.* (1989). This probe gives a spectrum similar to that for pdDTBN.

pleted in ^{13}C, have been employed in studies of phase transitions and hydration of the SC (Rehfeld *et al.*, 1988, 1990a,b). The advantage of using these spin probes is that not only phase transitions, but also the polarity of the microenvironment in which the probe resides, as well as acyl chain motion, can be determined. The EPR spectra of these molecules in the SC provide clues about the location and dynamics of exogenous chemicals of similar polarity and size in the tissue. A major reason for the stability of spin probes in the SC is the depletion of antioxidant systems in this tissue. For example, compared to the viable epidermis, the level of ascorbic acid in the SC is relatively low (Giroud and Leblond, 1951).

C. Experimental Considerations

1. *Labeling Tissue Samples with the Spin Probe*

Only approximately 1 mg of SC is required for most spin-labeling studies. In most cases the nitroxide free radicals can be detected at concentrations $>10^{-5}M$. The spin probe-labeled biological lipid shown in Fig. 1 was added directly to the SC (Gay *et al.*, 1989) or fractioned cow snout epidermal cells (2.5 x 10^{-6} cells) (Tanaka *et al.*, 1986) in a minimum quantity of ethanol. The spin probes shown in Fig. 2 were all coated onto the EPR quartz tube in isopentane and the isopentane was allowed to evaporate. The SC samples were then placed in the tube and the labeling proceeded by vapor transfer (Rehfeld *et al.*, 1988, 1990a,b; Plachy *et al.*, 1989). After 24 hours, the contents of the tube were swept with argon or nitrogen to remove oxygen (paramagnetic), which would cause line broadening in the EPR signal. The effect of this procedure is shown in Fig. 3. Because these small spin probes approach spherical symmetry, a different approach in data analysis was taken, revealing both the approximate spin probe mobility as a function of temperature, and the approximate polarity of the spin probe's environment.

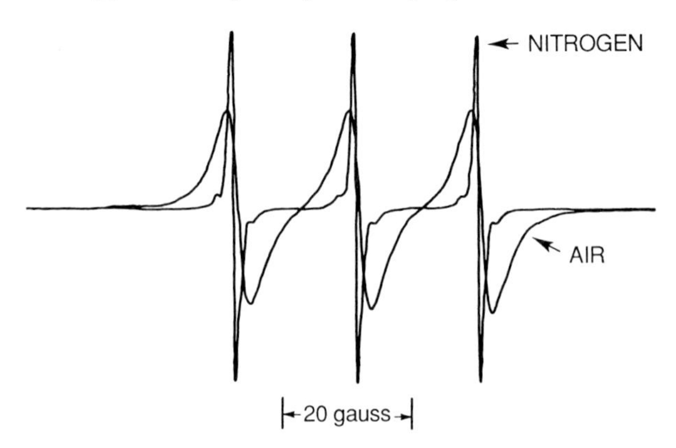

FIG. 3. Line broadening due to the presence of oxygen (paramagnetic molecules) for the EPR spectrum of pdDTBN dissolved in *n*-hexane.

2. *Interpretation of Spectra*

When pdDTBN is dissolved in a nonviscous isotropic solvent (i.e., water, ethanol, *n*-hexane, or acetone), extremely rapid random tumbling motion (reorientation correlation time, $\tau \ll 10^{-9}$ seconds) occurs equally about all probe axes, and the resulting EPR spectrum consists of three well-resolved lines, low (+1), center (0), and (-1) high field lines (Fig. 2). The line widths and intensities are dependent upon the dynamic motions of the pdDTBN, which in turn depend upon the structure of the probe's local molecular environment and temperature (Wertz and Bolton, 1986). In more viscous isotropic solvents ($\tau \geq 10^{-9}$ seconds), the spectral lines differ in widths and thus in their peak intensities (Wertz and Bolton, 1986). For example, the EPR spectra observed for pdDTBN dissolved in vacuum-dried SC (Fig. 2) show rapid motion averaging over all of the spin probe axes (from the probe in a fluid site) coexisting with a small amount of spin probe immobilization. Line separations (hyperfine splitting, a_0) depend on the polarity of the spin probe's environment (Reddoch and Konishi, 1969; Griffith *et al.*, 1974). The values of a_0 reported for pdDTBN in various solvents are listed in Table VI. In summary, a study of EPR line intensities, line widths, and line separations with temperature may provide the following information: (1) changes in the spin probe's molecular environment (phase transition), (2) relative preferential spin probe reorientation (isotropic or anisotropic motion), and (3) polarity of the spin probe's environment.

Unlike pdDTBN, when spin-labeled lipids are added to biological or model lipid membranes, immobilized EPR spectra are observed (Fig. 1). In this case, the spin probe mobility (or membrane fluidity) was determined from the order parameter (Hubbell and McConnell, 1971; Gaffney and McConnell, 1974; Gaffney, 1976). An example of the calculation is given in Fig. 1. The order parameter, S, is defined as the ratio of the observed anisotropy to the maximum anisotropy.

3. *Choice of Spin Probe: Type I Versus Type II*

Several factors can influence the choice of the spin probe. The spin-labeled lipid probes, e.g., doxyl fatty acids and doxyl phosphatidylcholine (type I), are believed to mimic lipid molecules native to the SC. In that sense they may be "less perturbing" to the SC lipid structure. However, their restricted anisotropic motion results in relatively broad line widths and thus reduced spectral resolution. The smaller, approximately spherical, type II spin probes offer greatly reduced line widths due to their rapid and often isotropic motion. This results in increased spectral resolution that increases our ability to detect multiple probe sites (Fig. 4), polarity of probe sites, and probe correlation times.

The SC is a tissue in which the fate of the exogenous molecules is of special interest. Thus the study of type II probes in the SC allows us to infer the dynamics and the location in the SC of exogenous molecules of similar size, shape, and po-

Table VI

pdDTBN HYPERFINE SPLITTING (a_0)[a]

Solvent	a_0[b] A	a_0[b] B	Solvent	a_0[b] A	a_0[b] B
Polar solvents			**Nonpolar solvents**		
Triethylamine	15.20	—	*n*-Pentane	15.12	—
Diethylether	15.27	—	*n*-Hexane	15.12	15.10
Di-*n*-propylamine	15.28	15.32	*n*-Hexene	15.17	—
Toleune	15.34	—	Heptane:pentane (1:1 by vol)		15.13
Piperidine	15.36	15.40	Cyclopentane	15.14	—
n-Butylamine	15.37	15.41	Cyclohexane	15.15	—
Tetrahydrofuran	15.38	—	Tetramethylsilane	15.15	—
Ethylacetate	15.41	15.45	1,5-Hexadiene	15.26	15.30
Hexamethylphosphoramide	15.44	—	Mesitylene	15.30	—
Isopropylamine	15.45	—	Carbon tetrachloride	15.31	—
2-Butanone	15.45	15.49	p-Xylene	15.33	—
Acetone	15.48	15.52	Benzene	15.38	—
i-Butylnitrile	15.50	—	Tetrachloroethylene	15.41	—
Pyridine	15.54	—	**Hydrogen-bonding solvents**		
Nitrobenzene	15.56	—	*N*-Methylpropionamide	15.72	—
Dimethylformamide	15.57	15.63	Dicholoromethane	15.73	—
Benzonitrile	15.58	—	Trichloromethane	15.86	—
1-Nitropropane	15.58	—	*N*-Methylformamide	15.87	15.91
Propionitrile	15.59	—	1-Decanol	—	15.87
Nitroethane	15.62	—	1-Octanol	—	15.89
Acetonitrile	15.64	—	2-Propanol	—	15.94
Dimethyl sulfoxide	15.65	—	1-Hexanol	—	15.97
Propylene carbonate	15.71	—	Ethanol	16.03	16.06
Nitromethane	15.75	—	1-Propanol	—	16.05
Solvents with two conformations			Methanol	16.16	16.21
p-Dioxane	15.48	—	Formamide	16.29	16.33
1,2-Dibromoethane	15.58	—	Acetic acid[c]	16.43	—
1,2-Dichloroethane	15.63	—	Ethanol:water (1:1 by vol)	—	16.69
1,2-Ethanediol	16.36	16.40	Water	17.12	17.16
			1.0 *M* LiCl (aq)	—	17.52
			Anomalous solvents		
			Carbon disulfide	15.30	—
			Hexafluorobenzene	15.44	—
			Chlorobenzene	15.46	—
			Bromobenzene	15.47	—
			α-Chloronaphthalene	15.49	—
			Iodobenzene	15.50	—
			α-Bromonaphthalene	15.51	—

[a]Values given are in gauss, at 25°C.

[b]Data in column A are from Reddoch and Konishi (1969); data in column B are from Griffith *et al.* (1974) unless otherwise noted. S. J. Rehfeld (unpublished observations) and Rehfeld *et al.* (1988; 1990b) are the source of a_0 values for the following lipids: fatty acids, hexadecanoic (65°C) and *cis*-9-octadecanoic acids (15.75); the triglyceride triolein (15.31); bovine brain ceramides, stearic/nervonic acid (≈50:50 mix) and α-hydroxy fatty acid (≈90% α-hydroxy) (15.18 and 15.21, respectively); bovine brain cerebrosides, ≈98% nonhydroxy and ≈50:50 nonhydroxy:hydroxy fatty acids (15.21 and 15.18, respectively).

[c]From S. J. Rehfeld (unpublished observations).

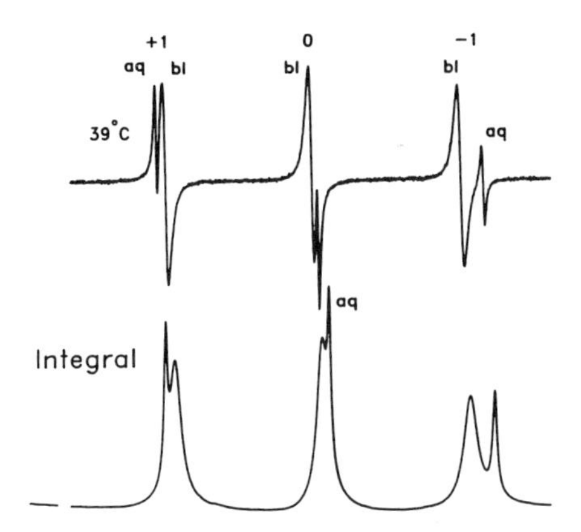

FIG. 4. EPR spectrum for ^{13}C-depleted pDTBN in 40% hydraged murine SC at 39°C. The upper trace is the first-derivative spectrum whereas the lower one is the absorption (integrated) spectrum. The peaks labeled "aq" and "bl" are due to pDTBN dissolved in the aqueous and lipid (bilayer) domains, respectively, in the SC.

larity. The possible perturbing effect of spin probes in lipid domains may be less important in SC, where lipid domains are hundreds of angstroms thick, than in a single bilayer membrane, which is typically 40 Å thick.

D. Membrane Fluidity Studies Using 5-Doxyl Stearic Acid

An EPR study of skin penetration enhancer effect on the membrane fluidity in 5-doxyl stearic acid-labeled model lipid liposomes (multilayer dipalmitoyl phosphatidylcholine) and human SC was conducted by Gay *et al.* (1989). The EPR spectrum observed for human SC was similar to the spectrum for the same spin probe in epidermal cells (Fig. 1), observed by Tanaka *et al.* (1986). The EPR spectra observed for both the model ipid membranes and human SC were similar and so were the effects on the order parameters upon treatment with the selected penetration enhancers. However, no data on thermal transitions in the SC were reported.

E. Phase Transitions in Mammalian Stratum Corneum

T he more spherical spin probe, pdDTBN, shown in Fig. 2A, gives both approximately isotropic EPR spectra when partitioned into a fluid environment, and anisotropic spectra when immobilized in a crystalline domain or when "bound" (by weak hydrogen bonding) in protein domains. Furthermore, the partitioning of the probe between different environments can give rise to two or more EPR

Table VII
LIPID PHASE TRANSITIONS IN MAMMALIAN STRATUM CORNEUM

Temperature (°C) at phase transition						
[1]	[2]	[3]	[4]	Water (%)	Method[a]	Reference
				Human		
40	75	85	107	20,60	DSC	Van Duzee (1975)
—	60–65	70–75	95–100	20	DSC	Knutson *et al.* (1985)
35	65	75	95	50	DSC	Golden *et al.* (1986)
—	65	85	—	60	IR	Golden *et al.* (1986)
42	77	91	—	0	DSC	Goodman and Barry (1989)
41	73	86	113	0–20	DSC	Goodman and Barry (1989)
39	72	85	99	20–40	DSC	Goodman and Barry (1989)
37	71	83	98	41–60	DSC	Goodman and Barry (1989)
38	70	83	95	>60	DSC	Goodman and Barry (1989)
				Porcine		
—	60–65	70–75	95–100	20	DSC	Knutson *et al.* (1985)
—	62	75	105	30	DSC	Golden *et al.* (1987)
				Murine		
38	62	68	—	54	DSC	Rehfeld and Elias (1982)
42	52, 67	82	107	0	DSC	Knutson *et al.* (1985)
—	55–65	70–85	—	20	IR	Knutson *et al.* (1985)
≈ 35	≈ 60	—	—	0	EPR	Rehfeld *et al.* (1985)
≈ 35	≈ 60	—	—	15–90	EPR	Rehfeld and Plachy (1991)
30–40	60–70	—	—	0–70	^{1}H NMR	Rehfeld and Plachy (1991)

[a]Abbreviations: DSC, differential scanning calorimetry; IR, infrared spectroscopy; EPR, electron paramagnetic resonance; NMR, nuclear magnetic resonance.

signals, e.g., one originating from the aqueous phase and the other from a more hydrophobic phase, as shown in Fig. 4. The application of EPR with the spin probe pdDTBN for measuring phase transitions, polarity, and hydrogen-bonding effects on lipid in SC have been described by Rehfeld *et al.* (1988, 1990a,b). Table VII compares thermal phase transition temperatures determined in these studies with those determined with DSC and other spectroscopic methods for human, porcine, and murine SC at various hydration levels. The polarities (hyperfine splitting, a_0) of the microenvironment of pdDTBN in various SC samples at 37°C from these studies are listed in Table VIII.

1. Murine Stratum Corneum

The recent study of Rehfeld *et al.* (1990b) determined the domains in the whole SC, which the spin probe, pdDTBN, partitions into and reports from, and whether the thermal phase transitions observed in the tissue could be ascribed to one or more specific lipid component(s). For this purpose, both unextracted and solvent-

Table VIII
POLARITY OF MICROENVIRONMENT OF pdDTBN IN VARIOUS SC SAMPLES AT 37°C

SC sample	Polarity a_0 (gauss)
Hairless mouse whole SC	15.42[a]
Hairless mouse SC membrane complexes	15.37[a]
Human sunburn scale	15.55[b]
Recessive X-linked ichthyosis scale	15.42[b]
Human cadaver whole SC	15.33[b]

[a]Rehfeld *et al.* (1990b). Samples are vacuum dried. Values ± 0.05 gauss.
[b]Rehfeld *et al.* (1988). Values ± 0.02 gauss.

extracted SC whole sheets and membrane complexes were compared as a function of temperature. Two spin probe reorientation correlation times were defined for the mobilities of the spin probe in the samples: $\tau_{e\pm1}$, which is derived from the line width of the low field line and the intensities of both the low and high field lines, and $\tau_{e\text{-}1}$, which is derived from the line width of the center field line and the intensities of both the center and high field lines. If the spin probe is in an anisotropic environment, these values differ from each other. However, if the environment is isotropic, the values are approximately equal. Also, in the latter case, an Arrhenius plot of the τ values (logarithm of τ versus inverse temperature) should be a straight line within certain limits of τ. The polarity of the spin probe's environment is indicated by the hyperfine splitting, which remains constant with temperature if the spin probe remains highly mobile. Finally, the amount of spin probe remaining mobile, or conversely immobile, can be estimated from the ratio of the sum of the areas of the three peaks to the total area in the EPR absorption spectrum obtained by computer integration of the first derivative spectrum normally obtained from EPR spectrometers.

In the physiological temperature range (~20–40°C), the spin probe was found to be largely dissolved in isotropic domains in the SC. A thermal transition was observed in this temperature range (midpoint, ~33°C), which was comparable to the previously described transition in DSC studies of neonatal mouse SC (38°C; Table VII) (Rehfeld and Elias, 1982). At 60°C, it was found that $\tau_{e\pm1} \sim \tau_{e\text{-}1}$, indicating rapid isotropic motion of the spin probe. This transition also coincided with a supraphysiological lipid phase transition previously found in neonatal mouse SC with DSC (60°C; Table VII). Between 40 and 60°C, the τ values remained essentially constant, suggesting melting of the lipid domains with constant viscosity.

Both of the above phase transitions are less well-resolved in ESR data obtained from SC membrane complexes. Further, the Arrhenius plots did not coincide exactly with those obtained from whole SC. From 23 to ~-20°C, the motion of the spin probe in SC membrane complexes became significantly more anisotropic.

This decrease in mobility is undoubtedly due to the removal of the lower melting-point unsaturated lipids, such as oleic acid and linoleic acid (Rehfeld *et al.*, 1990b), during the preparation of the SC membrane complexes. With increasing temperature, the spin probe partitioned predominantly into the "melted" lipid domains. The environment of the spin probe became more polar in SC than in SC membrane complexes with increasing temperature. The spin probe's mobility and polarity values in model lipids, e.g., cholesterol, free fatty acids, and ceramides, were significantly different from values observed for SC and SC membrane complexes (Rehfeld *et al.*, 1988, 1990b). In conclusion, both these data and the data from X-ray diffraction (see above) show that the properties of the lipid domains in SC membrane complexes and single-model lipids differ from the properties of the lipid domains in whole SC.

2. *Effects of Hydration*

Rehfeld et al. (1990b) examined the effect of hydration on murine SC thermal phase transitions using the spin probe pdDTBN (^{13}C depleted). When dried trypsin-isolated SC sheets were hydrated to various levels (10% to excess), the spin probe partitioned into an aqueous site and a hydrophobic site. In a heterogeneous system such as this, the resulting spectrum is a superposition of spectra of the probe in the separate phases, and if the polarities of the phases differ sufficiently, the resultant spectrum should split, particularly in high fields (Likhtenshtein, 1976). The composite spectrum for 40% hydrated SC is shown in Fig. 3. The mobility of the spin probe in the hydrophobic site was found to increase with increasing hydration, indicating increases in lipid acyl chain mobility. This has recently been confirmed by ^{1}H NMR (Rehfeld and Plachy, 1991). However, the polarity of the hydrophobic site remained constant for all levels of hydration. Between -70 and 60°C, the phase transitions determined from the signal of the probe in the hydrophobic site also were independent of the degree of hydration. These results indicate that the SC intercellular bilayers are highly impermeable to water, and hence constitute the major contributing phase to the water barrier of the SC.

3. *Human Stratum Corneum in Disease States*

Rehfeld *et al.* (1988) compared the EPR spectra of pdDTBN as a function of temperature in normal human scale (NHS) (from sunburn desquamation) and recessive X-linked ichthyosis scales (RXLIS), which has a higher cholesteryl sulfate content than normal. The spectra for both samples (which are solids) showed isotropic motion of the spin probe. Whereas DSC did not reveal any phase transitions in RXLIS, EPR spectra showed comparable phase transitions in NHS and RXLIS using Arrhenius plots of reorientation correlation times (only a single τ

calculated, which is defined like τ_{e-1} but with a different coefficient and for a different time limit) (Rehfeld *et al.*, 1988). However, the spin probe mobility in RXLIS was much higher than the mobility of the probe in NHS at all temperatures. Moreover, the spin probe's environment was found to be more polar in NHS in comparison to that in RXLIS. In NHS, the environment is similar to that in hydrogen bond acceptor solvents, such as ketones. SC isolated from cadaver skin using trypsin showed a more fluid and less polar (similar to that in triolein) environment than both of the above scales. Model lipid systems consisting of binary mixtures of cholesterol or cholesteryl sulfate and palmitic acid or ternary mixtures of cholesterol, palmitic acid, and oleic acid with or without cholesteryl sulfate, in the same proportions as those found in the scales, were also studied. The results on mobility and polarity paralleled those for scales. However only the mixtures with oleic acid gave mobilities similar to those for the scales. Overall, the date suggest that there is a difference in the degree of hydrogen bonding in NHS versus RXLIS. In NHS, cholesterol hydrogen forms bonds with esterified and nonesterified fatty acids, but in RXLIS, excess cholesteryl sulfate cannot hydrogen bond with fatty acids unless the pH is low (Rehfeld *et al.*, 1986, 1988). This model is consistent with observations on the spin probe's mobility and polarity in the two types of scales. In general, hydrogen-bonding lipids (e.g., cholesterol) would provide a more polar environment of less mobility to the spin probe in comparison to non-hydrogen-bonding lipids (e.g., cholesterol sulfate). How hydrogen-bonding characteristics of lipid species actually influence normal and pathological desquamation still has to be established (see Williams, this volume).

F. Spin Labeling of Stratum Corneum Cyanoacrylate Glue Strippings

To circumvent the use of trypsin for obtaining normal SC samples, Plachy *et al.* (1990) obtained intact sheets of SC from normal human subjects by bonding the skin surface to the surface of a small quartz plate spread with a small amount of cyanoacrylate glue (Super Glue; Duro, Loctite Corp., Cleveland, Ohio). This sampling method was first utilized by Marks and Dawber (1971) and Marks and Pearse (1975) to obtain SC sheets. The adherent SC samples can be labeled with the spin probe perdeuterated di-*tert*-amyl nitroxide (5,5-NO) by vapor-phase transfer as usual, and the entire quartz plate, with the bonded SC, is placed in the EPR spectrometer. Once the glue has solidified, no signals arise from the cured resin or from the spin probe dissolved in the resin; the only signals observed arise from the spin probe in the attached SC material. This method not only has the advantage of avoiding prior exposure of the SC to enzymes, but also, depending on how thick a sample is removed by each stripping, it could potentially be used to study SC properties as a function of depth within the SC, i.e., successive strippings.

G. Oxygen Transport in Stratum Corneum Using Spin Probes

Plachy and Hatcher (1990) determined the diffusion coefficient of dioxygen in trypsin-isolated SC using the spin probe technique. Due to its paramagnetism, dissolved dioxygen increases the EPR line width of spin probes (as shown in Fig. 3), depending on the product of the dioxygen diffusion coefficient and solubility in the tissue. The intensity of the spin probe signal was monitored as a function of time in the second-derivative mode of the spectrometer. The dioxygen diffusion coefficient in whole-sheet SC was found to be 3 x 10^{-8} cm^2 sec^{-1}. This is 50–100 times lower than the diffusion coefficient of dioxygen in an SC lipid mixture. Therefore, dioxygen appears to diffuse through the SC via a lipid pathway (intercellular lipids), with its effective length being about eight times the thickness of the SC, due to the tortuosity of the pathway. This is reasonable because the solubility of oxygen in hydrocarbons such as hexane or nonane is 10 times its aqueous solubility (Battino and Clever, 1966). Thus, oxygen concentration is expected to be higher in the lipid domain than in the corneocyte interior or the rest of the tissue. Moreover, the amount of water present in the tissue studied was very small because it was vacuum dried.

IV. Conclusions

EPR has provided information about the SC that is both parallel and complementary to information provided by X-ray diffraction. Both techniques provide information on a molecular level about temperature-dependent phase behavior that correlates with DSC determinations. However, X-ray diffraction provides more direct information about structure, whereas EPR provides more direct information about the physical properties of membranes, including polarity, microviscosity, and phase transitions. Unlike X-ray diffraction, EPR requires the introduction of spin probes into the system. Thus one is observing the behavior of the probe and thereby making indirect interpretations about its microenvironment. The use of spin probes that are more selective than pdDTBN regarding partitioning into either hydrophilic or hydrophobic domains of the SC would provide further information. Due to its much shorter time scale compared to X-ray diffraction in the study of lipids, EPR is useful for elucidating kinetic aspects of the properties of the SC. This is evident in the oxygen diffusion studies described above, and further applications are to be expected. Further applications of the spin probe technique for understanding the basis of exogenous chemicals in affecting the properties of the SC should also yield fruitful information.

Spectroscopic methods are powerful techniques that can provide detailed molecular information about biological and model lipid systems. However, to make valid interpretations on observations, parallel detailed information about the chemical composition of the system is required. In this article we have surveyed available information on the SC lipid domains obtained with X-ray diffraction and electron paramagnetic resonance spectroscopy. The aforementioned need in the study of SC is very much evident and is critical to further success in using these and other spectroscopic methods, e.g., nuclear magnetic resonance. Most studies performed to date have been on whole tissue. Isolation and identification of components would also allow the construction of model systems wherein spectroscopic techniques could in turn provide more interpretable information, which would further our understanding of the actual biological system. Most studies about the SC are motivated by a desire to learn about the system, as perturbed by chemicals or diseases. Indeed, such studies can also provide insights into normal structure and function, but actual information about normal SC is still very limited.

ACKNOWLEDGMENTS

This work was supported by the Veterans Administration Medical Research Service and NIH Grant AR 19098 (both to Peter M. Elias) and NIH Grant G-5305500-250161 (to William Z. Plachy).

References

Abrahamsson, S., Dahlen, B., Lofgren, H., and Pascher, I. (1978). *Prog. Chem. Fats Other Lipids* **16,** 125–143.

Andersen, H. C. (1978). *Annu. Rev. Biochem.* **47,** 359–383.

Battino, R., and Clever, H. L. (1966). *Chem. Rev.* **66,** 395–463

Berliner, L. J., ed. (1976). "Spin Labeling." Academic Press, New York.

Berliner, L. J., and Reuben, J., eds. (1989). "Spin Labeling." Plenum, New York.

Blaurock, A. E. (1982). *Biochim. Biophys. Acta* **650,** 167–207.

Bouwstra, J. A., Peischier, L. J. C., Brussee, J., and Bodde, H. E. (1989). *Int. J. Pharmacol.* **52,** 47–54.

Bouwstra, J. A., de Vries, M. A., Bras, W., and Gooris, G. S. (1990a). *Int. Symp. Controlled Release Bioact. Mater., Reno, Nev.* Abstr. p. 53.

Bouwstra, J. A., de Vries, M. A., Bras, W., and Gooris, G. S. (1990b). *Proc. Int. Symp. Control Release Bioact. Mater.* **17,** 33–34.

Cadenhead, D. A., and Muller-Landau, F. (1973). *Biochim. Biophys. Acta* **307,** 279–285.

Caffrey, M. (1989). *Annu. Rev. Biophys. Biophys. Chem.* **18,** 159–186.

Campbell, I. D., and Dwek, R. A. (1984). "Biological Spectroscopy." Benjamin/Cummings, Menlo Park, California.

Cantor, C. R., and Schimmel, P. R. (1980). "Biophysical Chemistry," pp. 687–791. Freeman, San Francisco, California.

Chapman, D. (1965). "The Structure of Lipids," pp. 221–315. Wiley, New York.

Cowley, J. M. (1981). "Diffraction Physics." North-Holland Publ., Amsterdam.

Elias, P. M., and Friend, D. S. (1975). *J. Cell Biol.* **65,** 108–191.

Elias, P. M., Goerke, J., and Friend, D. S. (1977). *J. Invest. Dermatol.* **69,** 535–546.

Elias, P. M., Bonar, L., Grayson, S., and Baden, H. (1983). *J. Invest. Dermatol.* **80,** 213–214.

Elliott, A. (1965). *J. Sci. Instrum.* **42,** 312–316.

Franks, A. (1955). *Proc. Phys. Soc. London, Sect. B* **68,** 1054–1064.

Franks, A. (1958). *Br. J. Appl. Phys.* **9,** 349–352.
Franks, N. P., and Levine, Y. K. (1981). *In* "Membrane Spectroscopy" (E. Grell, ed.), pp. 437–487. Springer-Verlag, Berlin.
Franks, N. P., and Lieb, W. R. (1981). *In* "Liposomes: From Physical Structure to Therapeutic Applications" (C. G. Knight, ed.), pp. 243–272. Elsevier/North-Holland, Amsterdam.
Fraser, R. D. B.,and MacRae, T. P. (1973). "Conformation in Fibrous Proteins," pp. 5–59. Academic Press, New York.
Friberg, S. E., and Osborne, D. W. (1985). *J. Dispersion Sci. Technol.* **6,** 485–495.
Friberg, S. E., Osborne, D. W., and Tombridge, T. L. (1985). *J. Soc. Cosmet. Chem.* **36,** 349–354.
Friberg, S. E., Kayali, I., Beckerman, W., Rhein, L. D., and Simion, A. (1990). *J. Invest. Dermatol.* **94,** 377–380.
Gaffney, B. J. (1976). *In* "Spin Labeling: Theory and Application" (L. J. Berliner, ed.), pp. 567–571. Academic Press, New York.
Gaffney, B. J., and McConnell, H. M. (1974). *J. Magn. Res.* **16,** 1–28.
Garson, J. C., Doucet, J., Lêvêque, J. L., and Tsoucaris, G. (1991). *J. Invest. Dermatol.* **96,** 43–49.
Gay, C. L., Murphy, T. M., Hadgraft, J., Kellaway, I. W., Evans, J. C., and Rowlands, C. C. (1989). *Int. J. Pharm.* **49,** 39–45.
Giroud, A., and Lebond, C. P. (1951). *Ann. N.Y. Acad. Sci.* **53,** 613–626.
Golden, G. M., Guzek, D. B., Harris, R. R., McKie, J. E., and Potts, R. O. (1986). *J. Invest. Dermatol.* **86,** 255–259.
Golden, G. M., McKie, J. E., and Potts, R. O. (1987). *J. Pharm. Sci.* **76,** 25–28.
Goldsmith, L. A., and Baden, H. (1970). *Nature (London)* **225,** 1052–1053.
Goodman, M., and Barry, B. W. (1989). *In* "Percutaneous Absorption" (H. I. Maibach and R. L. Bronaugh, eds.), 2nd Ed., pp. 567–593. Dekker, New York.
Grayson, S., and Elias, P. M. (1982). *J. Invest. Dermatol.* **78,** 128–135.
Griffith, H. O., Dehlinger, P. J., and Van, S. P. (1974). *J. Membr. Biol.* **15,** 150–192.
Gruner, S. M. (1987). *Science* **238,** 305–319.
Guinier, A. (1963). "X-Ray Diffraction in Crystals, Imperfect Crystals, and Amorphous Bodies." Freeman, San Francisco, California.
Harrison, S. C., Winkler, F. K., Schutt, C. E., and Durbin, R. M. (1985). *Methods Enzymol.* **114,** 211–237.
Hosemann, R., and Bagchi, S. N. (1962). "Direct Analysis of Diffraction by Matter." North-Holland Publ., Amsterdam.
Hou, S. Y. E., Mitra, A. K., White, S. H., Menon, G. K., Ghadially, R., and Elias, P. M. (1991). *J. Invest. Dermatol.* **96,** 215–223.
Hubbell, W. L., and McConnell, H. M. (1971). *J. Am. Chem. Soc.* **93,** 314–326.
James, R. W. (1965). "The Optical Principles of the Diffraction of X-Rays." Ox Bow Press, Woodbridge.
Keana, J. F. W. (1984). *In* "Spin Labeling in Pharmacology" (J. L. Holtzman, ed.), pp. 1–85. Academic Press, New York.
Knutson, K., Potts, R. O., Guzek, D. B., Golden, G. M., McKie, J. E., Lambert, W. J., and Higuchi, W. I. (1985). *J. Controlled Release* **2,** 67–87.
Laggner, P. (1988). *Top. Curr. Chem.* **145,** 173–202.
Levine, Y. K. (1973). *Prog. Surf. Sci.* **3,** 279–352.
Likhtenshtein G. I. (1976). "Spin Labeling Methods in Molecular Biology" (P. S. Shelnitz, transl.), p. 196. Wiley, New York.
Makowski, L., Li, J. (1983). *In* "Biomembrane Structure and Function" (D. Chapman, ed.), pp. 43–166. Macmillan, New York.
Marks, R., and Dawber, R. P. R. (1971). *Br. J. Dermatol.* **84,** 117–123.
Marks, R., and Pearse, A. D. (1975). *Br. J. Dermatol.* **92,** 651–657.

McElhaney, R. N. (1982). *Chem. Phys. Lipids* **30,** 229–259.
Michaels, A. S., Chandrasekaran, S. K., and Shaw, J. E. (1975). *AIChEJ.* **21,** 985–996.
Mitsui, T. (1978). *Adv. Biophys.* **10,** 97–135.
Pape, E. H. (1985). *In* Bengha G, ed. "Structure and Properties of Cell Membranes" (G. Bengha, ed.), Vol. 3, pp. 167–193. CRC Press, Boca Raton, Florida.
Philips, W. C. (1985). *Methods Enzymol.* **114,** 300–316.
Philips, W. C., and Rayment, I. (1985). *Methods Enzymol.* **114,** 316–329.
Plachy, W. Z., and Hatcher, M. E. (1990). *J. Invest. Dermatol.* **94,** 567. (Abstr.)
Plachy, W. Z., Viereck, J. C., and Yee, C. W. (1989). *J. Labelled Compd. Radiopharm.* **28,** 99–110.
Plachy, W. Z., Rehfeld, S. J., and Elias, P. M. (1990). *J. Invest. Dermatol.* **94,** 567 (Abstr.)
Potts, R. O. (1989). *In* Hadgraft J, Guy RH, "Transdermal Drug Delivery" (J. Hadgraft and R. H. Guy, eds.), pp. 23–57. Dekker, New York.
Reddoch, A., and Konishi, S. (1969). *J. Chem. Phys.* **70,** 2121–2130.
Rehfeld, S. J., and Elias, P. M. (1982). *J. Invest. Dermatol.* **91,** 499–505.
Rehfeld, S. J., and Plachy, W. Z. (1991). *Biophys. J.* **59,** 643a. (Abstr.)
Rehfeld, S. J., Williams, M. L., and Elias, P. M. (1986). *Arch. Dermatol. Res.* **278,** 259–263.
Rehfeld, S. J., Plachy, W. Z., Williams, M. L., and Elias, P. M. (1988). *J. Invest. Dermatol.* **91,** 499–505.
Rehfeld, S. J., Plachy, W. Z., and Elias, P. M. (1990a). *J. Invest. Dermatol.* **94,** 571. (Abstr.)
Rehfeld, S. J., Plachy, W. Z., Hou, S. Y. E., and Elias, P. M. (1990b). *J. Invest. Dermatol.* **95,** 217–223.
Rosenbaum, G., and Holmes, K. C. (1980). *In* "Synchrotron Radiation Research" (H. Winick and S. Doniach, eds.), pp. 533–564. Plenum, New York.
Ruocco, M. J., Atkinson, D., Small, D. M., Skarjune, R. P., Oldfield, E., and Shipley, G. G. (1981). *Biochemistry* **20,** 5957–5966.
Shipley, G. G. (1973). *In* "Biological Membranes," Vol. II (D. Chapman and D. F. H. Wallach, eds.), pp. 1–89. Academic Press, New York.
Small, D. M. (1986). "The Physical Chemistry of Lipids." Plenum, New York.
Stout, G. H., and Jensen, L. H. (1989). "X-Ray Structure Determination," pp. 7–17. Wiley (Interscience), New York.
Swanbeck, G. (1959). *Acta Derm.-Venereol.* **39,** Suppl., p. 43.
Swanbeck, G., and Thyresson, N. (1961). *Acta Derm.-Venereol.* **41,** 289–296.
Swanbeck, G., Thyresson, N. (1962). *Acta Derm.-Venereol.* **42,** 445–457.
Swartzendruber, D. C., Wertz, P. W., Madison, K. C., and Downing, D. T. (1987). *Invest. Dermatol.* **88,** 709–713.
Swartzendruber, D. C., Wertz, P. W., Kitko, D. J., Madison, K. C., and Downing, D. T. (1989). *J. Invest. Dermatol.* **92,** 251–257.
Sweet, R. M., and Woodhead, A. D. (1989). "Synchrotron Radiation in Structural Biology." Plenum, New York.
Tanaka, T., Sakanashi, T., Kaneko, N., and Ogura, R. (1986). *J. Invest. Dermatol.* **87,** 745–747.
Tinoco, J., Ghosh, D., and Keith, A. D. (1972). *Biochim. Biophys. Acta* **274,** 279–285.
Van Duzee, B. F. (1975). *J. Invest. Dermatol.* **65,** 404–408.
Welte, W., and Kreutz, W. (1979). *Adv. Polym. Sci.* **30,** 161–225.
Wertz, J. E., and Bolton, J. R. (1986). "Electron Spin Resonance." Chapman & Hall, New York.
Wertz, P. W., and Downing, D. T. (1986). *Biochem. Biophys. Res. Commun.* **137,** 992–997.
White, S. H., Mirejovsky, D., and King, G. I. (1988). *Biochemistry* **27,** 3725–3732.
Wilkes, G. L., Nguyen, A. L., and Wildnauer, R. (1973). *Biochim. Biophys. Acta* **304,** 267–275.
Witz, J. (1969). *Acta Crystallogr., Sect. A.* **A25,** 30–42.
Worthington, C. R. (1973). *Curr. Top. Bioenerg.* **5,** 1–39.
Yoshimatsu, M., and Kozaki, S. (1977). *In* "X-Ray Optics" (H. J. Queisser, ed.), pp. 9–33. Springer-Verlag, Berlin.

ADVANCES IN LIPID RESEARCH, VOL. 24

Strategies to Enhance Permeability via Stratum Corneum Lipid Pathways

RUSSELL O. POTTS,* VIVIEN H. W. MAK,[†] RICHARD H. GUY,[‡] AND MICHAEL L. FRANCOEUR[†]

*Cygnus Therapeutic Systems
Redwood City, California 94063
[†]Dermal Therapeutics Group
Central Research Division
Pfizer, Inc.
Groton, Connecticut 06340
[‡]Departments of Pharmacy and Pharmaceutical Chemistry
University of California, San Francisco
San Francisco, California 94143

I. Introduction

Stratum corneum (SC) lipids, like those of other biomembranes, can be studied by a variety of biophysical techniques. Methods that have been used to investigate lipid membrane biophysics include X-ray diffraction (Shipley, 1981; Franks and Levine, 1981), differential scanning calorimetry (DSC) (Donovan, 1984), nuclear magnetic resonance (NMR) (Chan *et al.*, 1981; Bloom and Smith, 1985), electron spin resonance (ESR) (Gordon and Curtain, 1988; Marsh, 1981), and infrared (IR) (Casal and Mantsch, 1984; Fringelli and Gunthard, 1981) spectroscopy. These techniques have been particularly useful in establishing the phase behavior of a number of lipid membrane systems. In addition, the results of these studies have led to a number of interesting conclusions about the role of membrane lipids in cell and liposome permeability. The reader is referred to several reviews on the use of these biophysical tools to evaluate lipid membranes (Chapman, 1975; Small, 1981).

In order to understand more completely the barrier properties of the SC, we have used the techniques of IR spectroscopy and DSC. These results, which are summarized in this article, lead one to the conclusion that, in spite of profound morphological and compositional differences among SC lipids and those commonly found in other biological systems, there is a remarkable similarity in many biophysical properties. Thus, insight can be gained from studies of SC lipids by comparing the results to those obtained for simpler and more conventional systems.

II. *In Vitro* Studies

The technique of Fourier transform IR (FTIR) spectroscopy has been extensively used to study the phase behavior of lipid membranes. Casal and Mantsch (1984) provide a comprehensive survey of the use of FTIR spectroscopy to evaluate phospholipid membranes.

The lipids of the SC can also be studied by FTIR techniques. However, unlike other membrane systems in which samples are suspended in water, SC samples were evaluated as intact sheets. Consequently, the intense IR absorption bands associated with water do not significantly compromise the spectra of SC, even for highly hydrated samples. In the experiments described here, SC sheets were obtained by enzymatic digestion (Golden *et al.*, 1986). Other isolation techniques may be used, however, to produce SC suitable for IR study. The sheet of SC is treated under the desired experimental conditions and is sealed between two IR-transparent windows. This sandwiched sample is then mounted in a temperature-controlled heating mantle in the IR spectrometer. Due to the thin (10–20 μm) cross-section of the SC, a significant amount of IR radiation passes through the sample though some is absorbed due to molecular vibrations. Analysis of the absorbed radiation provides a spectrum with characteristic peaks associated with each vibrational mode.

A. Effect of Temperature on SC Lipids

A typical FTIR spectrum of porcine SC is shown in Fig. 1. This spectrum looks like most other hydrated biological samples, with peaks due to lipids, proteins, and water. Of particular interest for the study of SC lipid biophysics are the peaks between 2800 and 3000 cm^{-1}, due to C–H stretching vibrations primarily associated with lipid alkyl chains. These absorbances occur near 2920 and 2850 cm^{-1} for the asymmetric and symmetric C–H vibrations, respectively. Upon heating the

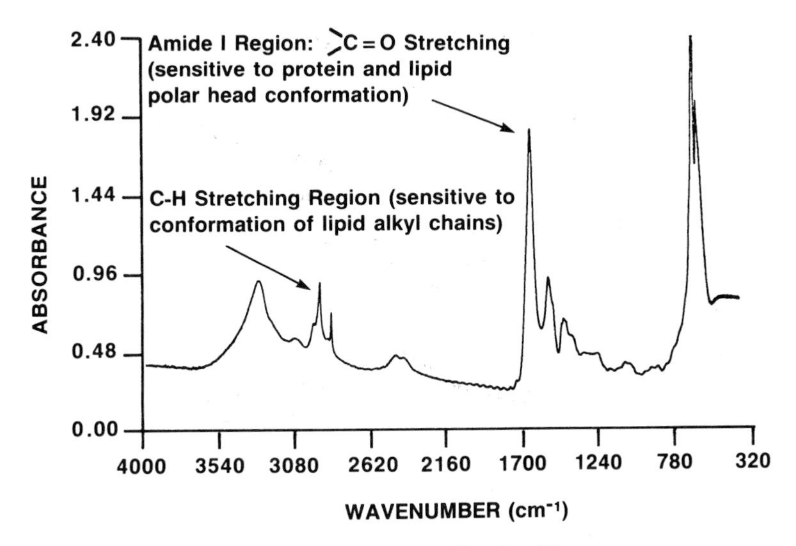

FIG. 1. The IR spectrum of porcine SC.

sample, both C–H stretching peaks broaden and shift to a higher wavenumber. A plot of the shift in C–H symmetric stretching frequency [$\nu_s(CH_2)$] over a continuous range of temperature is shown in Fig. 2A. These results show that there is a modest increase in $\nu_s(CH_2)$ between about 30 and 60°C and above about 80°C. In contrast, there is an abrupt increase in $\nu_s(CH_2)$ between about 60 and 80°C. Identical results were obtained for the asymmetric stretching frequency [$\nu_a(CH_2)$]. Similar sigmoidal curves of $\nu_s(CH_2)$ and $\nu_a(CH_2)$ versus temperature have been obtained for a variety of lipid systems and are associated with lipid phase transition(s). For example, dipalmitoyl phosphatidylcholine (DPPC) undergoes a similar, although even more abrupt, increase in $\nu_s(CH_2)$ at 41°C, corresponding to a gel to liquid-crystal phase transition in this lipid (Casal and Mantsch, 1984). Note that for a highly homogeneous sample such as DPPC, the transition occurs over a very narrow range of temperature. In contrast, the transition in SC is very broad, primarily due to the heterogeneous nature of SC lipids.

The assignment of the C–H stretching peaks to SC lipids was further substantiated by the results of DSC experiments that were performed with a biological microcalorimeter using as little as 10 mg of sample (Golden *et al.*, 1986). The SC was prepared in a manner identical to preparation of samples used for the FTIR experiments. Simply stated, DSC measures the excess heat liberated (exothermic) or absorbed (endothermic) from a sample when compared to a reference, as both

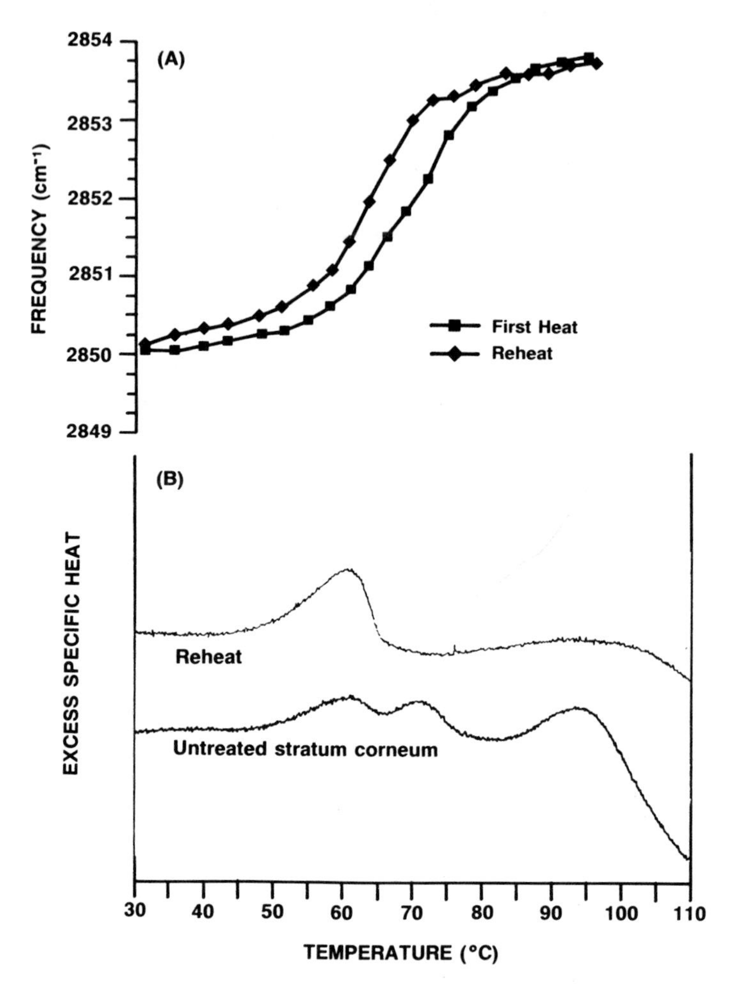

FIG. 2. (A) The change in the C–H symmetric stretching frequency [$\nu_s(CH_2)$] as a function of temperature for porcine SC. (B) The DSC thermal profile for identically treated porcine SC.

are heated at a uniform rate. A more complete description of lipid membrane microcalorimetry has been published (Donovan, 1984).

The IR and DSC thermal profiles for porcine SC between about 20 and 110°C are compared in Fig. 2A and B, respectively. The DSC results in Fig. 2B show that intact SC has three thermal transitions: two lipid-associated transitions with midpoints (T_m) near 60 and 70°C and a keratin-associated transition near 100°C (Golden *et al.*, 1986). In contrast, upon reheat, only a single endotherm at about

60°C was observed; the enthalpy of this endotherm was equal to the sum of the two lipid transitions. This IR data for an identical sample (Fig. 2A) show that the initial heat resulted in a transition that spans the range seen for the two lipid transitions in DSC. In addition, the IR results show a slight inflection at about 70°C, suggestive of two transitions in the range of 50–80°C. Upon reheat, the transitions seen with both IR and DSC occurred over a narrower temperature range with a lower T_m. The parallel between the IR and DSC data (Fig. 2A and B) at temperatures below about 80°C is striking. Qualitatively, the DSC data appear to be a first derivative of the IR results.

Further evidence for the assignment of the C–H stretching peaks comes from the results of solvent extraction experiments (Golden *et al.*, 1986). Removal of extracellular SC lipids via chloroform–methanol extraction led to a dramatic decrease in the intensity of the C–H stretching peaks. In addition, the lipids extracted from the SC showed a temperature dependence similar to that of the intact sample. The data obtained for extracted lipid again showed a parallel between DSC and IR results. Taken together, these results suggest that both techniques are sensitive to the lipid transitions that occur in the extracellular lipids of the SC.

The T_m values associated with SC lipid transitions (60 and 70°C) are much higher than those seen in cell membranes and are consistent with both the long, saturated lipid alkyl chains and the high concentration of ceramides in these samples (see Schurer and Elias, this volume). The area under the lipid thermal peaks (as determined by DSC) gives an estimate of the enthalpy (ΔH) of the transition. Results obtained for both intact porcine SC and extracted lipids show that ΔH is around 5 cal per gram of lipid. A similar value has been reported for human SC (Bouwstra *et al.*, 1989). Assuming an average molecular weight near 500 for these lipids (see Schurer and Elias, this volume), the molar enthalpy is around 2.5 kcal/mol. This value is low by comparison to many other membranes (Small, 1986), but may reflect the enthalpy-lowering effect of cholesterol or cholesterol sulfate, which comprise a significant fraction of SC lipids. In addition, the low value of ΔH could reflect the fact that some of the SC lipids are in a liquid state at the transition temperature, consistent with the results of White *et al.* (1988), who showed the coexistence of solid and liquid SC lipids at physiological temperatures.

In addition to providing information on phase transitions, IR results can also provide details of molecular rearrangements. The molecular basis for the shift in the C–H stretching frequencies has been well described (Casal and Mantsch, 1984). A qualitative explanation of the effect is offered here. The shift to higher frequency results when methylene groups along the alkyl chain adopt a gauche conformation. In the minimum free-energy state, lipid alkyl chains conform to an all-trans configuration. As shown in Fig. 3, when adjacent carbons are in such a configuration, every fourth methylene along the chain is trans across the intervening carbon–carbon bond. As a result, each methylene group experiences minimal steric hindrance from neighboring groups. In contrast, as the temperature is

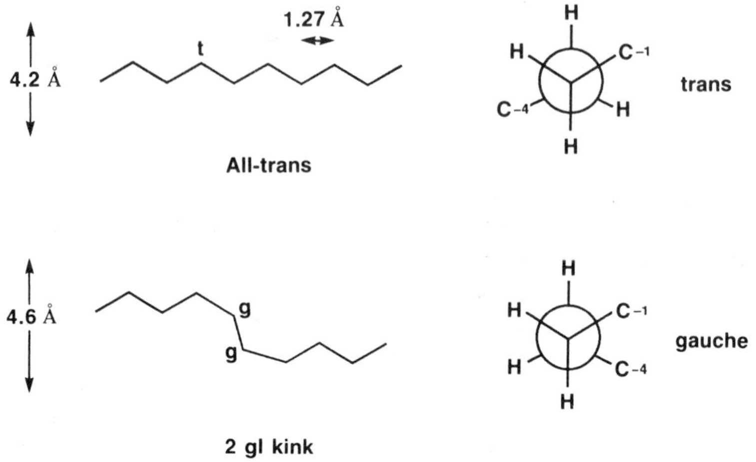

FIG. 3. A schematic representation of the trans and gauche conformers in an alkyl chain.

increased, gauche conformers begin to occur where C-1 and C-4 along the chain are no longer maximally separated. As a consequence, C–H stretching is sterically hindered and, thus, more energy is required to stretch this bond. Thus, the vibration occurs at higher frequency.

It has been demonstrated that the magnitude of the shift in $\nu_s(CH_2)$ is directly related to the ratio of the number of gauche to trans conformers in the lipid alkyl chain (often called rotamer disordering) (Casal and Mantsch, 1984). Thus, frequency shifts provide a measure of the lipid alkyl chain conformational disorder in general, and the number of gauche conformers, in particular. The phase transition from gel to liquid crystal results in significant alkyl chain disorder, which is reflected in a shift in $\nu_s(CH_2)$. The remarkable similarity between IR data shown here for SC lipids and those of other biomembranes suggests that, even though the lipids of the SC are compositionally different from other biomembranes, they all contain long alkyl chains that undergo rotamer disordering upon heating. Clearly, the precise temperature and width of the transition depends critically upon the particular lipid; however, the underlying biophysics remains the same.

In a number of lipid membrane systems, permeability and biophysical properties have been correlated (Stein, 1986). Other results have shown that the permeability of small nonelectrolytes and ions increased as lipids with short or unsaturated alkyl chains were incorporated into the membranes, or as the temperature was increased (Fettiplace, 1978; Blok *et al.*, 1976). Results obtained from temperature-dependent increases in water vapor permeability suggest that a similar correlation exists for porcine SC (Golden *et al.*, 1987a; Potts and Francoeur, 1990). Water vapor permeability was measured for a number of reasons. First,

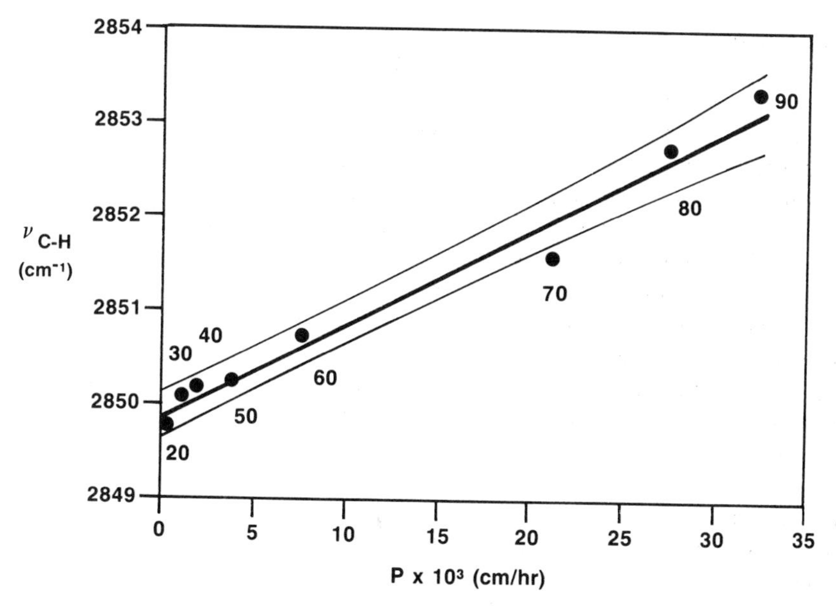

FIG. 4. The water permeability coefficient (P) of SC versus the $\nu_s(CH_2)$ value derived for an identically treated sample, each obtained over a temperature range from 20 to 90°C.

though unstirred layer effects can be appreciable in liquid-phase diffusion experiments, they are negligible in vapor studies. Thus, temperature-dependent changes in permeability are not due to changes in unstirred layer effects. Second, vapor experiments were performed using saturated aqueous salt solutions in the donor and receptor compartments. In these experiments, NaCl was used to provide a constant 75% relative humidity over the temperature range from 10 to 90°C. Changes in permeability, therefore, could not be associated with altered water activity.

The water permeability coefficient (P) and $\nu_s(CH_2)$ for porcine SC samples were measured at 10°C intervals from 20 to 90°C (Potts and Francoeur, 1990). These data are plotted in Fig. 4. A linear regression analysis reveals a high correlation ($r > 0.999$) between P and $\nu_s(CH_2)$, suggesting a functional relationship between water permeability and the IR spectral shift. Though it is possible that this correlation reflects a fortuitous relationship between two activated processes, several lines of evidence argue against this. First, as described above, water permeability through porcine SC appears to be dependent upon lipid biophysical properties. More importantly, the results show that P and $\nu_s(CH_2)$ are correlated at temperatures above and below the SC lipid phase transitions. This requires both processes to share a common T_m and to display a similar temperature dependence on either side of the transitions, an unlikely chance event.

As described above, the shift in $\nu_s(CH_2)$ results from an increased number of gauche conformers in the lipid alkyl chain. Thus, the results shown in Fig. 4 suggest that water permeability through SC and the number of gauche conformers in the SC lipids are functionally related. This conclusion is consistent with two related mechanisms involving the formation of free volume defects, which have been postulated for the permeation of water through lipid membranes. Lieb and Stein (1971) proposed that permeant molecules were transported by randomly "jumping" from donor to acceptor "holes" within the membrane. Alternatively, Trauble (1970) proposed that water permeation through lipid bilayers involves the propagation of water-carrying "kinks" along the alkyl chain. Furthermore, he postulated that the kinks occurred due to the random (e.g., thermal) occurrence of gauche conformers in an otherwise primarily trans lipid alkyl chain. The data presented in Fig. 4 suggest that water permeation through the SC is highly correlated with the number of gauche conformers [as measured by $\nu_s(CH_2)$]. While not distinguishing between the water "jump" or "kink" propagation hypotheses, these results nevertheless provide compelling evidence that water permeation through a lipid membrane is associated with the formation of gauche conformers, or free-volume defects.

The possible origin of these holes is illustrated in Fig. 3. Upon formation of a gauche–trans–gauche (g-t-g´) kink, the lateral spacing between the hydrocarbon chains increases by about 10% relative to the all-trans configuration. As a consequence, free-volume holes of about 50 $Å^3$ are created, especially in the region immediately adjacent to the gauche–trans–gauche linkage (Trauble and Haynes, 1971). As described above, this leads to increased membrane permeability. In addition, the average chain length decreases as gauche conformers are formed. Once again membrane permeability should increase because permeability and membrane thickness are inversely related. Results obtained with DPPC membranes, for example, suggest that at the gel to liquid-crystal transition the bilayer thickness decreased by about 10%, whereas the free volume increased by about 2% (Trauble and Haynes, 1971). As described by Small (1986), however, the free-volume change in a gel to liquid crystal transition could be as large as 15%. Changes in free volume and thickness also occur at other temperatures, albeit with a lower temperature dependence. Thus, the occurrence of gauche conformers could directly increase the membrane permeability via the formation of defects and/or the reduction in membrane thickness. Finally, the permeability of compounds through the SC decreases exponentially with increasing permeant molecular volume, consistent with transport through lipid, free-volume holes (Potts and Guy, 1991).

In conclusion, the results of FTIR experiments show that the extracellular lipids of porcine SC undergo phase transitions between about 60 and 80°C. Furthermore, temperature-dependent increases in water vapor permeability and the C–H stretching frequency are highly correlated. The spectral shift is due to increased gauche conformers along the lipid alkyl chain. Thus, these results suggest that water permeability through the SC is dependent upon conformational changes in the lipid alkyl chain.

B. Effect of Oleic Acid on SC Lipids

Free fatty acids (FFAs) have been shown to induce a number of membrane-associated effects at the cellular level, including lymphocyte mitogenesis (Meade and Mertin, 1978), stabilization of erythrocytes (Raz and Levine, 1973), and changes in membrane-bound enzyme function (Anderson and Jaworski, 1977). Similarly, certain FFAs have been found to enhance the permeability of topically applied drugs across the skin (Golden *et al.*, 1987b; Wickett *et al.*, 1981; Cooper, *et al.*, 1985). The ability of FFAs to increase skin permeability appears to be related to perturbation of the intercellular lipid bilayers present in the SC. Several studies (Golden *et al.*, 1987b; Francoeur and Potts, 1988; Francoeur *et al.*, 1990) have correlated FFA-induced changes in lipid properties with enhanced permeability of a coapplied drug (piroxicam). For example, the results of calorimetric studies with porcine SC have demonstrated that the T_m values of the two lipid transitions that occur between 60 and 70°C are significantly reduced following oleic acid (OA) treatment (Francoeur *et al.*, 1990). The decrease in T_m also correlated with the amount of OA taken up by the SC and the extent of *in vitro* flux enhancement. Together, these correlations stress the importance of the lipid pathway for drug transport across the skin. The enhancement was further found to depend on the concentration of ionized permeant (piroxicam) within the applied vehicle, contrary to the classical pH-partition hypothesis (Francoeur and Potts, 1988; Francoeur *et al.*, 1990). The enhanced flux of charged permeants is consistent with reports in the phospholipid literature relating increased ion transport to lipid phase separation (Blok *et al.*, 1976; Wu and McConnell, 1973; Shimshick *et al.*, 1973; Papahadjopoulos *et al.*, 1973). Consequently, a mechanism was proposed whereby OA causes the formation of a separate liquid phase within the endogenous SC lipids, increasing the number of permeable defects at liquid–solid lipid interfaces (Francoeur *et al.*, 1990). Such a mechanism is also consistent with reports of increased transepidermal water loss (TEWL) observed in a number of skin disorders characterized by lipid abnormalities, which also may involve phase separation (Rehfeld *et al.*, 1986; Elias and Williams, 1985). Grubauer *et al.* (1989a) have, in fact, experimentally correlated TEWL to the presence of separate polar and nonpolar lipid phases in mouse skin. The experimental results from our laboratories supporting these conclusions are summarized below.

1. Oleic Acid Uptake

The *in vitro* uptake of OA by isolated porcine SC and silastic membranes was measured from a series of ethanol:H_2O vehicles (Francoeur *et al.*, 1990). As depicted in Fig. 5A and B, the maximum amount of OA taken up by SC and silastic

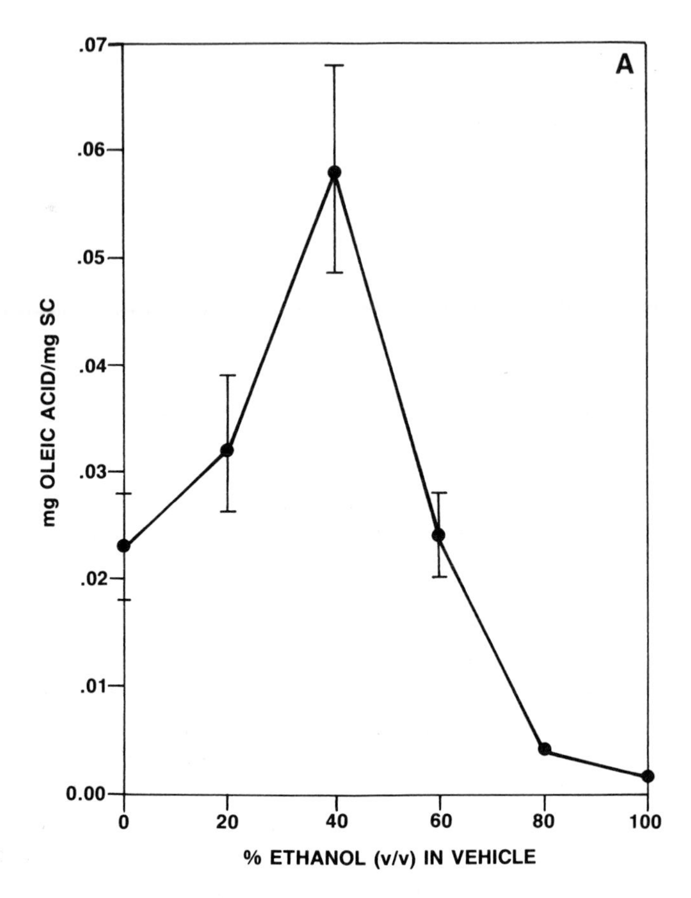

FIG. 5. (A) The uptake of oleic acid by isolated porcine stratum corneum as a function of the vehicle ethanol concentration. All vehicles contained 0.25% (v/v) oleic acid. Error bars depict SEM (n = 3). (B) The uptake of oleic acid by silastic membranes as a function of the vehicle ethanol concentration. All vehicles contained 0.25% (v/v) oleic acid. Error bars depict SEM (n = 2).

membranes was achieved with the vehicle containing 40% ethanol. The bell shape of these two curves can be partially explained by the fact that, above 40% ethanol, OA is entirely solubilized by the vehicle. Therefore, increasing the ethanol concentration has the effect of reducing the effective distribution coefficient for OA from the vehicle into the lipophilic SC and silastic membranes. Below 40% ethanol concentration, however, the OA vehicles were not homogeneous solutions, but existed as dispersions. Without defining the complete phase diagram for the OA–ethanol–water system, it is not possible to completely describe a mechanism responsible for low ethanol uptake. Nevertheless, the similarity in the shape of these two curves suggests that the uptake into each

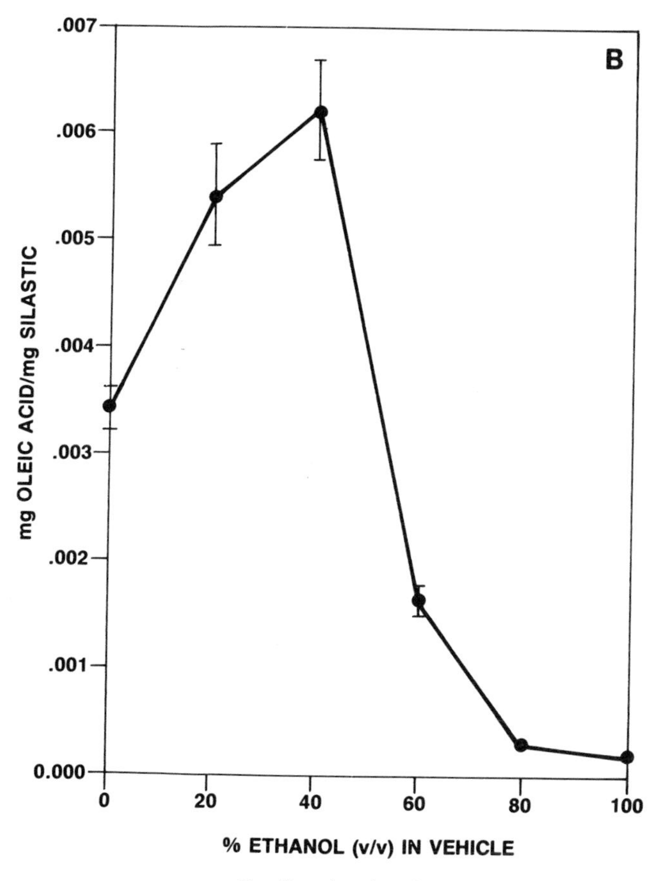

FIG. 5. (*continued*)

membrane was governed by the thermodynamic activity of OA in the applied vehicle.

2. *Differential Scanning Calorimetry*

Porcine SC, when heated, undergoes three apparent phase transitions at 60, 70, and 100°C (Fig. 2B). In order to study the effects of oleic acid uptake on these thermal transitions, SC samples were treated with vehicles of different ethanol:H_2O concentrations, all containing 0.25% (v/v) OA. As shown by the data in Table I, treatment of SC with OA shifts the T_m of the first two endotherms to lower values. In contrast, there was no effect on the third transition. The decrease in T_m was greatest for the first lipid peak. The plot shown in Fig. 6 indicates that T_m is most extensively reduced by exposure to the vehicles containing 40–50% ethanol, the same vehicles that delivered the greatest amount of OA to

Table I

EFFECTS OF OLEIC ACID ON THE THERMAL PROPERTIES OF PORCINE STRATUM CORNEUM

Treatment	ΔH (cal/g SC)[a]	T_m (°C)[b]
None ($n = 8$)	0.40 [0.07][c]	60.4 [0.3]
	0.32 [0.03]	72.5 [0.2]
40% ethanol:H_2O ($n = 4$)	0.25 [0.11]	63.2 [0.9]
	0.26 [0.12]	73.6 [2.8]
40% ethanol:H_2O, 0.25% oleic acid ($n = 4$)	0.27 [0.05]	53.6 [1.1]
	0.18 [0.04]	69.8 [2.5]

[a]Based on dry weight of stratum corneum.
[b]Temperature range over which both lipid transitions occur.
[c]Numbers in brackets represent SEM.

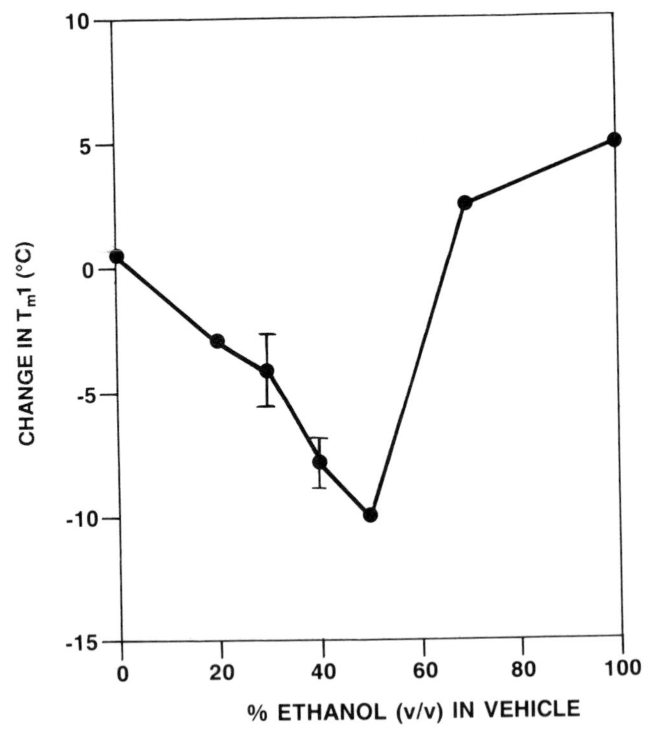

FIG. 6. The change in the transition temperature (T_{m1}) for the lipid endotherm ($T_{control}$ - $T_{treated}$) in porcine stratum corneum as a function of the vehicle ethanol concentration. All vehicles contained 0.25% (v/v) oleic acid. Error bars depict SEM ($n = 3$); for those points with no error bars, $n = 1$.

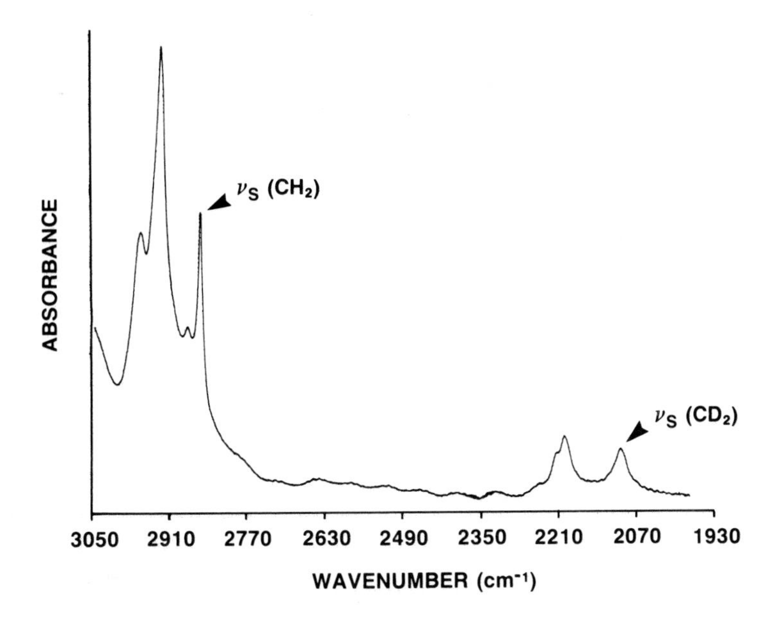

FIG. 7. Partial infrared spectrum of porcine stratum corneum treated with perdeuterated oleic acid. Note the separate C–H and C–D stretching bands at about 2850 and 2096 cm^{-1}.

the SC. The lack of effect on the 100°C transition implies that OA perturbs primarily the lipids of the SC. The enthalpy associated with the two lipid transitions is significantly decreased following OA treatment (Table I). However, a similar reduction in enthalpy is also observed with the ethanol:H_2O control, suggesting that the changes are not due solely to OA. It should also be recognized that the precise enthalpy for such noncooperative transitions is difficult to calculate due to baseline uncertainties. The correlation between OA uptake and ΔT_{m1} ($r = 0.99$) indicate that (1) OA markedly affects the physical properties of SC lipids and (2) a constant fraction of OA is distributed into the lipid bilayers.

3. *Infrared Spectroscopy*

Figure 7 shows an infrared spectrum, spanning from about 3000 to 1900 cm^{-1}, for a porcine SC sample that had been treated with perdeuterated oleic acid, [^{2}H]OA. Of particular interest for these studies are the C–H and C–D symmetric stretching frequencies at approximately 2850 and 2090 cm^{-1}, respectively. The significance of these particular bands is that changes in width and frequency reflect, on a molecular level, conformational changes of the alkyl lipid chains (Mendelsohn *et al.*, 1981; Cameron and Mantsch, 1982; Huang *et al.*, 1982). A

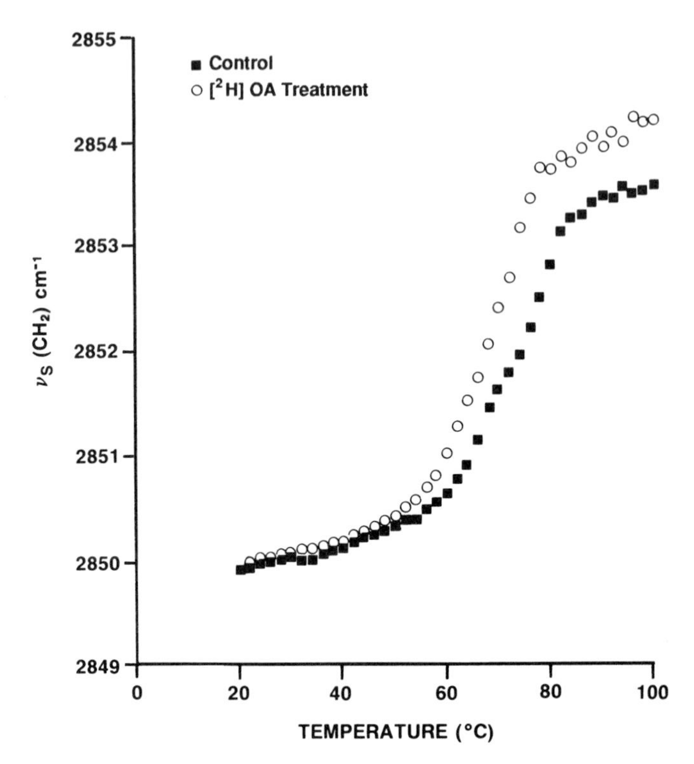

FIG. 8. The change in C–H symmetric stretching frequency, $\nu_s(CH_2)$, for porcine stratum corneum as a function of temperature. Included are the oleic acid-treated and the untreated (control) samples.

further advantage of analyzing C–H and C–D signals is that the SC lipids and OA can be separately evaluated.

Figure 8 illustrates the changes in $\nu_s(CH_2)$ for the OA-treated and the untreated SC as a function of temperature. In these experiments, the average concentration of OA taken up by the SC was determined to be 60 mg/mg SC (SEM = 4.0, $n = 6$). The incorporation of OA had no significant effect on the conformational order of the endogenous lipids below the T_m (Table II). Treatment with OA did, however, lower the T_m of the inherent SC lipids by 7.7°C (SEM = 1.3, $n = 4$), in agreement with results obtained by differential scanning calorimetry (Table I). At temperatures above the T_m, higher values ($p > 0.95$) for $\nu_s(CH_2)$ were found in the treated samples, indicating that melted SC lipids are further disordered in the presence of OA. Similar effects were observed in the extracted SC lipid samples, where the T_m of the extracted lipids was shifted to a lower value (ΔT_m = -9.5°C, SEM = 1.25, $n = 3$).

The decrease in the SC lipid T_m following treatment with OA can be partially explained on the basis of thermodynamic principles. In fact, a similar explanation

Table II
THE SYMMETRIC STRETCHING FREQUENCIES, $\nu_s(CH_2)$ AND $\nu_s(CD_2)$[a]

	Stratum corneum		Extracted lipids	
	30°C	90°C	20°C	75°C
Control				
$\nu_s(CH_2)$	2850.1 [0.1][b] ($n = 8$)	2853.4 [0.1] ($n = 7$)	2849.7 [0.0] ($n = 2$)	2853.2 [0.2] ($n = 2$)
Oleic acid treated				
$\nu_s(CH_2)$	2850.1 [0.1] ($n = 4$)	2854.2 [0.2] ($n = 4$)	2849.7 [0.1] ($n = 2$)	2854.2 [0.2] ($n = 2$)
$\nu_s(CD_2)$	2097.2 [0.6] ($n = 2$)	2097.6 [0.6] ($n = 2$)	2097.1 [0.0] ($n = 2$)	2097.2 [0.0] ($n = 2$)

[a]Values are for the stratum corneum and extracted lipids treated with [^{2}H]oleic acid. The selected temperatures represent values above and below the transition.
[b]Numbers in brackets represent SEM.

was offered for a downward shift in T_m following the incorporation of cis-unsaturated FFA into a phospholipid membrane (Klausner *et al.*, 1980). At the SC lipid phase transition, $\Delta G = 0$, because the respective lipid phases are in equilibrium. Accordingly, T_m will be equal to $\Delta H/\Delta S$, where $\Delta S = S_{final} - S_{initial}$. The calorimetric experiments have indicated that the ΔH of SC lipids was only slightly altered by OA, suggesting that an increase in ΔS may, in part, account for the reduction in T_m. Indeed, the increase of $\nu_s(CH_2)$ above the T_m suggests that OA increases S_{final} with no effect on $S_{initial}$. This result implies that the mechanism by which the transport of polar molecules is enhanced at physiological temperatures may be due to subtle microphysical changes within the lipid bilayers, and cannot be ascribed to gross perturbation or fluidization.

The increase in conformational disorder of the endogenous SC lipids above their T_m may reflect at least two effects. Deuterium NMR and FTIR studies have shown for fluid phospholipids that the greatest conformational disorder exists toward the middle of the bilayer (Seelig and Seelig, 1974; Snyder *et al.*, 1983), whereas the methylene groups closest to the polar head region remain somewhat ordered, even in the liquid-crystalline state. If this analogy holds for SC lipids, then one may speculate that OA exerts its primary effect(s) on the more ordered alkyl carbons nearest to the polar region of the bilayer. Alternatively, OA may increase the fraction of total lipids that have undergone a fluid-phase transition. The latter possibility, though, would suggest that there are highly ordered lipids present in the untreated SC above the observed phase transition. Though infrared spectroscopic analysis cannot distinguish between the two effects, X-ray diffraction studies with hairless mouse SC suggest that it is unlikely that ordered lipids exist at 90°C (White *et al.*, 1988). Hence, we suggest that OA primarily disorders the alkyl chain near the polar region of the bilayers. Results obtained for the

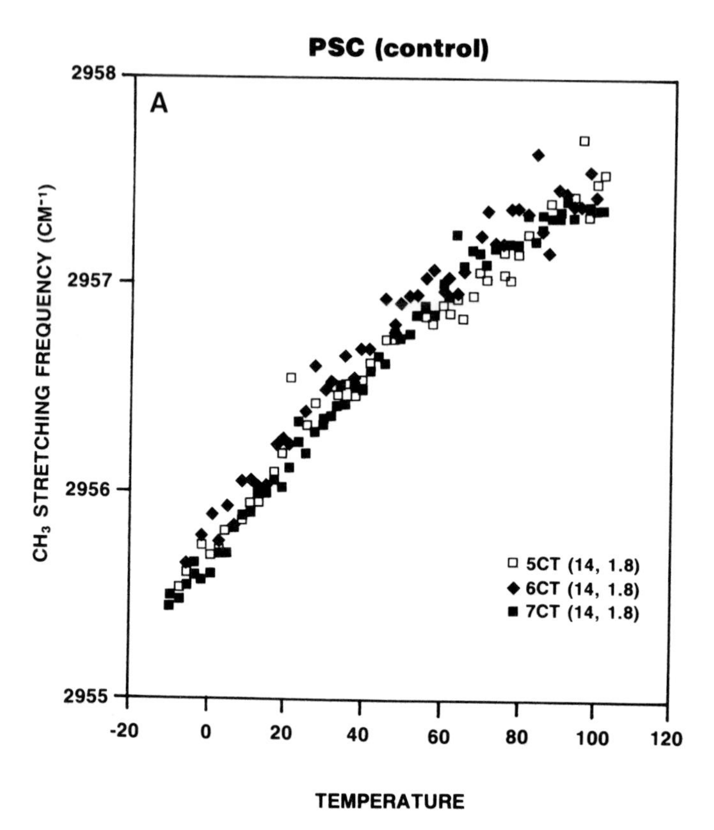

FIG. 9. The methyl asymmetric stretching vibration $\nu_a(CH_3)$ of porcine SC (PSC) before (A) and after (B) treatment with OA. The results shown in Fig. 6–9 were obtained with identically treated samples.

methyl asymmetric stretching vibrations $\nu_a(CH_3)$ (Ongpipattanakul, Ph.D. Thesis) are in agreement with the hypothesis that OA primarily affects the alkyl chain near the head group area. The results, shown in Fig. 9A and B, compare the temperature dependence of $\nu_a(CH_3)$ before and after treatment of porcine SC with [^{2}H]OA, respectively. Two interesting observations can be made from these data. First, in contrast to CH_2 stretching (Fig. 2A), there is no cooperative transition observed for CH_3 stretching. Similar results were obtained by Umemura *et al.* (1980) for DPPC, where $\nu_a(CH_2)$ showed an abrupt increase at 41°C and $\nu_a(CH_3)$ showed a gradual increase over the temperature range studied. Umemura *et al.* suggested that the difference between the CH_2 and CH_3 results reflected the decoupling of the terminal methyl group from the rest of the alkyl chain, due to greater motional freedom near the center of the bilayer. The results presented here similarly suggest that the center of the SC lipid bilayers is inherently more disor-

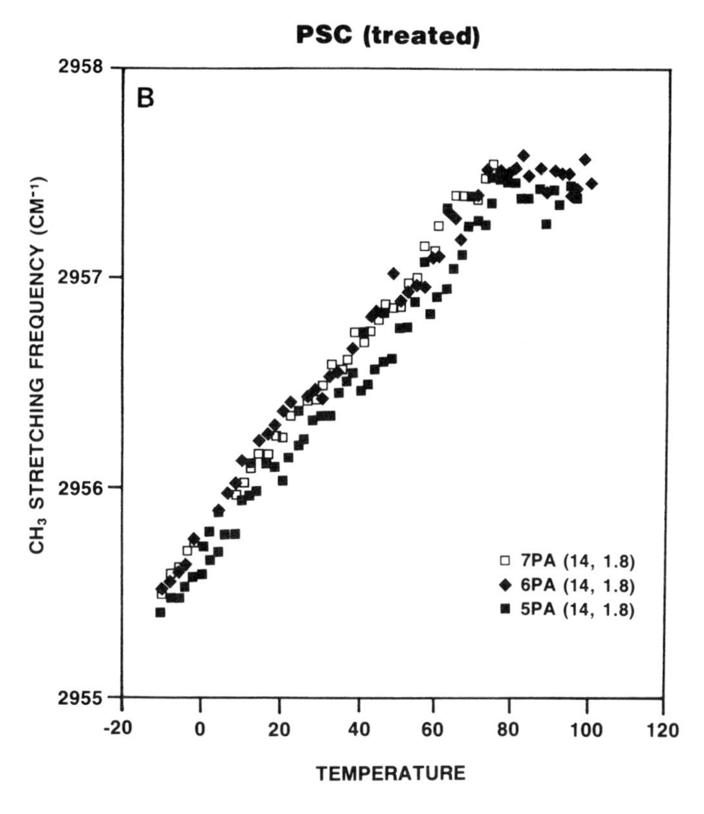

FIG. 9. (*continued*)

dered than the rest of the chain. Second, OA has no effect on $\nu_a(CH_3)$ while causing a significant shift in $\nu_a(CH_2)$ (Fig. 8), especially at elevated temperatures. Thus, these results strongly support the fact that OA has little effect on the methyl terminus of the alkyl chain, but rather exerts its primary disordering effect on the more ordered regions near the polar head of the lipids.

The net frequency changes for the [^{2}H]OA and the SC lipids are compared in Table II. The $\nu_s(CH_2)$ for the SC lipids increased by 4–5 cm^{-1}, but the corresponding change in $\nu_s(CD_2)$ was negligible. Thus, it is apparent that the OA present in the SC or extracted lipids does not undergo a highly cooperative phase transition. The actual value of $\nu_s(CD_2)$ at 32°C corresponds to [^{2}H]OA, which is almost fully disordered (i.e., melted). Therefore, the alkyl chains of the [^{2}H]OA within the SC are, on average, in the liquid state. It should be mentioned that FTIR analysis cannot measure the long-range order of the liquid OA molecules to establish whether the fatty acid is contained in the same plane as the endogenous lipid bilayer, or if it induces the formation of more exotic domains, such as hexagonal II phase. Further, the precise composition of the lipid phase containing the liquid OA is not known. It is conceivable that OA is heterogeneously dispersed among one or more other SC lipid components, and, as such, would not be expected to manifest a separate cooperative phase transition. Nevertheless, it is unequivocal that the OA molecules in the SC at physiological temperatures are in the liquid state and are, therefore, phase separated from the endogenous solid lipids.

4. *Effect of Oleic Acid on Transport*

The effect of OA on transport is illustrated here with the drug piroxicam (Francoeur *et al.*, 1990). Similar results have been obtained for a number of other ionized permeants (Francoeur and Potts, 1988). Shown in Fig. 10 are the piroxicam flux data as a function of the vehicle ethanol content. Enhanced transport was observed for both human and hairless mouse skin (HMS), albeit of significantly different magnitudes, reaching 200-fold in the case of HMS. Surprisingly, there is a marked flux "window" for the ethanol vehicle concentration, which has also been observed for several other hydrophilic permeants tested in conjunction with either OA or Azone (Francoeur and Potts, 1988). In fact, this parabolic flux dependence has also been reported for the transdermal delivery of insulin from aqueous vehicles containing Azone and propylene glycol (Priborsky *et al.*, 1988).

The plots in Figs. 6 (ΔT_m versus ethanol) and 10 (flux versus ethanol) are nearly mirror images of one another. Plotting the change in T_{m1} (i.e., control minus treated) versus the flux of piroxicam across human skin and HMS does, indeed, yield highly linear correlations ($r = 0.95$) between flux and ΔT_m. A similar correlation was also observed for ΔT_m versus the flux of salicylic acid for a series of different FFAs (Golden *et al.*, 1987b). Therefore, the enhanced transport appears to be directly related to OA perturbation of the SC lipids. While thermodynamic considerations may explain the OA-induced decrease in SC lipid T_m, the relevance of this effect to the actual mechanism of enhanced diffusion is not straightforward, because at physiological temperatures (viz. ~32°C), the SC lipids are largely unmelted even in the presence of OA. Whatever the mechanism, it appears that the extent of enhancement may be predicted with the appropriate DSC experiments. Further, these correlations between flux, uptake, and ΔT_m again reinforce the conclusion that the intercellular lipids are a primary element of the microenvironment through which enhanced diffusion occurs.

Additional experiments were conducted to ascertain the effect of pH on the enhanced transport of piroxicam (Table III). As expected, increases in the pH of the vehicle led to an increase in total piroxicam solubility. Because each donor solution was saturated with respect to total drug concentration, the amount of unionized drug in the vehicle did not change appreciably over this pH range. However, the flux was significantly altered by changes in pH. If permeation through the skin was only a function of neutral drug concentration (i.e., the pH-partition hy-

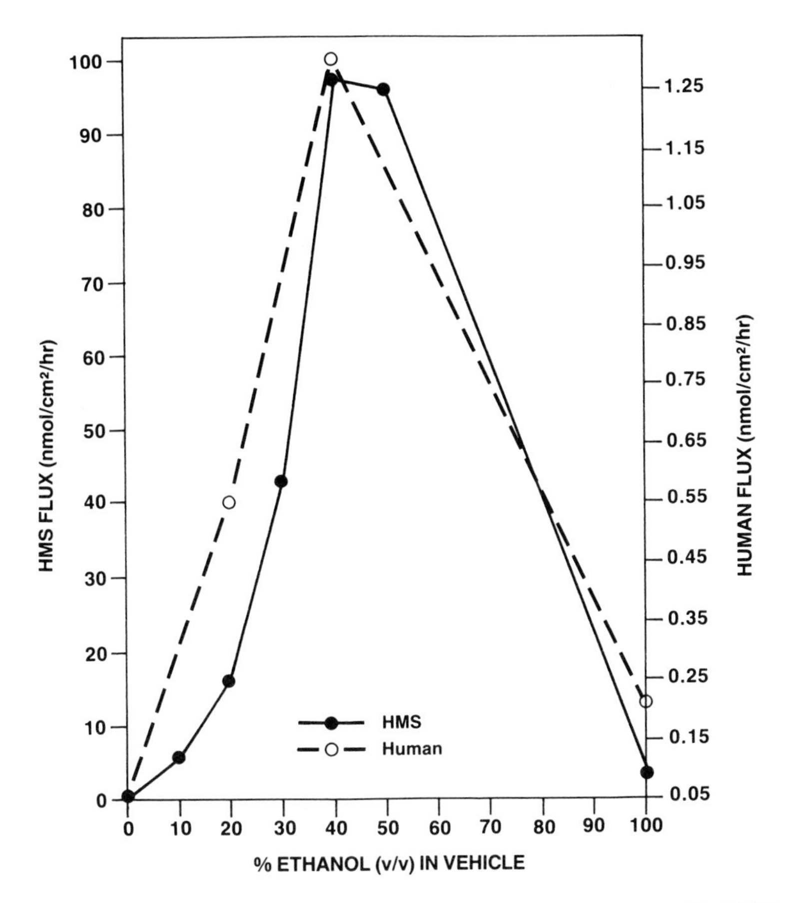

FIG. 10. The *in vitro* transport of piroxicam across human and hairless mouse skin (HMS) as a function of the vehicle ethanol concentration. All vehicles contained 0.25% (v/v) oleic acid present as a penetration enhancer.

pothesis applies), then changes in vehicle pH (at drug saturation) should produce little or no effect on transport, and the diffusion of piroxicam across untreated HMS should be independent of pH over the range of 4 to 7. The data obtained with OA, though, show that the enhanced transport is dependent on the pH and total piroxicam concentration (anion plus neutral drug species). Oleic acid would appear, therefore, to alter the properties of the SC such that the diffusion of ionized molecules is facilitated. The fact that both OA and piroxicam are negatively charged at this pH implies that the enhanced transport is not the result of ion pairing. Hence, the identification of highly soluble, hydrophilic salts in combination with OA could yield dramatic improvements in absorption (Francoeur and Potts, 1988).

Table III

EFFECT OF pH ON THE SOLUBILITY AND FLUX OF PIROXICAM[a]

pH	Apparent solubility (mg/ml)	Flux (μg/hr/cm^2)	Permeability x 10^3 (cm/hr)
4.2	1.1	3.8	3.4 (0.96)[b]
5.2	1.2	8.1	6.8 (3.63)
6.7	2.3	26.7	11.5 (1.07)

[a]Values are for a 50% ethanol vehicle containing 0.25% oleic acid. The transport studies were conducted *in vitro* with hairless mouse skin.

[b]Numbers in parentheses refer to the standard deviation.

5. *Mechanism of Enhancement*

The possible significance of OA-induced phase separation to increased SC transport can be inferred from the phospholipid literature. Anomalously high diffusion rates for small ions such as Na^+ and K^+ have been reported for various phospholipid systems that were heated to their phase-transition temperature (Blok *et al.*, 1976; Papahadjopoulos *et al.*, 1973). The enhanced flux of these ions was thought to be related to the formation of permeable defects at the fluid–solid interface of lipids at their T_m. In addition, enhanced ion transport has been observed in two-component lipid systems that exhibit lateral phase separation (Wu and McConnell, 1973; Shimshick *et al.*, 1973). Given the likely existence of separate liquid and solid phases within OA-treated SC lipids, and dramatic increase in the permeability of charged compounds, it seems reasonable to propose a similar mechanism of enhanced skin transport. For any molecule to diffuse across the SC it must, at some point, encounter lipid bilayers, because they constitute the only continuous domain. In addition, several independent lines of evidence strongly support the idea that lipids alone account for SC permeability properties (Potts and Francoeur, 1991; Potts and Guy, 1991). The formation of permeable interfacial defects within the bilayer, therefore, could result in enhanced transport by reducing either the diffusional path length and/or resistance.

The effects of OA on SC and phospholipid thermal-phase properties provide additional evidence for this phase separation mechanism. As indicated previously, OA selectively perturbs the inherent lipid structure of the SC, reducing the transition temperatures and decreasing the alkyl chain order above the T_m. Results from DSC and Raman spectroscopy have shown that, when incorporated into the lipid domain, cis-unsaturated fatty acids (viz. oleic and linoleic) induced gel–fluid phase immiscibility and reduced the T_m of phospholipid vesicles (Ortiz and Gomez-Fernandez, 1987; Verma *et al.*, 1980). Furthermore, these effects appear to be specific for cis-unsaturated fatty acids, as they were not observed for either saturated ($C > 12$) or trans-unsaturated FFAs. The specificity for cis-unsaturated FFAs is consistent with the proposed mechanism, because transport studies have shown that the saturated (C > 12) or trans-unsaturated fatty acids have little, if any, effect on SC permeability (Golden *et al.*, 1987b).

The enhanced transport of polar or ionized molecules through phase-separated "defects" may require water to be associated with these interfacial regions. Indeed, the presence of water in such regions has been implied by the results of Klausner *et al.* (1980), who studied mixed-lipid, phase-separated systems using a fluorescent probe. The results of fluorescence lifetime analyses indicated that three distinct lipid domains may exist in these systems, which were assigned to the gel, fluid, and gel–fluid interface regions of the bilayer. Subsequent experiments showed that the fluorescence assigned to the interfacial domain was quenched by D_2O, suggesting that water is associated with this region.

The electrical impedance of porcine SC was also measured (Burnette *et al.*, 1990 unpublished observations). These results showed that the SC resistance was dramatically reduced from 20,000 to 200 ohms following OA treatment. This decrease in electrical impedance, together with enhanced drug permeability, both obtained under closely similar conditions, strongly support the idea that OA treatment forms permeable, interfacial defects within the SC bilayer. In conclusion, the FTIR, DSC, electrical impedance, and flux results, along with the phospholipid literature, are all consistent with a phase-separation mechanism of enhanced transport across the skin.

C. Effect of Water on SC Lipids

Increased hydration of the SC by occlusion of the skin generally increases the transdermal delivery of topically applied drugs (Barry, 1983). Because SC lipids play an important role in the skin's barrier function, we have investigated the effect of hydration on SC structure by FTIR spectroscopy, with particular focus on lipid organization.

Human SC (HSC) or porcine SC (PSC) samples were prepared as described previously (Golden *et al.*, 1986) and were dried *in vacuo* overnight in order to remove bound water. The dehydrated samples were incubated in D_2O or H_2O for approximately 8 hours and then were once again vacuum dried overnight. Following this second dehydration, the IR spectrum of one sample was obtained. Other samples of identically treated SC were then placed in a hydrated atmosphere over a saturated solution of potassium sulfate in D_2O or H_2O (95% relative humidity). At 30-minute intervals, a sample was removed from the hydration chamber, weighed, and an IR spectrum taken. In this way the change in $\nu_a(CH_2)$ was determined as a function of D_2O or H_2O uptake.

The results showed that $\nu_a(CH_2)$ increased when the SC was hydrated with H_2O (Fig. 11). In contrast, when hydrated with D_2O, the C–H asymmetric stretching frequency was unchanged for either HSC or PSC. The difference can be explained as follows. When the lipids are hydrated with H_2O, there is an increase

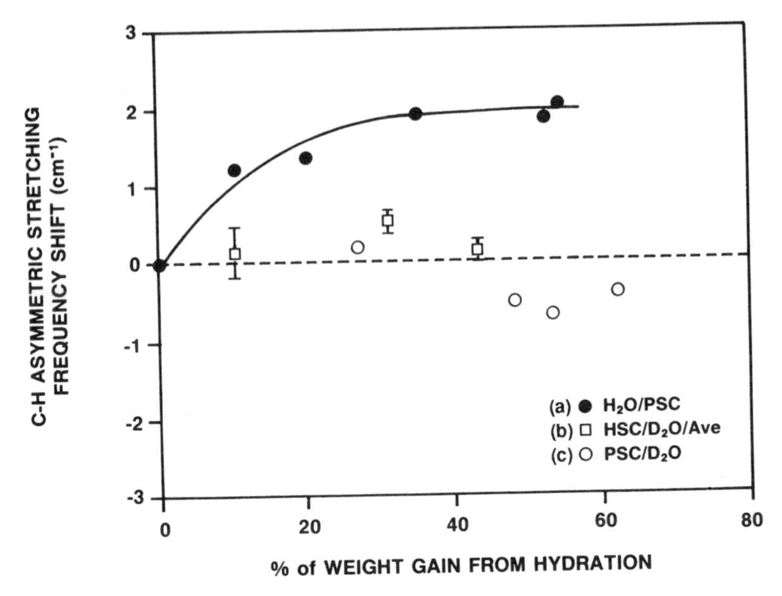

FIG. 11. The effect of hydration at 95% relative humidity on the maximum frequency of the C–H asymmetric stretching frequency of (a) porcine SC (PSC) hydrated in H_2O and (b) PSC and (c) human SC (HSC) hydrated in D_2O.

in the OH stretching band centered around 3200 cm^{-1}. The adjacent C–H stretching frequencies may be shifted by this increased absorbance because they partially overlap. However, when hydrated with D_2O, the major absorbance is the OD stretching frequency that occurs around 2500 cm^{-1}, and thus does not interfere with the spectral regions of interest. As a result, the C–H stretching frequency can more accurately reflect the influence of hydration on SC alkyl chain disorder.

The lack of changes in $\nu_a(CH_2)$ for HSC or PSC with D_2O suggests that the organization of the SC lipids was not altered significantly by hydration. Similar results have recently been obtained by Bouwstra *et al.* (1990), who investigated the small-angle X-ray scattering of HSC as a function of water content. Their results showed that the lamellar d-spacing was unaffected by hydration from 0 to greater than 40% (w/w). Similarly, Hou *et al.* (1991) found no hydration-dependent changes in d-spacing for mouse SC. Because alkyl chain packing can strongly influence the lamellar spacing, the combined IR and X-ray data show that hydration has little effect on chain disorder. This is in sharp contrast to other lipid systems, in which hydration has a large effect on d-spacing and alkyl chain disorder (Small, 1981). Furthermore, whereas hydration caused a decrease in SC lipid T_m (Golden *et al.*, 1986), the change was rather small (5–8°C) compared to changes of greater than 80°C commonly seen with phospholipid systems (Small, 1981). One possible explanation for the relatively small effect of hydration on SC lipids may be the low solubility of water in the lamellae. Direct measurement of water uptake in SC lipids and analysis of the water-dependent freezing-point depression of SC lipid transitions (Potts and Francoeur, 1991) suggest that SC lipids are maximally hydrated at less than one water molecule per lipid. In contrast, lamellar phospholipid systems are maximally hydrated at more than 10 water molecules per lipid, even in the solid phase (Small, 1981). Given that the head groups of SC lipids are significantly less polar than phospholipids, it is likely that SC lipids have less affinity for water. The question remains, "How does increased hydration lead to enhanced drug penetration?" The results presented here strongly suggest that lipid alkyl chain disorder is not involved. However, in analogy to the effects of OA on SC lipids, water may induce lipid phase separation. Alternatively, water may affect SC permeability via other mechanisms, such as increased drug uptake by the tissue, or by a direct action on other regions of the SC (e.g., protein).

III. *In Vivo* Studies: Use of Attenuated Total Reflectance Infrared Spectroscopy to Evaluate the SC

In addition to providing information on the biophysical properties of the SC, FTIR spectroscopy can be used to measure the amount of drug in the skin, provided the compound absorbs IR radiation. More importantly, this can be done *in vivo* (at least semiquantitatively) for the outer few micrometers of the SC using attenuated total reflectance infrared (ATR-IR) techniques. In brief, the ATR effect occurs when radiation propagating through a medium of refractive index n_1 strikes an interface with a medium of lower refractive index, n_2 (Harrick, 1979). If the incident beam strikes the interface at an angle greater than the critical angle, defined as $\Theta_c = \sin^{-1}(n_2/n_1)$, then the beam will be totally reflected. In addition, an evanescent wave is established at the interface and propagates into the medium of lower refractive index. If this medium has an absorption band at the frequency of the incident radiation, the reflected radiation will lose energy due to absorbance. Energy losses due to nonabsorptive processes (e.g., scattering) may also occur.

The ability of ATR spectroscopy to detect IR absorbances depends upon a number of factors, including the intensity, wavelength, and entry angle of the incident radiation, the absorptivity of the sample, the degree of contact between the two media, and the depth of penetration of the evanescent wave into the sample. The term *depth of penetration* is somewhat misleading, because the energy of the evanescent wave decreases exponentially with increasing depth into the sample. Energy coupling can be increased by matching the index of refraction of the sample (n_2) and the internal reflection element (IRE) (n_1), whereas the depth of penetration can be increased by choosing an incident angle close to, but greater than, Θ_c.

The absorbance measured by the ATR technique varies with the degree of contact between the IRE and sample (skin in the experiments described here). Because the degree of contact can change from one experiment to the next, it is necessary to ratio the absorbance being studied (i.e., that from the drug) to one due to the SC alone. This ratio is independent of the degree of contact and therefore can be used in quantitative analysis. Due to individual variations in the optical properties of the SC, however, the precise depth of penetration remains ill-defined, especially at incident angles near Θ_c. Thus, the results obtained by ATR techniques should be considered a semiquantitative estimate of drug concentration in the outer regions of the SC. The ATR technique has been extensively used to measure water concentration in the SC. The water content was inferred from the ratio of the amide I absorbance peak due to water and SC proteins, divided by the absorbance of the amide II peak due to SC protein alone. These results were particularly qualitative, however, because (1) though water absorbs IR radiation primarily in the amide I region, it can influence both amide I and II absorbances due to changes in protein hydrogen bonding, and (2) the ceramide lipids of the SC absorb IR radiation in the amide I region, further complicating the interpretation of the amide I:II ratio. Nevertheless, the results of these experiments (for review see Potts, 1986) showed that the ratio increased with increasing water content of the SC.

The water concentration of the SC was measured via ATR techniques using FTIR spectroscopy (Potts *et al.*, 1985). The IRE used in these experiments was made of ZnS (n_1 = 2.24), because this material provides high-energy coupling with the skin (n_2 = 1.6). The critical angle for the ZnS/skin interface is about 46°. Using samples from human volunteers, FTIR spectra were obtained with the following advantages over previous experiments. First, spectra were obtained rapidly (about 1–2 minutes), resulting in minimal occlusion of the skin by the optics. Second, the ATR-FTIR system, coupled with a dry nitrogen purge of the optical pathway (to prevent IR absorption by atmospheric water vapor), allowed for the detection of a weak water absorbance near 2100 cm^{-1} that was distinct from all absorbances of the SC. The results of this study provided an *in vivo* estimate for the water concentration in the outer region of human SC (0.1 g/cm^3) that was in excellent agreement with values obtained by other noninvasive techniques.

A. Effect of Ethanol on Human SC *in Vivo*

Ethanol is believed to act as a penetration enhancer in the commercially available estradiol transdermal delivery system (Estraderm; Ciba-Geigy, New York) (Pershing *et al.*, 1990; Knutson *et al.*, 1987; Good *et al.*, 1985; Ghanem *et al.*, 1987). However, the mechanism by which ethanol increases drug flux across the

SC is poorly understood. It is known that ethanol disorders the alkyl chain regions of phospholipid bilayers, and that this effect causes an increase in permeability (Lyon *et al.*, 1981; Rowe, 1985). Consequently, it has been suggested that ethanol causes lipid disordering in the intercellular domains of the SC. This hypothesis has been tested *in vivo*, in normal human volunteers, using ATR-IR. SC treatment with ethanol in both liquid and vapor phases was examined.

The volunteers were healthy adults (aged 23–30 years); experiments were conducted on the ventral forearm. A baseline spectrum was first recorded and the measurement site was tape stripped four times to prevent contamination of the IR signal originating from the SC intercellular lipids with that originating from sebaceous lipids (Bommannan *et al.*, 1990). The examination site (about 20 cm^2) was then treated either (1) for 30 minutes with 10 ml of spectroscopic-grade absolute ethanol or (2) for 1 hour with saturated ethanol vapor. After the treatment period, spectra were recorded periodically over the next 4 hours and at 24 hours post-treatment. The ethanol treatment liquid was preserved for spectroscopic analysis. The following spectroscopic features were examined post-ethanol treatment:

1. The change in $\nu_a(CH_2)$.
2. The integrated intensity under the C–H asymmetric stretching absorbance (over the frequency range 2945–2875 cm^{-1}), which is directly proportional to the amount of the absorbing species (primarily the SC lipids). Different degrees of contact between the skin and the reflection element were correctly taken into account, as reported elsewhere (Bommannan *et al.*, 1990).
3. The absorbances due to dissolved ethanol at 880 [CC skeletal vibration (Herzberg, 1954)] and 1050 cm^{-1} (CO stretching).

Skin pretreatment with ethanol resulted in significant levels of the solvent in the SC. Treatment with ethanol vapor elicited similar findings. Figure 12 shows the shift in $\nu_a(CH_2)$ during and following the treatment protocol with liquid ethanol. The baseline (pretreatment) spectrum was first recorded. The site was then tape stripped four times, causing $\nu_a(CH_2)$ to decrease significantly (by ~2 cm^{-1}). The skin was then treated for 30 minutes with ethanol, following which another IR spectrum was obtained. There was a further net decrease (by ~1 cm^{-1}) in the peak frequency of $\nu_a(CH_2)$, indicating a slight ordering (not disordering) of the lipid alkyl chains. Furthermore, because the ethanol that partitioned into the SC also contributed to the C–H absorbances, and $\nu_a(CH_2)$ for ethanol was about 10 cm^{-1} higher than that from the SC lipids, the change observed following ethanol treatment may represent an underestimate of the ordering effect. Note also the DSC results in Table I showing that the T_m values of porcine SC lipid transitions are increased following treatment with aqueous ethanol. A possible explanation for this apparent lipid "ordering" is that ethanol promotes interdigitation of the hydrocarbon chains by displacement of bound water molecules at the lipid head group/membrane interface region (Lewis *et al.*, 1989). Alternatively,

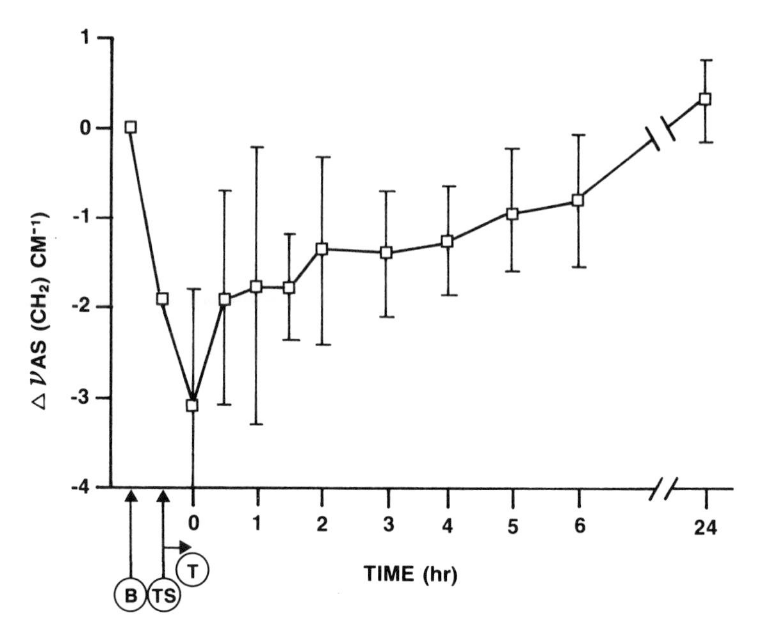

FIG. 12. Average values ($n = 6$, error bars = SD) of the shift in the peak maximum of the C–H asymmetric stretching absorbance plotted as a function of time during the ethanol liquid treatment protocol. ANOVA followed by Scheffe's F test revealed a statistically significant difference ($p < 0.05$) between the location of the peak maximum before and after the 30-minute ethanol treatment (T). There was no statistically significant difference between the baseline (B) value and and that at 24 hours; TS, after four tape strippings.

ethanol could extract disordered, liquid-like SC lipids, leaving behind the more ordered lipids, of higher T_m.

The possibility that lipid removal could explain ethanol's action on skin barrier function was next considered. Figure 13 shows the IR-determined, relative SC lipid content during and following the liquid ethanol treatment protocol. The results suggest that ethanol indeed extracts an appreciable amount of SC lipid. Again, as the ethanol present in the SC contributes to the C–H stretching absorbance, the perception of reduced lipid content may be an underestimate of the total effect. This conclusion was confirmed by evaporation of the ethanol treatment liquid and subsequent IR analysis of the residue. Spectra from all volunteers showed large lipid-associated absorbances in the 3000- to 2750-cm^{-1} region.

The time course of lipid recovery is consistent with earlier results measuring SC barrier function perturbation by solvents (Grubauer *et al.*, 1989a,b; Imokawa *et al.*, 1986). The lipids recover quickly in the initial phase, and completely within 24 hours. In summary, therefore, we conclude that *in vivo*, in humans, ethanol does not disorder SC lipid domains. Short-term exposure to the solvent,

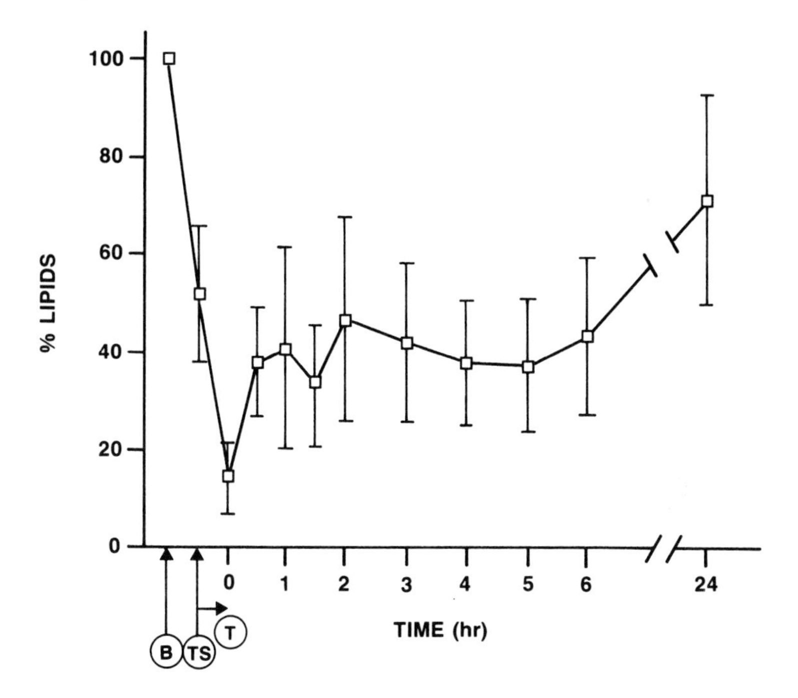

FIG. 13. Average values (n = 6, error bars = SD) of the amount of ATR-IR-detected SC lipids plotted as a function of time during the ethanol liquid treatment protocol. ANOVA followed by Scheffe's F test revealed a statistically significant difference ($p < 0.05$) between the amount of lipids before and after the 30-minute ethanol treatment (T). There was no statistically significant difference between the baseline (B) value and that at 24 hours; TS, after four tape strippings.

however, causes lipid extraction, which may lower skin barrier function and render the membrane more permeable.

B. ALKANOL-ENHANCED TRANSPORT ACROSS HAIRLESS MOUSE SKIN *IN VITRO*

Subsequent to our study with ethanol, we further examined the interaction of alkanols with mammalian skin (Kai *et al.*, 1990) by (1) measuring the flux of a model permeant across hairless mouse skin *in vitro* following pretreatment of the membrane with a series of *n*-alkanols (the C_2–C_6, C_8, C_{10}, and C_{12} homologues), and (2) using FTIR spectroscopy to examine the molecular interaction(s) between alkanols and SC.

Hairless mouse (SKH-hr-1) skin was mounted in vertical flow-through diffusion chambers and exposed to 0.5 ml of pure alkanol for 3 or 6 hours. At the end of this pretreatment period, the alkanol was removed and the epidermal surface dried and washed with a small volume of distilled water. Then, the donor compartment was charged with 0.5 ml of an aqueous nicotinamide solution (0.1 mg/ml), spiked with radiolabeled solute. The flux of penetrant was monitored for the next 24 hours. All alkanols enhanced nicotinamide flux compared to the no treatment and water pretreatment controls. The degree of enhancement varied parabolically with alkanol chain length and was only slightly greater when the exposure time was extended from 3 to 6 hours.

In order to investigate the mechanism by which the alkanols were compromising the murine barrier, hairless mouse skin was pretreated as above and then examined immediately by ATR-FTIR. The reflectance approach was used to obviate the possibility of artifacts associated with SC isolation. Figure 14A–C shows the spectra (in the region of the C–H stretching absorbances) obtained from (A) untreated SC, (B) SC pretreated with ethanol, hexanol, and octanol, and (C) the solvent extract after ethanol exposure. The results again indicate that ethanol extracts SC lipids. There is an apparent augmentation, however, of the C–H absorbances following hexanol and octanol treatments. The reason for this observation becomes apparent when the octanol experiment is repeated using the perdeuterated alkanol (Fig. 14D and E). Now C–H absorbances from the SC lipids are separated from C–D absorbances due to octanol, which partitioned into the SC (and which caused the enhanced C–H peaks in Fig. 14B). The spectra also reveal that SC lipids are extracted efficiently by octanol; the extraction is sufficiently extensive that one cannot reliably assess whether the order of the remaining SC lipid hydrocarbon chains has been altered by the alkanol treatments.

Overall, these experiments reveal that alkanols substantially compromise hairless mouse SC through (principally) lipid extraction. The effect of ethanol on the murine barrier appears to be considerably greater than that on human SC.

C. Oleic Acid Concentration and Effect in Human SC *in Vivo*

Pursuing the *in vitro* findings using porcine stratum corneum and the effects of unsaturated fatty acids on the intercellular lipid domains, we then designed a study to investigate the mechanism of action of these agents on human SC *in vivo* (Mak *et al.*, 1990a).

Human ventral forearm was occluded with 0.5 ml of an OA solution (0.5, 1.0, 5, and 10% w/v in ethanol) for 30 minutes. Prior to OA treatment, the application site was cleaned and an IR spectrum was recorded. At the end of 30 minutes, the occlusive pad was removed and the skin wiped clean of residual material. An IR spectrum was recorded immediately, and further spectra were obtained during the next 6 hours and at 24 hours postdosing. It was found that exposure of human SC to 1.0% OA for 30 minutes causes a significant increase in $\nu_a(CH_2)$, indicating that the presence of the fatty acid causes an overall increase in lipid hydrocarbon chain disorder. Note, however, that $\nu_a(CH_2)$ reflects contributions from both SC

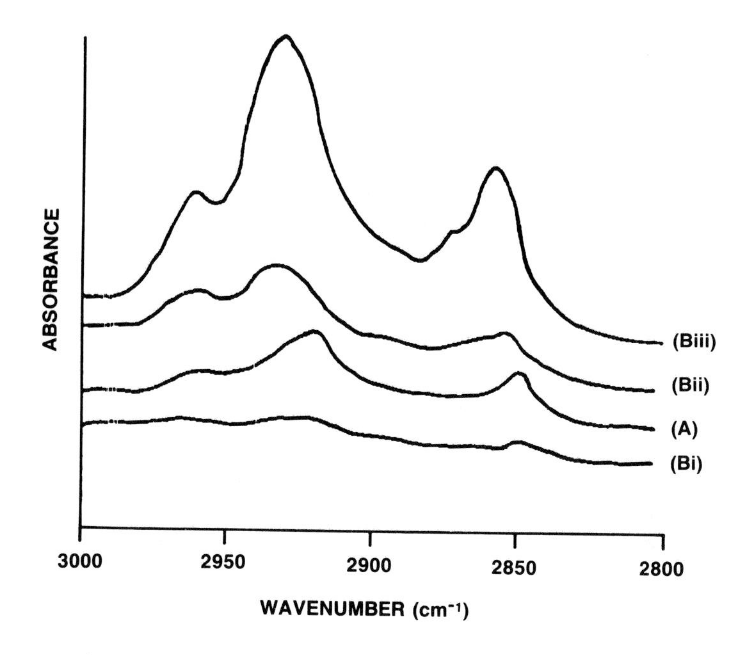

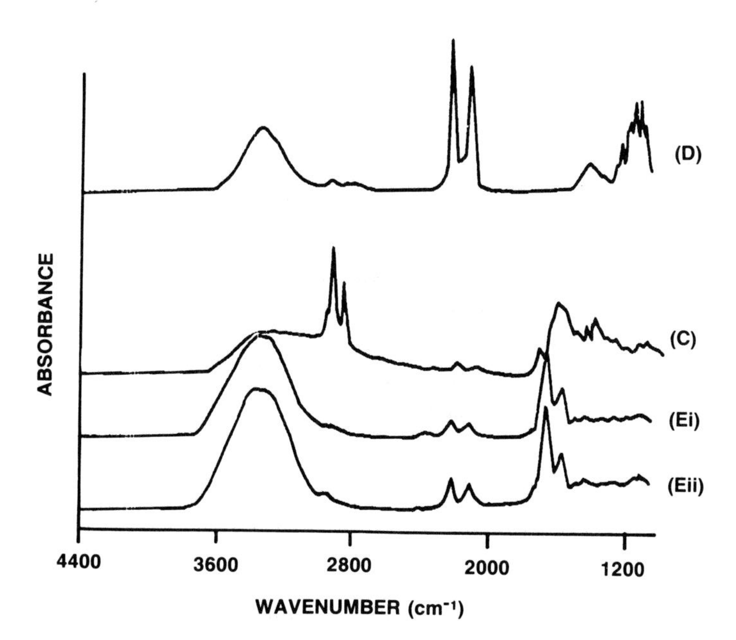

FIG. 14. IR spectra of (A) untreated hairless mouse SC, (B) mouse SC following exposure for 6 hours to (i) ethanol, (ii) hexanol, and (iii) octanol, (C) material extracted from the SC by treatment with ethanol, (D) perdeuterated n-octanol, and (E) SC following 3 (i) and 6 (ii) hours of exposure to perdeuterated *n*-octanol.

lipids and OA. The effect of OA lasted for a substantially longer period than the time of enhancer application. Longer exposures (or higher concentrations of OA) maintained the maximal perturbation at about +3 cm^{-1} for more than 6 hours. Control experiments, in which the SC was dosed with ethanol alone, caused no shift in the C–H absorbance frequencies. The lack of an ethanol effect in this case (as compared to that discussed above) is probably due to differences in the experimental protocol. We have shown that the outer SC layers contain both epidermal and sebaceous lipids (Bommannan *et al.*, 1990). The sebaceous lipids are relatively disordered compared to the epidermal material, and they are present in large amounts in the outer SC. The ordering effect due to ethanol treatment was seen in experiments wherein the outer SC (and, hence, sebaceous lipids) was removed prior to treatment. The lack of effect here may reflect a property of sebaceous lipids that were not removed.

To understand the saturable nature of the OA perturbation, we measured (radiolabeled) OA uptake into excised human SC *in vitro*. The results indicated a linear uptake and suggest that, in the concentration range studied, there is no solubility limitation on the partitioning of OA into SC. The *in vivo* IR data were then examined more carefully to see if the enhancer level could be obtained from a unique OA-derived absorbance. An absorbance at 1710 cm^{-1}, which originated primarily from the enhancer, was utilized. To quantify OA uptake using this absorbance, the 1710-cm^{-1} peak height was normalized using an SC lipid peak at 1743 cm^{-1} (thereby allowing different degrees of skin–IRE contact to be appropriately taken into account). These spectroscopic measurements showed that enhancer uptake is not solubility limited *in vivo*. Figure 15 demonstrates the excellent correlation between *in vivo* and *in vitro* results.

In vitro results presented above showed that OA exists in a "liquid" state when incorporated into the SC. The IR spectrum of a treated sample reflects a weighted contribution of both the "solid" SC lipids [low relative $\nu_a(CH_2)$] and "liquid" OA [high relative $\nu_a(CH_2)$]. As progressively more OA was incorporated into the SC, the measured value of $\nu_a(CH_2)$ would approach that of the pure OA component. Thus, the "saturable" nature of the shift in $\nu_a(CH_2)$ could simply reflect the uptake of OA by the SC, but is consistent with separate lipid domains for each. As discussed above, such lipid phase separation has important consequences for SC permeability.

In conclusion, this study demonstrates that (1) the effect of penetration enhancers on SC barrier function can be followed *in vivo* by ATR-FTIR, and (2) in the case of OA, the technique can assess the level of chemical within the SC as a function of time.

D. *In Vivo* Measurement of Penetration Enhancement by ATR-FTIR

To illustrate the potential outlined above, we have recently undertaken a series of experiments in which human volunteers were dosed topically with either a

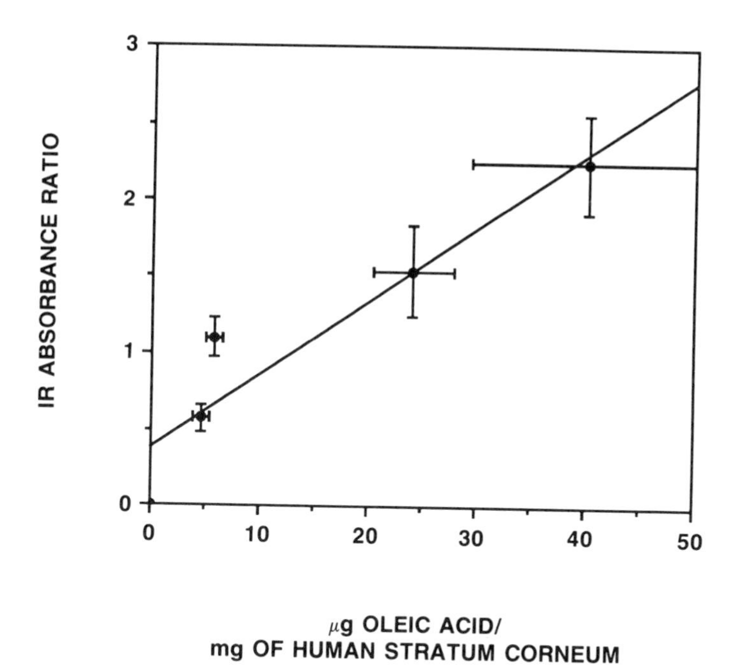

FIG. 15. Correlation ($r = 0.95$) between *in vivo* IR measurements of oleic acid in human SC and the uptake of the labeled enhancer into excised SC *in vitro*.

10% w/v solution of 4-cyanophenol (CP) in propylene glycol or the same 10% CP solution containing 5% v/v of the penetration enhancer, oleic acid (Mak *et al*., 1990b; Higo *et al*., 1990). The selection of CP as a model penetrant was based primarily upon the fact that the C≡N functional group provides an intense IR absorbance at 2230 cm^{-1}, a spectral region in which the SC is transparent. Experimentally, human volunteers (aged 23–40 years) were treated on the ventral forearm with either (1) 10% CP in propylene glycol or (2) the same solution containing 5% OA. Treatment was administered via a saturated gauze pad for 1, 2, or 3 hours. At the end of the treatment period, the delivery system was removed, the skin surface was cleaned, and serial IR spectra were then acquired from the treated skin sites over the following 9-hour period.

The representative spectrum in Fig. 16 shows a strong and unique absorbance at 2230 cm^{-1} due to the C≡N group of the topically absorbed CP. Relative amounts of the penetrant in the SC, as a function of time of posttreatment, were evaluated from the ratio of the 2230 cm^{-1} absorbance to that at 1743 cm^{-1} due to the C=O stretch of the endogenous SC lipids. The relative CP levels, in the outer SC, versus time profiles following 1-, 2-, and 3-hour pretreatments are shown in Fig. 17. Because drug input has ceased once IR measurements commence, the CP levels decay progressively. This clearance process is first order for all treatment times.

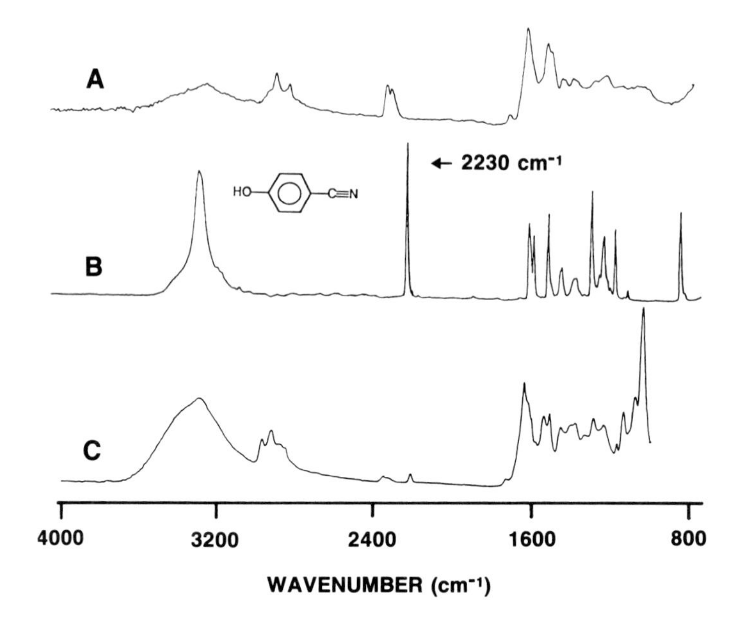

FIG. 16. IR spectra of (A) human forearm SC, (B) 4-cyanophenol, and (C) human forearm SC after treatment with 10% w/v 4-cyanophenol in propylene glycol.

The results in Fig. 17 offer insight into drug uptake and elimination by the SC. In the absence of enhancer, as the treatment time increases, the relative amount of CP in the SC when the first IR spectrum is obtained (i.e., at the end of the treatment period) reaches an apparent plateau. In the presence of OA, on the other hand, the relative amount is always less and decreases with increasing exposure time (Table IV). The decay of the CP absorbance following removal of the delivery system is faster, however, following treatment with the enhancer. For 1-, 2-, and 3-hour treatments, the relative elimination rates (with enhancer/without) are 1.7, 2.0, and 2.8, respectively. These observations are believed to be consistent with the hypothesis that, though CP is delivered into the SC at comparable rates from the vehicles with and without enhancer, the presence of OA promotes percutaneous transport and facilitates throughput of the penetrant.

It should also be noted that the same spectra, which have already revealed the effect of the enhancer on SC lipid domains and the promotion of penetrant transport, also contain information about the relative amounts of OA and propylene glycol as a function of time. Unique absorbances at 1710 cm^{-1}, due to the OA carboxyl C=O stretch (see above), and at 1040 cm^{-1}, from the propylene glycol C–O stretch, have been identified and quantified (Mak *et al.*, 1990b).

Finally, we have used ATR-FTIR to evaluate the effect of OA on the distribution of cyanophenol in human SC following topical application as (1) a 10% w/v solu-

Table IV

EFFECT OF 5% (v/v) OLEIC ACID ON THE NORMALIZED CYANOPHENOL (CP) LEVEL IN THE OUTER SC AT THE END OF THE TREATMENT PERIOD

	Treatment time (hours)		
	1	2	3
Relative CP amount without enhancer (mean ± SE)	0.82 ± 0.11	1.27 ± 0.12	1.22 ± 0.09
Relative CP amount with enhancer (mean ± SE)	0.56 ± 0.05	0.56 ± 0.04	0.15 ± 0.03

tion (spiked with [^{14}C]CP) in propylene glycol and (2) in propylene glycol containing a penetration enhancer (5% w/v OA) (Higo *et al.*, 1990). The formulations (0.5 ml) were applied to the ventral forearms of human volunteers for 1, 2, or 3 hrs. At the end of application period, the dosing site was cleaned and an ATR-FTIR spectrum was recorded. Then, the treatment area was tape stripped once and a second IR spectrum was recorded. These latter two procedures were repeated up to a maximum of 20 times. The weight of each tape was measured before and after stripping. The amount of CP in the skin was quantified spectroscopically by measuring the normalized integrated intensity between 2244.6 and 2199.8 cm^{-1}. The absolute amount of CP in the different layers of skin was determined by subjecting the SC tape strips to liquid scintillation counting. The key findings were as follows:

1. The weight of SC removed by tape stripping after 1 hour of treatment was not influenced by the presence of OA in the formulation. However, the weight of SC removed following 2 or 3 hours of OA treatment was significantly greater than the corresponding controls.
2. Treating the skin for 2 and 3 hours with the fatty acid formulation caused frank irritation and severely undermined SC integrity.
3. The integrated intensity under the CP-associated absorbance at 2230 cm^{-1} was plotted as a function of the cumulative SC weight removed by the repeated tape-stripping process (which is, in turn, proportional to the depth into the SC) (Weigand and Gaylor, 1973). Treatment for 1 hour in the presence of OA does not alter the distribution of CP in the SC compared to the control. However, the 2- and 3-hour treatments with OA significantly enhanced the delivery of CP into and across the SC (i.e., CP was found in higher amounts throughout the SC and at greater SC depth following OA treatment for 2 or 3 hours).
4. The [^{14}C]CP distribution in the sequentially removed tape strips paralleled closely the spectroscopic observations (Fig. 18). When the radioactivity data were plotted against the average integrated intensity multiplied by the weight of the corresponding SC strip, there was a high degree of correlation between the absolute level of CP (measured by ^{14}C) and the spectroscopically determined values.

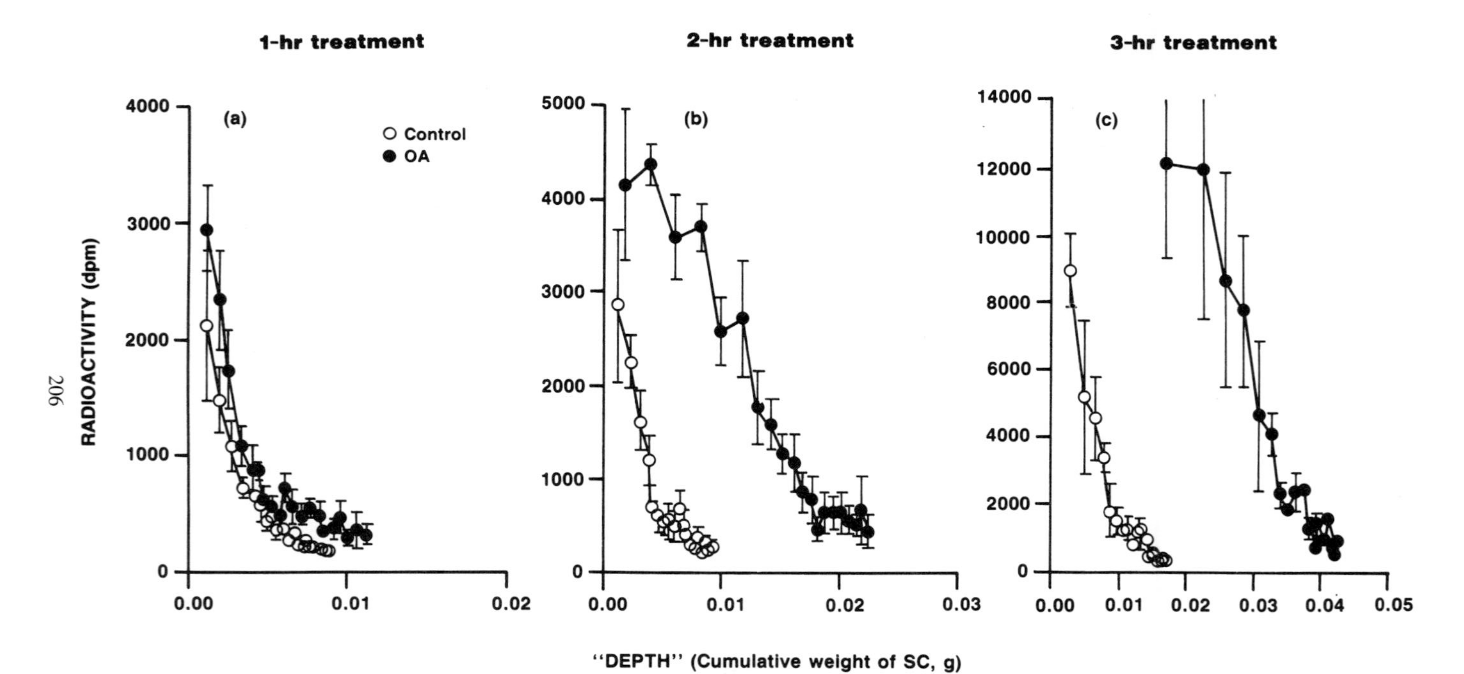

FIG. 17. Semilog plots of the relative amounts of 4-cyanophenol (CP) remaining in human forearm SC after 1, 2, and 3 hours of treatment with 10% w/v CP in (○) propylene glycol or (●) propylene glycol containing 5% oleic acid ($n = 5$).

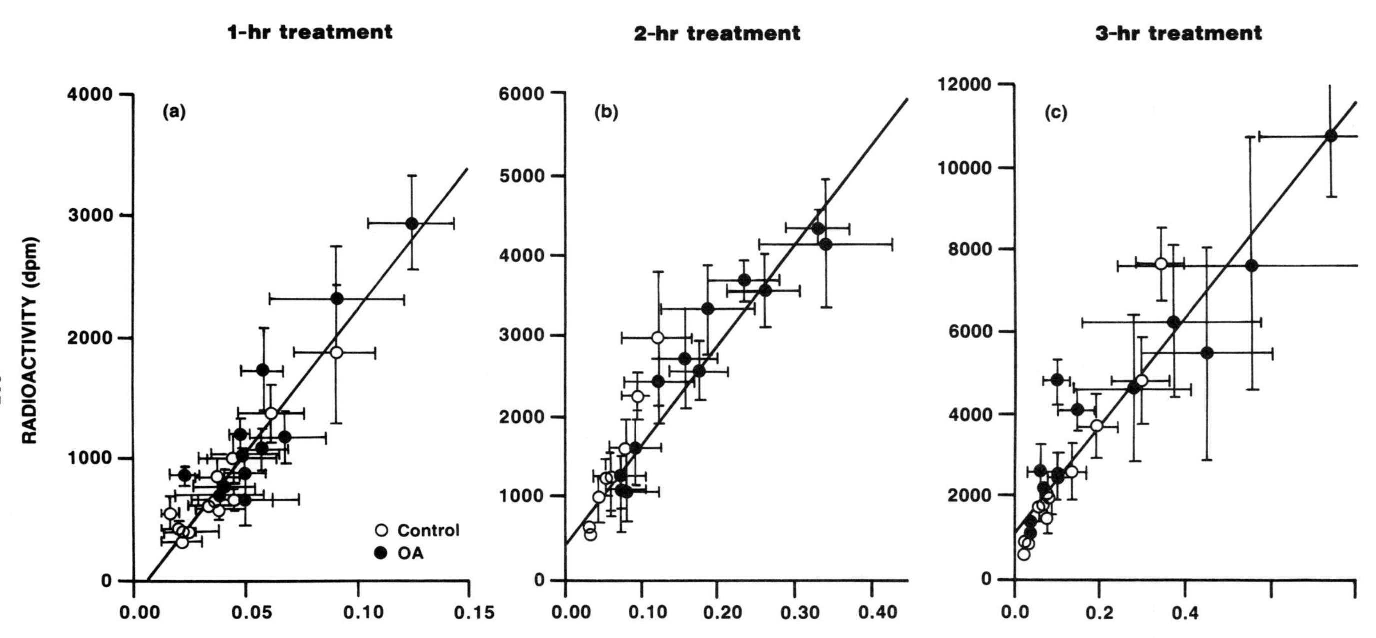

FIG. 18. Correlation between the absolute, radiochemical measurement of 4-cyanophenol in the SC and the corresponding spectroscopically determined level, i.e., the integrated intensity (mean + SE; $n = 4$ or 5).

The results indicate that the noninvasive technique of ATR-FTIR has the potential to quantify accurately, *in vivo*, the distribution of transdermally delivered drugs in the SC. Facile evaluation of formulation changes and of the effects of putative penetration enhancers should now prove possible.

IV. Conclusion

The results presented here strongly support the hypothesis that the SC lipid pathway is important to the permeability of water and numerous drugs. This conclusion was derived through the use of biophysical tools commonly used to study lipid membranes. More importantly, the mechanisms of drug and water transport through the SC deduced from these studies have precedent in lipid membrane biophysics. Thus, one is left with the interesting observation that despite differences in lipid composition and distribution, the mechanisms of molecular permeation through the SC and other biomembranes are similar.

Acknowledgments

Supported in part by the U.S. National Institutes of Health (HD-23010). We thank Bommi Bommannan, Naruhito Higo, Alane Kennedy, Stephen McNeill, Boonsri Ongpipattanakul, and Guia Golden for their intellectual and experimental contributions.

References

Anderson, W., and Jaworski, C. (1977). *Arch. Biochem. Biophys.* **180,** 374–383.
Barry, B. W. (1983). "Dermatological Formulations." Dekker, New York.
Blok, M. C., Van Der Neut-Ko, E. C. M., Van Deenen, L. L. M., and DeGier, J. (1976). *Biochim. Biophys. Acta* **406,** 187–196.
Bloom, M., and Smith, I. C. P. (1985). *In* "Progress in Protein–Lipid Interactions" (A. Watts and J. J. H. H. M. DePont, eds.), pp. 61–88. Elsevier/North-Holland, Amsterdam.
Bommannan, D., Potts, R. O., and Guy, R. H. (1990). *J. Invest. Dermatol.* **95,** 408–418.
Bouwstra, J. A., Peschier, L. J. C., Brusse, J., and Boddé, H. E. (1989). *Int. J. Pharm.* **52,** 47–54.
Bouwstra, J. A., de Vries, M. A., Bras, W., Brussee, J., and Gooris, G. S. (1990). *Proc. Int. Symp. Controlled Release Bioact. Mater.* **17,** 33–34.
Cameron, D. G., and Mantsch, H. H. (1982). *Biophys. J.* **38,** 175–184.
Casal, H. L., and Mantsch, H. H. (1984). *Biochim. Biophys. Acta* **779,** 381–401.
Chan, S. I., Bocian, D. F., and Petersen, N. O. (1981). *In* "Membrane Spectroscopy" (E. Grell, ed.), pp. 1–46. Springer-Verlag, New York.
Chapman, D. (1975). *Q. Rev. Biophys.* **8,** 185–235.
Cooper, E., Loomans, M., and Fauzi, M. (1985). U.S. Pat. 4,552,872.
Dluhy, R. A., Moffatt, D. J., Cameron, D. G., Mendelsohn, R., and Mantsch, H. H. (1985). *Can. J. Chem.* **63,** 1925–1932.
Donovan, J. W. (1984). *Trends Biochem. Sci. (Pers. Ed.)* **9,** 340–344.

Elias, P. M., and Williams, M. L. (1985). *Arch. Dermatol.* **121,** 1000–1008.
Fettiplace, R. (1978). *Biochim. Biophys. Acta* **513,** 1–10.
Francoeur, M. L. and Potts, R. O. (1988). Eur. Pat. Appl. 87309459.3.
Francoeur, M. L., Golden, G. M., and Potts, R. O. (1990). *Pharm. Res.* **7,** 621–627.
Franks, N. P., and Levine, Y. K. (1981). *In* "Membrane Spectroscopy" (E. Grell, ed.), pp. 437–488. Springer-Verlag, New York.
Fringeli, U. P., and Gunthard, H. H. (1981). *In* "Membrane Spectroscopy" (E. Grell, ed.), pp. 270–327. Springer-Verlag, New York.
Ghanem, A., Mahmoud, H., Higuchi, W. I., Rohr, U. D., Borsadia, S., Liu, P., Fox, J. L., and Good, W. R. (1987). *J. Controlled Release* **6,** 75–83.
Golden, G. M., Guzek, D. B., Harris, R. R., McKie, J. E., and Potts, R. O. (1986). *J. Invest. Dermatol.* **86,** 255–259.
Golden, G. M., Guzek, D. B. Kennedy, A. H., McKie, J. E., and Potts, R. O. (1987a) *Biochemistry* **26,** 2382–2388.
Golden, G. M., McKie, J. E., and Potts, R. O. (1987b). *J. Pharm. Sci.* **76,** 25–28.
Good, W. R., Powers, M. S., Campbell, P., and Schenkel, L. (1985). *J. Controlled Release* **2,** 89–97.
Gordon, L. M., and Curtain, C. C. (1988). *In* "Advances in Membrane Fluidity" (R. C. Aloia, C. C. Curtain, and L. M. Gordon, eds.), pp. 25–88. Alan R. Liss, New York.
Grubauer, G., Elias, P. M., and Feingold, K. R. (1989a). *J. Lipid Res.* **30,** 323–333.
Grubauer, G., Feingold, K. R., Harris, R. M., and Elias, P. M. (1989b). *J. Lipid Res.* **30,** 89–96.
Harrick (1979). "Internal Reflection Spectroscopy." Harrick Sci. Corp., Ossining, New York.
Herzberg, G. (1954). "Infrared and Raman Spectra of Polyatomic Molecules," p. 361. Van Nostrand, Princeton, New Jersey.
Higo, N., Bommannan, D., Potts, R. O., and Guy, R. H. (1990). *Proc. Int. Symp. Controlled Release Bioact. Mater.* **17,** 413–414.
Hou, S. Y. E., Mitra, A. K., White, S. H., Menon, G. K., Ghadially, R., and Elias, P. M. (1991). *J. Invest. Dermatol.* (in press).
Huang, C., Lapides, J. R., and Levin, I. W. (1982). *J. Am. Chem. Soc.* **104,** 5926–5930.
Imokawa, G., Akasaki, S., Hattori, M., and Yoshizuka, N. (1986). *J. Invest. Dermatol.* **87,** 758–761.
Kai, T., Mak, V. H. W., Potts, R. O., and Guy, R. H. (1990). *J. Controlled Release* **12,** 103–112.
Klausner, R., Kleinfeld, A., Hoover, R., and Karnovsky, M. (1980). *J. Biol. Chem.* **255,** 1286–1295.
Knutson, K., Krill, S. L., Lambert, W. J., and Higuchi, W. I. (1987). *J. Controlled Release* **6,** 59–74.
Lewis, E. N., Levin, I. W., and Steer, C. J. (1989). *Biochim. Biophys. Acta* **986,** 161–166.
Lieb, W. R., and Stein, W. D. (1971). *Nature (London)* **234,** 220–221.
Lyon, R. C., McComb, J. A., Schreurs, J., and Goldstein, D. B. (1981). *J. Pharmacol. Exp. Ther.* **218,** 669–675.
Mak, V. H. W., Potts, R. O., and Guy, R. H. (1990a). *J. Controlled Release* **12,** 67–75.
Mak, V. H. W., Potts, R. O., and Guy, R. H. (1990b). *Pharm. Res.* **7,** 835–841.
Marsh, D. (1981). *In* "Membrane Spectroscopy" (E. Grell, ed.), pp. 51–137. Springer-Verlag, New York.
Meade, C., and Mertin, J. (1978). *Adv. Lipid Res.* **16,** 127–165.
Mendelsohn, R., Dluhy, R. H., Taraschi, J., Cameron, D. G., and Mantsch, H. H. (1981). *Biochemistry* **20,** 6699–6706.
Ortiz, A., and Gomez-Fernandez, J. (1987). *Chem. Phys. Lipids* **45,** 75–91.
Papahadjopoulos, D., Jacobson, K., Nir, S., and Isac, T. (1973). *Biochim. Biophys. Acta* **311,** 330–348.
Pershing, L. K., Lambert, W. J., and Knutson, K. (1990). *Pharm. Res.* **7,** 170–175.
Potts, R. O. (1986). *J. Soc. Cosmet. Chem.* **37,** 9–33.
Potts, R. O., and Francoeur, M. L. (1990). *Proc. Natl. Acad. Sci. U.S.A.* **87,** 3871–3873.
Potts, R. O., and Francoeur, M. L. (1991). *J. Invest. Dermatol.* **96,** 495–499.
Potts, R. O., and Guy, R. H. (1991). *Pharm. Res.* (submitted).

Potts, R. O., Guzek, D. B., Harris, R. R., and McKie, J. E. (1985). *Arch. Dermatol. Res.* **277,** 489–495.
Priborsky, J., Takayama, K., Nagai, T., Waitzova, D., and Elis, J. (1988). *Drug Des. Delivery.* **2,** 91–97.
Raz, A., and Livine, A. (1973). *Biochim. Biophys. Acta* **311,** 222–229.
Rehfeld, S. J., Williams, M. L., and Elias, P. M. (1986). *Arch. Dermatol. Res.* **278,** 259–263.
Rowe, E. S. (1985). *Biochim. Biophys. Acta* **813,** 321–330.
Seelig, A., and Seelig, J. (1974). *Biochemistry* **13,** 4839–4845.
Shimshick, E. J., Kleeman, W., Hubbell, W. L., and McConnell, H. M. (1973). *J. Supramol. Struct.* **2,** 285–295.
Shipley, G. G. (1986). *In* "The Physical Chemistry of Lipids" (D. M. Small, ed.), pp. 97–143. Plenum, New York.
Small, D. M. (1986). "The Physical Chemistry of Lipids." Plenum, New York.
Snyder, R. G., Maroncelli, M., Strauss, H. L., Elliger, C. A., Cameron, D. G., Casal, H. L., and Mantsch, H. H. (1983). *J. Am. Chem. Soc.* **105,** 133–134.
Stein, W. D. (1986). "Transport and Diffusion Across Cell Membranes." Academic Press, New York.
Trauble, H. (1970). *J. Membr. Biol.* **4,** 193–208.
Trauble, H., and Haynes, D. H. (1971). *Chem. Phys. Lipids* **7,** 324–335.
Umemura, J., Cameron, D. G., and Mantsch, H. H. (1980). *Biochim. Biophys. Acta* **602,** 32–44.
Verma, S., Wallach, D., and Sakura, F. (1980). *Biochemistry* **19,** 574–579.
Weigand, D. A., and Gaylor, G. R. (1973). *J. Invest. Dermatol.* **60,** 84–86.
White, S. H., Mirejovsky, D., and King, G. I. (1988). *Biochemistry* **27,** 3725–3732.
Wickett, R., Cooper, E., and Loomans, M. (1981). Eur. Pat. Appl. 81303128.3.
Wilkinson, D. A., and Nagle, J. F. (1981). *Biochemistry* **20,** 187–192.
Wu, S. H. W., and McConnell, H. M. (1973). *Biochem. Biophys. Res. Commun.* **55,** 484.

Lipids in Normal and Pathological Desquamation

MARY L. WILLIAMS

Departments of Dermatology and Pediatrics
University of California, San Francisco
San Francisco, California 94143

I. Introduction

Normal desquamation is an orderly and "invisible" process whereby individual corneocytes (squames) at the skin surface detach from their adjacent and underlying neighbors and are swept away (Fig. 1) (for review see McGuire, 1982). In pathological desquamation, the thickness of the stratum corneum compartment is increased (hyperkeratosis), giving a "dry," scaly appearance to the skin. In these pathological states, squames usually do not detach as single cells, but are shed in clusters, forming visible scales (Fig. 2). Scaling is seen in a variety of clinical settings; it may be focal or generalized to the entire skin surface, and it may be either genetically determined or acquired. Ichthyosis is a clinical term, derived from the Greek root *ichthys* for fish, and indicates a generalized scaling disorder. Because of the negative connotations to patients, the more neutral term, disorders of cornification, has been proposed for this group of disorders (Williams and Elias, 1987), where normal cornification may be broadly defined as those processes resulting in the production and maintenance of a normal stratum corneum.

These disorders can be pathogenically divided into two groups (Frost and Van Scott, 1966; Frost *et al.*, 1966) (Fig. 3): (1) *hyperproliferative states*, in which expansion of the proliferative cell population in the basal epidermal cell layers, in

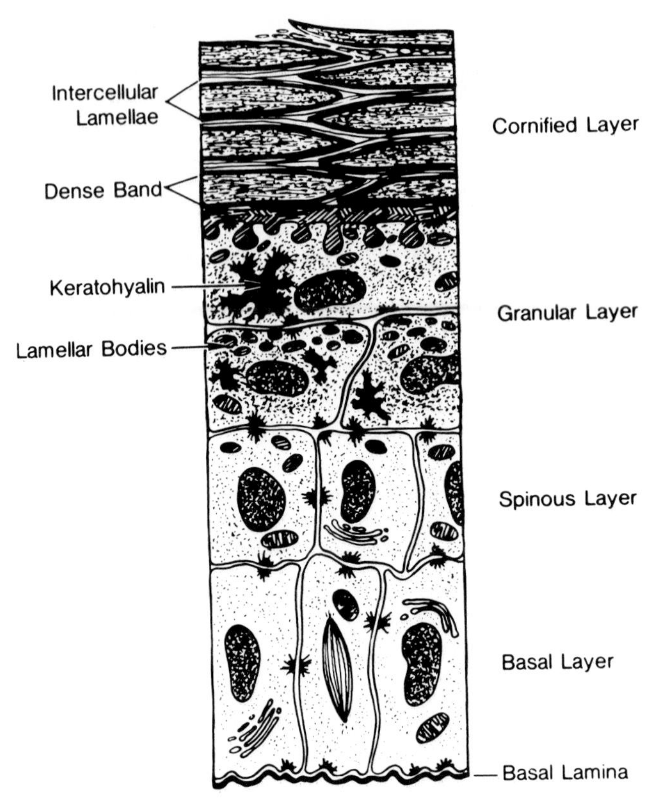

FIG. 1. Diagrammatic representation of the epidermis. [Reproduced from Williams, M. L. (1983). *Pediatr. Dermatol.* **1,** 1–24, with permission of Blackwell Scientific Publications, Inc.]

addition to more rapid transit of cells through the differentiating cell compartments (spinous and granular cell layers), result in an expanded population of incompletely differentiated cells in the outermost, normally anucleated cell layers (stratum corneum); and (2) *retention hyperkeratoses*, where increased thickness of the stratum corneum is due solely to a failure of desquamation, because epidermal proliferation rates and viable epidermal transit times are normal.

This review will focus upon evidence that stratum corneum lipids are critical effectors of corneocyte cohesion and desquamation. Interest in the role of lipids in mediating desquamation has been fueled by the following considerations. First, lipids in the stratum corneum are in the right place to modulate desquamation; i.e., they are localized to the intercellular domain (see Elias and Menon, this volume). Because normal desquamation occurs through detachment of single, intact corneocytes, it is reasonable to examine constituents of the cell membrane and intermembrane domain for effectors of this process. Desmosomes are intercellular attach-

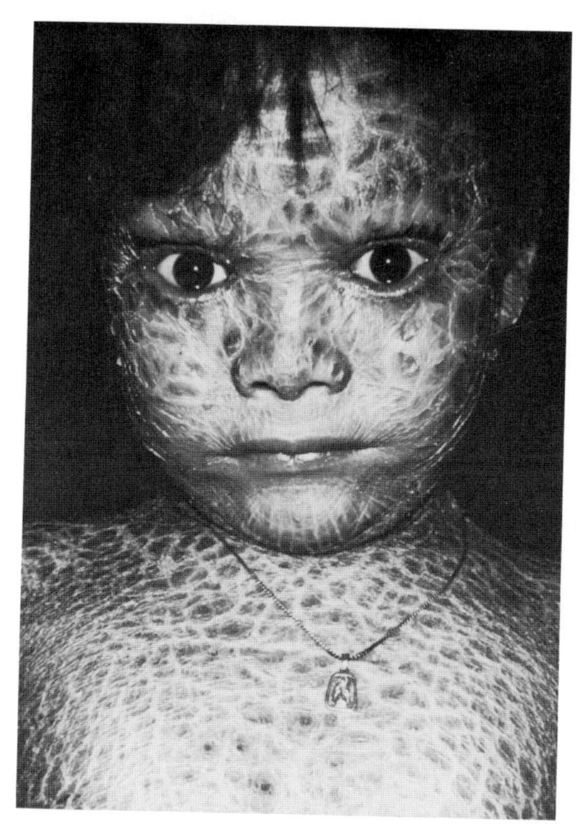

FIG. 2. Clinical example of an inherited disorder of cornification (type 4 disorder of cornification, lamellar type). [Reproduced from Williams, M. L., and Elias, P. M. (1985a). *Arch. Dermatol.* **121** 477–488, with permission of the American Medical Association.]

ment plaques that appear to function as spot welds in epithelial cell cohesion (see Section V). A progressive loss of desmosomal structures occurs during stratum corneum transit, such that these structures are fragmentary or absent in the outer layers of normal stratum corneum. Thus, it can be inferred that cohesion of at least these outermost corneocytes normally must be mediated by other factors. Second, a number of hypocholesterolemic agents induce scaling as a prominent side effect (Table I) (for review see Williams *et al.*, 1987a). Third, in most instances in which the underlying metabolic defect in genetic disorders of cornification has been defined, inborn errors of lipid metabolism are responsible (Table II). Both the drug-induced scaling disorders and the genetic ichthyoses associated with lipid metabolic defects have provided invaluable models for the examination of the role of specific lipids in stratum corneum cohesion/desquamation, and have been the

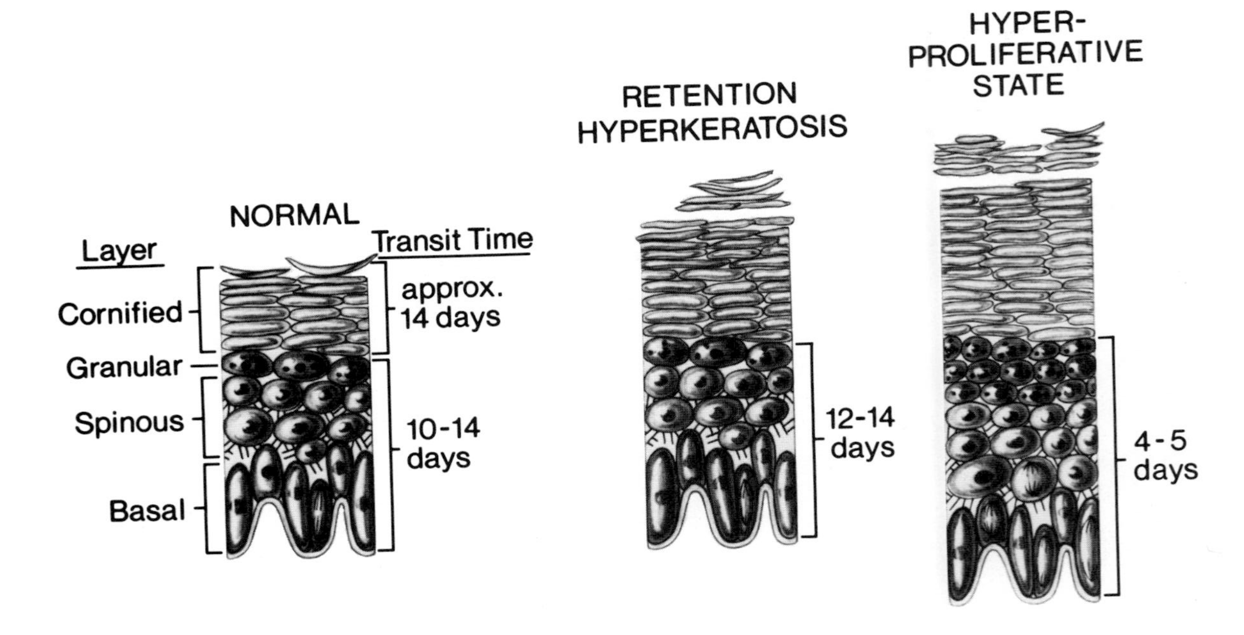

FIG. 3. Diagrammatic representation of transit times through epidermal compartments in normal skin, compared to retention hyperkeratoses and hyperproliferative states. [Reproduced from Williams, M. L. (1986). *Pediatr. Dermatol.* **3,** 476–497, with permission of Blackwell Scientific Publications, Inc.]

Table I

HUMAN HYPERKERATOTIC SKIN CONDITIONS ASSOCIATED WITH SYSTEMICALLY ADMINISTERED HYPOCHOLESTEROLEMIC AGENTS

Drug	Therapeutic use	Dermatologic side effects	Effect(s) on cholesterol metabolism
Nicotinic acid	Hypocholesterolemic agent	Generalized dry skin	Decreases hepatic VLDL synthesis; inhibits HMG-CoA reductase; increases cholesterol oxidation and excretion
Triparanol (Mer-29)	Hypocholesterolemic agent	Ichthyosis; hair loss and depigmentation	Inhibits Δ^{24} (side chain) reduction
WY-3457 (butyrophenone)	Antipsychotic agent	Ichthyosis; hair loss and depigmentation	?Inhibits postsqualene step(s)
Azacosterol (20,25-diaza-cholesterol	Hypocholesterolemic agent	Hyperkeratosis of palms and soles	Inhibits Δ^{24} (side chain) reduction
Gemfibrozil	Hypocholesterolemic agent	Exacerbation of psoriasis	Decreases hepatic VLDL secretion; increases peripheral VLDL hydrolysis; inhibits HMG-CoA reductase

Table II

ESTABLISHED OR PROPOSED METABOLIC DEFECTS IN SOME INHERITED DISORDERS OF CORNIFICATION

DOC[a]	Disorder	Inheritance[b]	Underlying defect
1	Ichthyosis vulgaris	AD	?Regulation of profilaggrin synthesis
2	Recessive X-linked ichthyosis	XR	Steroid sulfatase deficiency
7	Harlequin ichthyosis	AR	?Lamellar body deficiency
10	Sjögren–Larsson syndrome	AR	Fatty alcohol oxidoreductase deficiency
11	Refsum disease	AR	Phytanic acid oxidase deficiency
12	Neutral lipid storage disease	AR	?Intracellular lipase deficiency
13	Multiple sulfatase deficiency	AR	Deficiency of all sulfatases, including steroid sulfatase
16	CHILD syndrome	XD	Peroxisomal deficiency; ?Same as DIC 17a
17a	Conradi–Hünermann type chrondrodysplasia punctata	XD	Peroxisomal deficiency
17b	Rhizomelic type of chondrodysplasia puntata	AR	Peroxisomal deficiency

[a]DOC, Disorder of cornification, assigned types.

[b]AD, Autosomal dominant; XR, X-linked recessive; AR, autosomal recessive; XD, X-linked dominant.

subject of several previous reviews (Williams, 1983; Williams and Elias, 1985b, 1987; Epstein *et al.*, 1987; Melnick, 1989). Finally, a variety of epidermal lipid abnormalities have been reported in more common dermatologic disorders affecting the epidermis with altered cornification as part of the pathological process, e.g., senile xerosis, psoriasis, and atopic dermatitis. Whether these changes in lipid composition reflect the primary or initiating pathological event or arise from secondary events or represent an epiphenomenon is unclear, because the etiology of these disorders is not well understood at present. In view of space limitations, this review will focus upon more recent information concerning disorders in which there is a likely causal link between lipids and pathological desquamation, i.e., the drug-induced syndromes, essential fatty acid deficiency, and inherited disorders of lipid metabolism.

II. Cholesterol and Sterol Metabolites in Desquamation

A. Sterol Esterification and Keratinization

The concept that scaling skin disease and altered cholesterol metabolism are linked is not new even to this century (for reviews see Rothman, 1964; Yardley 1969). Most early work focused on a relationship between keratinization (i.e., the formation of the fibrillar network of keratins within the corneocyte) and sterol esterification. Increased sterol esterification was postulated to occur as cholesterol was liberated from the dissolution of organellar membranes during the transition from the nucleated epidermis to the stratum corneum (Freinkel and Aso, 1969). The observation that the esterified sterol content of fetal and neonatal rat increased coincident with the appearance of the stratum corneum during ontogeny provided support for this hypothesis (Freinkel and Fiedler-Weiss, 1974). A number of studies demonstrated alterations in the ratio of free to esterified cholesterol content of skin in psoriasis and other scaling skin disorders, and further supported a link between normal cornification and sterol esterification (for review see Yardley, 1969). However, in recent years the sterol esterification–keratinization hypothesis has fallen into disfavor. First, some of the earlier studies were flawed by inaccurate lipid identification techniques (for review see Yardlay, 1969). More recent studies of stratum corneum lipid composition have shown that free, and not esterified, sterols predominate in stratum corneum (Gray and Yardley, 1975; Elias *et al.*, 1979; Lampe *et al.*, 1983a,b). Second, it has been recognized that in many species, including rodents, sterol esters are major constituents of sebaceous gland-derived lipids (sebum) (Nikkari, 1974; Downing *et al.*, 1987; Grubauer *et al.*, 1989). The extent to which sterol esters in human stratum corneum or skin surface lipids may derive from sebaceous glands remains somewhat uncertain. In a widely quoted study, Kellum (1967) was unable to detect either cholesterol or sterol esters in isolated human sebaceous glands. Subsequently, Stewart *et al.* (1978) used more sensitive analytic techniques to detect cholesterol and cholesterol esters in isolated human sebaceous glands. Further support for a significant contribution of sebaceous glands to cholesterol esters in human skin surface or stratum corneum lipids has been inferred from their fatty acid composition, which shows a predominance of branched-chain species characterized of sebaceous-gland metabolism (Nicolaides *et al.*, 1972). In addition, the variation in sterol ester content in humans in relation to sebaceous gland density (Wilson, 1963; Nicolaides, 1963; Wilkinson, 1969; Lampe *et al.*, 1983a) and secretory activity (Stewart *et al.*, 1984) supports a sebaceous gland origin for at least some of these sterol esters. Finally, neonatal vernix caseosa (largely composed of sebum) is characterized by a particularly high sterol ester content (Karkainen *et al.*, 1965). In fetal epidermal preparations that contain nascent sebaceous glands, the appearance of lipidized glands morphologically correlates with a marked enrichment in sterol ester content (Williams *et al.*, 1988b). After the neonatal period, the relative contribution of sterol esters to sebum composition appears to depend upon sebum flow—a higher proportion of sterol esters occurring with lower flow rates (Stewart *et al.*, 1989). Thus, failure to control for potential sebaceous gland contributions through matching of age, sex, and body site in control and experimental groups may account for differences in sterol esterification in psoriatic skin that were observed by some investigators (see, e.g., Rothman, 1950; Gara *et al.*, 1964) but not others (Wilkinson and Farber, 1967).

The early interest in the sterol esterification/keratinization hypothesis seems also to have been fueled by X-ray diffraction studies of stratum corneum that attributed diffraction patterns to lipid–keratin filament interactions within the corneocyte (Swanbeck and Thyresson, 1962). Sterol esters were postulated to interact with keratin filaments to produce these patterns. The subsequent recognition of the predominant segregation of lipids to the extracellular domains of the stratum corneum (for review see Elias, 1983; see also Elias and Menon, this volume) and the reproduction of whole stratum corneum X-ray diffraction patterns in isolated stratum corneum membrane preparations (Elias *et al.*, 1983b) have cast serious doubt on this earlier interpretation. As the sterol esterification–keratinization linkage has fallen from favor in recent years, interest has been redirected to free sterols and another metabolite, cholesterol sulfate, in stratum corneum function.

B. Cholesterol Synthesis Inhibitors

1. *Introduction*

Nicotinic acid was the first hypocholesterolemic agent in which the side effect of generalized dry skin was reported (Parsons and Flinn, 1959; Ruiter and Meyler, 1960) (Table I). Though the mechanism whereby nicotinic acid therapy lowers serum cholesterol is not completely understood, it may be mediated both by decreased hepatic very low-density lipoprotein (VLDL) synthesis, as well as by inhibition of cholesterol synthesis through inhibition of hydroxymethylglutaryl CoA (HMG-CoA) reductase (for review see Holz, 1983). Three hypocholesterolemic drugs, which inhibit distal (postlanosterol) steps in sterologenesis—triparanol (MER-29), azacosterol (20,25-diazacholesterol), and a butyrophenone (WY-3457)—subsequently were associated with the appearance of scaling skin disorders in humans (Winkelmann *et al.*, 1963; Simpson *et al.*, 1964; Anderson and Martt, 1965). Horlick and Avigan (1963) also noted gross morphological changes in the skin of triparanol-treated rats, i.e., scaling, hair loss, and "atrophy," similar to those reported in humans. Both Clayton *et al.* (1963) and Horlick and Avigan (1963) observed an accumulation of a number of Δ^{24}-sterol intermediates in skin of rats fed triparanol, supporting the hypothesis that triparanol inhibits a single enzyme responsible for reduction of the Δ^{24} double bond in multiple sterol substrates (Avigan *et al.*, 1960; Steinberg and Avigan, 1960) (Fig. 4). In these studies, both sterol content and synthesis from radiolabeled acetate were determined in whole skin slices, thus, the lipid biochemical alterations may be presumed to reflect a combination of changes in sterol metabolism in epidermis, hair follicles, and sebaceous glands, as well as in other cellular constituents of the dermis.

2. *Azacosterol*

Azacosterol also inhibits distal steps in sterol biosynthesis, and some patients treated with this hypocholesterolemic drug develop hyperkeratosis of the palms and soles (Anderson and Martt, 1965). Like triparanol, the block is primarily localized to the Δ^{24} reduction of desmosterol to cholesterol (Ahrens *et al.*, 1965; Famagalli and Niemiro, 1964; Ranney and Daskalakis, 1964), but more proximal steps may also be inhibited (Emmons *et al.*, 1980) (Fig. 3). Azacosterol-fed hairless mice also develop a generalized hyperkeratosis (Elias *et al.*, 1983a, 1985; Geiger and Hartmann, 1986; Brod *et al.*, 1987). Elias *et al.* (1983a) examined the lipid composition of isolated stratum corneum from control versus azacosterol-fed mice. Though there were no differences in the total lipid weight percentage of stratum corneum sheets, azacosterol-fed mice showed a significantly decreased free sterol content (controls, 25.9 ± 1.5%; azacosterol, 14.6 ± 0.7% of the total lipid). Moreover, whereas in untreated mice cholesterol was the predominant sterol (89.2 ± 1.1% of total sterols), with 7-dehydrocholesterol accounting for the majority of the remainder, in azacosterol-treated mice desmosterol was the predominant sterol (76.7 ± 4.9%) and cholesterol accounted for only 15% of sterol species. These workers also observed the accumulation of other sterols, such as lathosterol and possibly zymosterol, in agreement with chick embryo fibroblast studies of Emmons *et al.* (1980), who suggested that azacosterol may inhibit more than one postlanosterol step.

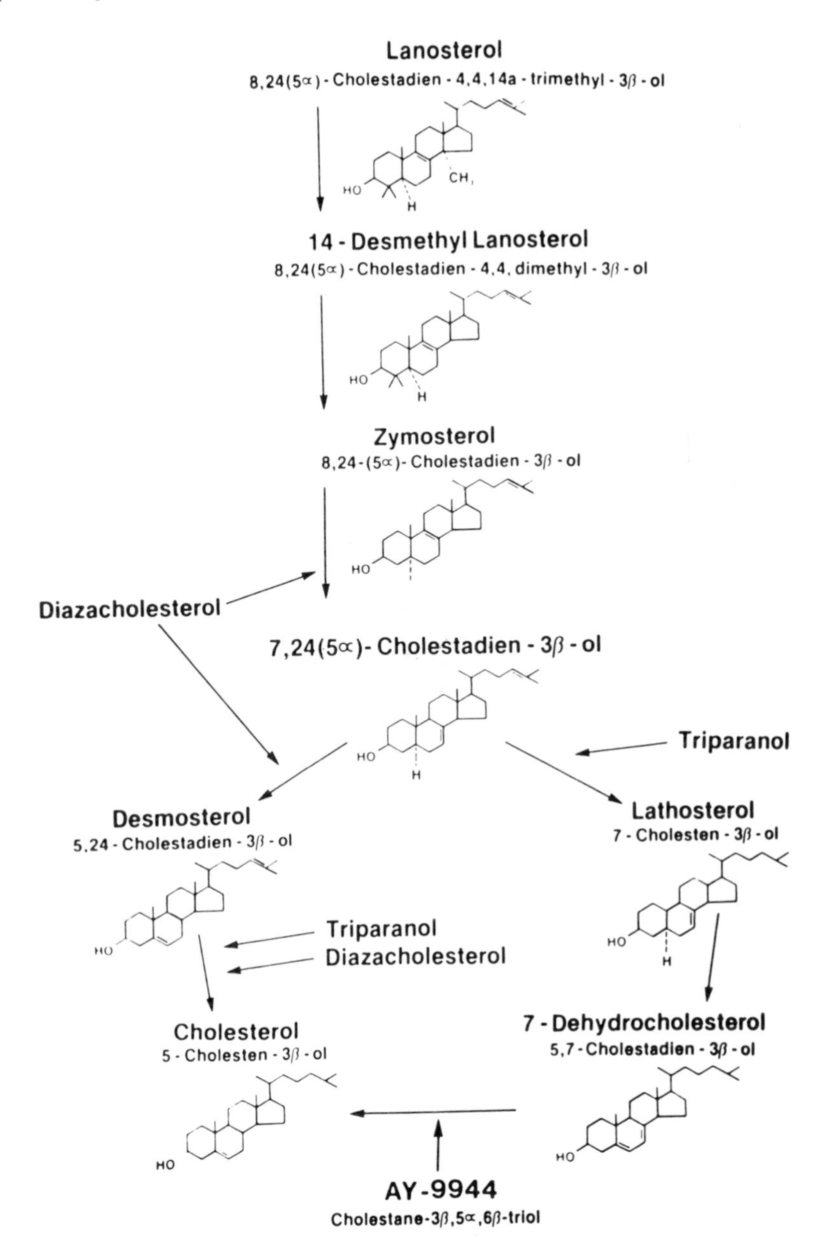

FIG. 4. Lanosterol to cholesterol biosynthetic pathway, with sites of inhibition by diazacholesterol and triparanol.

Azacosterol-treated mice also exhibit enrichment in stratum corneum sphingolipids (azacosterol treated, 150% of controls), particularly ceramides (Elias *et al.*, 1983a, 1985; Brod *et al.*, 1987). In contrast, the amount of free fatty acids and triacylglycerols in azacosterol-treated stratum corneum was unchanged. Because the total lipid weight percentage of stratum corneum in treated versus control animals was comparable, these data imply that epidermal sphingolipid synthesis may be specifically increased in response to azacosterol. Whether this represents a direct effect of this drug on one or more steps in the sphingolipid pathway or whether it is a consequence of altered stratum corneum cholesterol content is unclear. A link between sphingolipid and cholesterol biosynthetic pathways has been suggested by studies demonstrating that low-density lipoproteins (LDLs) down-regulate sphingolipid synthesis in cultured human fibroblasts (Verdery and Theolis, 1982), Chinese hamster ovary cells (Merrill, 1988), renal cells (Chatterjee *et al.*, 1986), and hepatocytes (Messmer, *et al.*, 1989). Further evidence for such linkage has been supported by recent studies in the Neiman–Pick type C disease in which sphingolipid storage accompanies a defect in the intracellular translocation of lipoprotein-derived cholesterol (Pentchev *et al.*, 1985, 1986; Sokol *et al.*, 1988; Thomas *et al.*, 1989; for review see Spence and Callahan, 1989). However, it should be kept in mind that epidermal cholesterol biosynthesis appears to be regulated largely independent of the LDL receptor pathway (for review see Williams *et al.*, 1987a) (see Ponec, this volume).

Alternatively, the increased sphingolipid content of stratum corneum in azacosterol-treated mice may be a consequence of altered stratum corneum cholesterol content. In this scenario, decreased stratum corneum cholesterol would result in disorganization of the lipid bilayers and/or altered membrane dynamics, and, consequently, increased transepidermal water loss (see below). As discussed by Feingold (this volume), disruption of barrier function could signal increased epidermal lipogenesis, including sphingolipid synthesis. It needs to be noted in this regard, however, that rates of transepidermal water loss were not increased in azacosterol-treated mice (Elias *et al.*, 1985; Brod *et al.*, 1987), although it could be argued that the compensatory increase in sphingolipid content effectively repaired the barrier defect.

Stratum corneum thickness is increased from 3- to 10-fold in azacosterol-treated mice, and membrane lipids are decreased as assessed in both oil red O-stained frozen sections and in freeze–fracture replicas (Elias *et al.*, 1983a). On thin-section electron microscopy, epidermal lamellar bodies often appear to be empty and in freeze–fracture replicas the stratum corneum intercellular domains have a fragmentary, "moth-eaten" appearance. The epidermal [^{3}H]thymidine labeling index is somewhat increased in azacosterol-treated animals, although the numbers of animals studied was small (n = 6), and these differences did not achieve statistical significance (Elias *et al.*, 1985). Acanthosis, indicative of epidermal hyperplasia in azacosterol-fed hairless mice, was also reported by Brod *et al.* (1987) and Geiger and Hartmann (1986). These observations are intriguing in light of recent evidence that links a defective permeability barrier to epidermal hyperplasia (Proksch *et al.*, 1991b).

In summary, azacosterol, an inhibitor of one or more distal steps in the cholesterol biosynthetic pathway, induces disordered cornification when systemically administered to both humans and experimental animals. In hairless mice, decreased stratum corneum free sterol content and altered sterol composition, with an accumulation of desmosterol and marked diminution in cholesterol content, are observed. The ceramide content is proportionally increased, such that the total quantity of lipid in the stratum corneum remains unchanged. Nonetheless, this perturbation in lipid composition may be sufficient to disrupt the lamellar organization of the stratum corneum membrane architecture, resulting in abnormal cohesion. Yet, epidermal acanthosis and an apparent increase in epidermal [^{3}H]thymidine labeling index suggest that the effects of azacosterol on the epidermis may be complex, and that increased stratum corneum thickness may arise both as a consequence of hyperproliferation and of delayed desquamation.

3. Lovastatin

As shown in Table I, other cholesterol-lowering drugs also can produce a scaling or hyperkeratotic skin disorder as a side effect (for review see Williams *et al.*, 1987a). Because epidermal cholesterol synthesis may be largely independent of regulation by circulating LDL cholesterol (see Ponec, this volume), and because all of these agents inhibit one or more steps in the cholesterol biosynthetic pathway, direct inhibition of epidermal cholesterol synthesis may be the common link to the scaling disorder produced by these agents. Yet surprisingly, cutaneous side effects have not been reported in association with systemic administration of competitive inhibitors of 3-hydroxy-3-methylglutaryl coenzyme A reductase (the rate-limiting enzyme in cholesterol synthesis), of which lovastatin (mevinolin) is an example (for reviews see Endo, 1985; Alberts, 1988). The hypocholesterolemic effects of lovastatin are ascribed to increased hepatic LDL receptor expression and LDL uptake as a consequence of inhibition of hepatic HMG-CoA reductase (Bilheimer *et al.*, 1983; Ma *et al.*, 1986; Reihner *et al.*, 1990). The epidermis may be able to maintain cholesterol biosynthetic rates during lovastatin therapy through a compensatory increase in epidermal HMG-CoA reductase enzyme content (see below). Alternatively, drug pharmacokinetics, such as predominant first-pass hepatic clearance (Alberts, 1988), may protect the epidermis from direct drug effects.

However, the failure to observe scaling skin disease in response to systemic lovastatin cannot be attributed to tissue insensitivity, because lovastatin profoundly inhibits cholesterol synthesis in cultured keratinocytes at concentrations comparable to other cell types (Ponec *et al.*, 1987). Moreover, a single topical application of lovastatin to the skin of hairless mice produces an approximately 90%

inhibition of epidermal cholesterol synthesis within 2 hours (Feingold *et al.*, 1990). However, only after five to seven daily applications of lovastatin does the skin become red and scaly (Feingold *et al.*, 1991). The epidermis appears hyperplastic, an observation confirmed by an increase in mitotic figures, [^{3}H]thymidine incorporation, and DNA content. A disturbance in barrier function, as measured by increased rates of transepidermal water loss, occurs coincident with these changes. Moreover, electron microscopy shows that lamellar bodies in the upper, nucleated cell layers appear abnormal in both size and internal structure, whereas in stratum corneum, abnormal lamellar body-derived contents are evident within the intercellular spaces, and many abnormal lamellar structures are retained within the corneocyte. Whereas initially (within a few hours) only cholesterol synthesis is inhibited, by 24 hours a significant increase in fatty acid synthesis is observed. Williams *et al.* (1991c) also have shown that lovastatin stimulates fatty acid synthesis and inhibits fatty acid oxidation in cultured keratinocytes, thereby inducing triacylglycerol accumulation. As in whole epidermis, the increased fatty acid synthesis does not occur until 12 to 24 hours after exposure to the drug. It is of interest that increased fatty acid synthesis is not observed with lovastatin in the presence of either LDL cholesterol or 25-hydroxycholesterol (Williams *et al.*, 1991c). With repeated topical applications of lovastatin coincident with the development of the morphological and functional abnormalities described above, epidermal cholesterol biosynthetic rates normalize and fatty acid biosynthesis remains elevated (>250% of control) (Feingold *et al.*, 1991). The normalization of cholesterol biosynthetic rates is probably achieved largely through increased cellular HMG-CoA reductase enzyme content, as has been described in liver (Edwards *et al.*, 1983; Singer *et al.*, 1988) and cultured UT-1 cells (Pathak *et al.*, 1986) in response to lovastatin and related drugs, because greater than 200% increase in enzyme activity can be demonstrated in epidermal enzyme preparations dialyzed free of inhibitor (Feingold *et al.*, 1991). In accordance with these biosynthetic data, epidermal cholesterol, 7-dehydrocholesterol, lanosterol, and lathosterol contents are unchanged, whereas fatty acid content is significantly increased.

Thus, while topical lovastatin treatment of hairless mice results in disordered desquamation and abnormal epidermal barrier function, these changes cannot be simply ascribed to a decrease in stratum corneum membrane cholesterol content, as one might have predicted. Indeed, stratum corneum cholesterol content is maintained during chronic lovastatin treatment, whereas fatty acid synthesis and content are increased. Feingold *et al.* (1991) have proposed that the disorganization of stratum corneum membrane bilayers they observed is due to an alteration in the membrane free sterol:free fatty acid ratio, resulting in phase separations. This membrane disorganization, in turn, is presumed to be the basis for the barrier defect. The barrier defect may in turn induce epidermal hyperplasia, because correction of the barrier through occlusion with a vapor-impermeable membrane returns epidermal DNA synthesis toward normal (Proksch *et al.*, 1991b). The abnormal desquamation in these chronically lovastatin-treated mice may arise as a consequence of epidermal hyperproliferation and/or as a consequence of altered stratum corneum membrane organization.

4. Summary

Disorders of cornification occur as side effects to a number of hypocholesterolemic agents, and it is likely that these cutaneous effects are due in each case to perturbation of epidermal cholesterol synthesis. Failure to observe cutaneous side effects with some cholesterol synthesis inhibitors is most likely due to insufficient epidermal exposure to the drug, rather than to organ resistance. Epidermal responses to inhibition of cholesterol biosynthesis are complex and include scaling, epidermal hyperplasia, increased transepidermal water loss, and perturbations in the metabolism of other lipid classes. Though the pathogenic sequences in these responses remain to be defined, they do strongly support the importance of sterols for epidermal homeostasis and normal desquamation.

C. Cholesterol Sulfate and Recessive X-Linked Ichthyosis

1. Recessive X-Linked Ichthyosis and the Steroid Sulfatase Gene

Since 1978 it has been recognized that recessive X-linked ichthyosis (RXLI) is caused by deficiency of a microsomal enzyme, steroid sulfatase (Koppe *et al.*, 1978; Shapiro and Weiss, 1978; Shapiro *et al.*, 1978). Placental steroid sulfatase deficiency was recognized nearly a decade earlier as the cause of abnormal maternal urinary estrogen profiles in some postterm pregnancies (France and Liggins, 1969; for review see Taylor, 1982). Though placental steroid sulfatase deficiency was initially linked to failure of labor to initiate or progress, this conclusion has been questioned because urinary estrogens may only be examined in postterm pregnancies (for review see Crawford, 1982). Moreover, Lykkesfeldt *et al.* (1984) reported no increased incidence of obstetrical complications in 23 placental sulfatase deficiency-associated pregnancies.

The steroid sulfatase gene has been the subject of considerable research in the past decade, in part because of its locus on the distal tip of the short arm of the X chromosome (Teipolo *et al.*, 1977, 1980; Mohandras *et al.*, 1979; Li *et al.*, 1990), a region that undergoes only partial X inactivation in humans (Migeon *et al.*, 1982; Willems *et al.*, 1986; for review see Shapiro, 1985). The steroid sulfatase gene has been cloned by several groups; most patients with RXLI have the disorder on the basis of gene deletion (Ballabio *et al.*, 1987, 1989; Bonifas *et al.*, 1987; Conary *et al.*, 1987; Gillard *et al.*, 1987; Yen *et al.*, 1987; Shapiro *et al.*, 1989). Genomic deletions may be large enough to detect by chromosomal banding (Tiepolo *et al.*, 1977; Curry *et al.*, 1984; Bick *et al.*, 1989) or, in the case of smaller deletions, by DNA flow cytometry (Cooke *et al.*, 1988). A number of patients have been described with ichthyosis due to steroid sulfatase deficiency and a variety of other organ system anomalies or defects (for review see Schnur *et al.*, 1989); these are considered to represent "contiguous gene syndromes" arising from disturbances in neighboring loci (Schmickel, 1986). Hence, the possibility of contiguous gene effects must be considered in all cases when unusual phenotypic features are ascribed to steroid sulfatase deficiency. To date the skin represents the only organ system wherein malfunction inevitably occurs in association with steroid sulfatase deficiency.

2. *Steroid Sulfatase and Arylsulfatase C*

Steroid sulfatase is a membrane-bound enzyme responsible for hydrolyzing the 3β-sulfate esters of cholesterol and the steroid hormones that exist in parallel to their hormonally active, nonsulfated counterparts (for review see Roberts and Liebermann, 1970). The term *steroid sulfatase* tends to be used interchangeably with arylsulfatase C. Arylsulfatase C is the term applied to the sulfohydrolase activity toward 4-methylumbelliferyl sulfate and other synthetic sulfate esters. Arylsulfatase C is distinguished from arylsulfatases A and B, active against some of the same substrates, by its nonlysosomal localization and nonacidic pH optima, as well as differences in activators and inhibitors (for reviews see Rose, 1982; Hobkirk, 1985; Shapiro, 1985). Recent studies have demonstrated considerable homology in the N-terminal regions of arylsulfatases A, B, and C (Robertson *et al.*, 1988; Stein *et al.*, 1989; Schuchman *et al.*, 1990).

Several groups have attempted to purify steroid sulfatase from human placenta (Noel *et al.*, 1983; Dibbelt and Kuss, 1986; Vaccaro *et al.*, 1987; Kawano *et al.*, 1989). The native protein, with an estimated molecular weight ranging from 238,000 to 530,000, is composed of identical subunits having a molecular weight of 62,000 to 78,000. Similar mass characteristics have been reported for the enzyme purified from rat and human livers (Moriyasu *et al.*, 1982; Kawano *et al.*, 1989). Differences in apparent molecular mass are likely to be due to incomplete purification, as well as differences in purification procedures. Moreover, detergents employed in these isolations may alter substrate kinetics (Kawano *et al.*, 1989). The enzyme is a glycoprotein of the high mannose type and is a transmembrane protein in microsomes, as assessed by its enzymatic activity, antibody binding, and susceptibility to transglutaminase in isolated microsomes in the presence or absence of detergent (Moriyasu and Ito, 1982). In human fibroblasts, steroid sulfatase is synthesized as a proenzyme with a high-mannose oligosaccharide chain possessing a molecular weight of 63,500 (Conary *et al.*, 1986). Processing of the oligosaccharide form results in the mature enzyme of 61,000 Da; the enzyme's half-life is 4 days. Using a human placental cDNA clone transfected into BHK-21 cells, an enzymatically active protein was expressed (Stein *et al.*, 1989). The expressed protein had a molecular weight of 62,000 with N-glycosylation at two of four asparagine residues. The deduced protein consisted of 583 amino acids with the two glycoyslated domains facing the lumen, linked by a membrane-spanning domain.

Whether one or more than one steroid sulfohydrolase occurs in mammalian tissues has been the subject of considerable debate (for reviews see Rose, 1982; Daniel, 1985; Shapiro, 1985). The earlier evidence for more than one steroid sulfatase was derived largely from studies using partially purified enzyme preparations, exhibiting differences in substrate specificities, pH optima, and thermal stability (see e.g., Zuckerman and Hagerman, 1966; Iwamori *et al.*, 1976). Shapiro (1985) has argued that there is a single gene for steroid sulfatase/arylsulfatase C because (1) antibodies against steroid sulfatase precipitate all enzyme activities (2) all activities map to the same region X chromosome, and (3) all activities are lost in patients with RXLI, all of whom belong to a single complementation group in cell fusion studies. Since Shapiro's review, several investigations have provided additional data consistent with a single steroid sulfohydrolase (Dibbelt and Kuss, 1986; Vaccaro *et al.*, 1987; Kawano *et al.*, 1989).

Yet, a number of investigators have provided evidence for more than one sulfatase enzyme (Meyer, *et al.*, 1982, 1984; Nelson *et al.*, 1983; Jobsis *et al.*, 1983; Gniot-Szulzycka and Januszewska, 1986; Dijkstra *et al.*, 1987; MacIndoe, 1988; Milewich *et al.*, 1990; van Diggelen *et al.*, 1989; Hobkirk *et al.*, 1989; Choi and Hobkirk, 1986; Chang *et al.*, 1986; Munroe and Chang, 1987; Chang *et al.*, 1990). When cultured fibroblasts or leukocytes were solubilized with detergents, the soluble proteins separated by polyacrylamide gel electrophoresis, and arylsulfatase activity demonstrated by a histochemical method, two distant bands were evident (Jobsis *et al.*, 1983). Only one of these bands exhibited activity against dehydroepiandrosterone sulfate (DHEAS), and in X-linked ichthyosis, only this DHEAS-active band was absent. However, the possibility that the DHEA-inactive band represented arylsulfatase A and/or B activities was not excluded. More recently, van Diggelen *et al.* (1989) have demonstrated differences in arylsulfatase C activity in a 50,000 *g* supernatant versus pellet fractions prepared from leukocytes. Though pellet-associated arylsulfatase activity was inhibited by excess DHEAS and was deficient in RXLI leukocytes, soluble activity was neither inhibited by DHEAS nor deficient in RXLI. The possibility that the soluble arylsulfatase activity was due to either arylsulfatase A or B was considered unlikely because 50 mmol of phosphate, which inhibits both these enzymes, was present in the assay, and because of the alkaline pH (8–8.5) optimum of the reaction. This second arylsulfatase C was postulated to account for the significant residual enzyme activities (up to 25% of control) reported by this group and others (Meyer *et al.*, 1979; Meyer and Grundmann, 1980; Dijkstra *et al.*, 1987; Herrmann, *et al.*, 1987), when arylsulfatase C, rather than steroid sulfatase, activity is assayed in some RXLI tissues. MacIndoe (1988) has described two steroid sulfohydrolases in a human breast carcinoma cell line. An estrone sulfate sulfatase and DHEAS sulfatase were distinguished by their differing substrate specificities and kinetics of inhibition. Though differing substrate kinetics alone would not distinguish between different catalytic sites on a single enzyme versus different enzymes, they observed that estrone sulfatase activity was down-regulated when cultures were incubated for 20 hours with physiologic concentrations of estrone sulfate or DHEAS, whereas the DHEAS sulfase activity was not down-regulated. These observations are most consistent with two isoenzymes. Hobkirk *et al.* (1989) observed sulfohydrolase activity toward estrone sulfate but not against DHEAS, pregnenolone sulfate, or androstenediol sulfate in porcine Leydig cells, an observation consistent with limited isoenzyme expression in this tissue; however, rat Leydig cells are active against DHEAS (Bedin *et al.*, 1988) and boar testes are active against pregnenolone sulfate (Ruokonen, 1978). Choi and Hobkirk (1986) solubilized estrone sulfatases from guinea pig uterus, testes, and brain microsomes and from rat liver and human placenta microsomes and subjected these preparations to chromatofocusing. Guinea pig tissues exhibited two (uterus and brain) or three (testes) chromatofocused peaks with activity, whereas liver and human placenta contained a single peak of activity. The rat liver and human placental enzymes showed a neutral p*I*, but the predominant sulfatase in all guinea pig tissues had an alkaline p*I*. These data suggest that there are both significant species and tissue differences in steroid sulfatase isoenzymes.

Most recently, Chang and associates (Chang *et al.*, 1986, 1990; Munroe and Chang, 1987) clearly demonstrated the existence of two arylsulfatase C isoenzymes in humans. Although both enzymes were membrane associated, they were distinguishable by their electrophoretic mobility, pH optima, K_m, heat stability, and tissue distribution. Moreover, one electrophoretic species, the "s" (slow migrating) form, exhibited steroid sulfatase activity, and this was the only form precipitated by antisera to steroid sulfatase. Though both isoenzymes localize to the same region of the X chromosome, they are nonallelic because both are expressed in male-derived tissues. Moreover, a full-length cDNA to the "s" isoenzyme did not hybridize with mRNA from somatic hybrid lines expressing only the other, faster migrating ("f") form. No RXLI fibroblasts expressed the "s" form, yet some, but not all, RXLI cell lines expressed the "f" form. These investigators therefore concluded that the "s" and "f" isoenzymes are products of separate but closely linked genes, the "s" isoenzyme gene being the same as steroid sulfatase (Chang *et al.*, 1990). Their studies would suggest that in some RXLI kindreds the genomic deletion may include not only the steroid sulfatase locus, but also the nearby "f" isoenzyme locus. Placenta expresses only the "s" form (Munroe and Chang, 1987), in agreement with several other studies describing a single arylsulfatase C/steroid sulfatase enzyme in this tissue (Vaccaro *et al.*, 1987; Dibbelt and Kuss, 1987; Choi and Hobkirk, 1986). Thus, the existence of more than one sulfohydrolase of the arylsulfatase C type seems established. Determination of the natural substrates of these isoenzymes and their tissue expression, as well as species variations, may serve to resolve much of the confusion of the preceding decades.

3. *Sulfated Steroids in RXLI*

Determination of the endogenous substrates of these isoenzymes may also shed light on the heretofore puzzling observation that serum sulfated hormone levels, with the exception of estrone sulfate (Lykkesfeldt *et al.*, 1985a), are not consistently elevated in RXLI males (Ruokonen *et al.*, 1980, 1986; Epstein and Leventhal, 1981; Lykkesfeldt *et al.*, 1985a). Attempting to account for these findings, Bergner and Shapiro (1988) administered [^{3}H]DHEAS to three RXLI males and measured excretion of the desulfation products as urinary [^{3}H]glucuronides. Though one patient failed to desulfate the radiolabeled substrate, in another patient desulfation was attributed to gastrointestinal bacterial enzymatic action, because it could be prevented by antibiotic (ampicillin) pretreatment. Yet, in the third patient, desulfation could not be prevented by antibiotic therapy and the source of this activity remained unexplained at the time. In the light of the recent demonstration of arylsulfatase C isoenzymes (Chang *et al.*, 1990; see above), one may speculate that the desulfation activity in this patient may have been due to residual arylsulfatase C isoenzyme activity.

In contrast to the relatively normal serum sulfated hormone levels, plasma and red blood cell membrane cholesterol sulfate content is increased 10- to 20-fold in RXLI (Bergner and Shapiro, 1981; Epstein *et al.*, 1981a). Cholesterol sulfate is carried by the β (or low density) lipoprotein fraction in RXLI serum, which alters the electrophoretic mobility of this fraction (Epstein *et al.*, 1981a; Ibsen *et al.*, 1986). In addition to increased cholesterol sulfate content of LDLs in RXLI, Nakamura *et al.* (1988) reported that the cholesterol ester content was significantly decreased but the triacylglycerol content and apoprotein B:cholesterol ratio were significantly increased. However, the size of the lipoprotein particle and its affinity for the LDL receptor were unchanged. The observation that cholesterol sulfate accumulates both in RXLI scale (Williams and Elias, 1981) and in blood (Bergner and Shapiro, 1981; Epstein *et al.*, 1981a) indicates that cholesterol sulfate is one of the natural substrates of steroid sulfatase.

4. *Cholesterol Sulfate and the Pathogenesis of Scaling in RXLI*

A generalized hyperkeratosis (ichthyosis) is the most constant phenotypic feature of steroid sulfatase deficiency. Whereas infant's skin is usually normal at birth, after a few weeks it undergoes a pronounced desquamation or peeling (Lykkesfeldt *et al.*, 1985b; Hoyer *et al.*, 1986); thereafter, the typical scale pattern is evident on the trunk and extremities. The hyperkeratosis in RXLI is attributed to delayed desquamation rather than hyperproliferation, because epidermal proliferation indices, i.e., the proportion of basal cells labeled by [^{3}H]thymidine (labeling index) and the time required for migration of labeled cells through the nucleated epidermal layers (epidermal transit time), are comparable to normal epidermis *in vivo* (Frost *et al.*, 1966; Frost, 1973). However, the number of patients studied was small ($n = 4$) and these studies were undertaken prior to delineation of steroid sulfatase deficiency as the cause of RXLI; diagnosis by clinical criteria alone is not invariably reliable (Yoshike *et al.*, 1985). Nonetheless, the absence of significant acanthosis (increased thickness of the nucleated epidermal compartment) in RXLI epidermis supports the concept of a normoproliferative, retention-type of hyperkeratosis. Despite the increased thickness of the stratum corneum in RXLI, rates of transepidermal water losses are modestly increased (normal range, 0.14–0.23 mg/cm^2/hour; mean, 0.18 mg/cm^2/hour; RXLI range, 0.19–0.30 mg/cm^2/hour; mean, 0.26 mg/cm^2/hour; $p < 0.05$), comparable to those observed in other cornification disorders (Frost *et al.*, 1968).

When the lipid content of the pathological scale in RXLI was examined, a five-fold increase in cholesterol sulfate and a 50% decrease in free sterol content was observed in comparison to those sterols in scales from normal, postsunburn, or postorthopedic cast desquamated scales (Williams and Elias, 1981). Yet the proportion of sterol esters and other lipid classes, as well as the total sterol content (i.e., sum of free sterols, sterol esters, and sterol sulfates) and the overall lipid weight percentage, are unchanged in RXLI scale (Table III).

To further examine the possibility that regulation of stratum corneum cholesterol/cholesterol sulfate content by steroid sulfatase may mediate normal cohesion and desquamation, Elias and associates (1984) isolated epidermal cell populations from normal murine and human, as well as RXLI, epidermis, and assayed these tissues for both steroid sulfatase activity and cholesterol sulfate content. In humans, cholesterol sulfate normally comprises 5.2% of the total lipid of human stratum granulosum, with somewhat lower proportions in the lower nucleated epidermal layers and stratum corneum (3.4 and 2.6%, respectively). Moreover, steroid sulfatase specific activity is 3 to 20 times higher in stratum granulosum in comparison to lower epidermal cell populations, activity that is substantially preserved in stratum corneum. Because cholesterol sulfotransferase activity predominates in the lower epidermal cell layers (Epstein *et al.*, 1984a), Epstein *et al.* (1984b) postulated the operation of an epidermal cholesterol sulfate cycle in which cholesterol first is sulfurylated in lower epidermal cell layers and later hydrolyzed, regenerating cholesterol in the outer layers. Thus, cholesterol sulfotransferase and steroid sulfatase activities, as well as cholesterol sulfate content, have been used as differentiation markers in cultured rabbit tracheal epithelial cells (Rearick and Jetten, 1986; Jetten *et al.*, 1989a,b) and cultured keratinocytes (Ponec *et al.*, 1989).

Grayson and Elias (1982) developed a technique whereby the plasma membranes of adjoining corneocytes, their intervening membrane bilayers, can be isolated as “sandwiches.” When the membrane regions of normal human stratum corneum are isolated, both steroid sulfatase and its substrate, cholesterol sulfate,

Table III
LIPID CONTENT AND COMPOSITIONS OF RXLI VERSUS NORMAL SCALE[a]

	RXLI	Normal	Significance
Sterols			
Total lipid content (% dry wt)	11.2 ± 1.3	11.7 ± 0.5	NS[c]
Polar lipids (% total lipid)	2.6 ± 0.3	2.0 ± 0.4	NS
Cholesterol sulfate	12.2 ± 1.4	2.3 ± 0.6	$p < 0.01$
Free sterols	9.3 ± 1.0	15.4 ± 1.9	$p < 0.02$
Sterol esters	4.2 ± 0.8	5.5 ± 0.5	NS
Total sterols[b]	21.6 ± 1.6	23.4 ± 1.4	NS
Neutral Lipids (% total lipids)			
Glycolipids and ceramides	25.8 ± 2.3	25.5 ± 1.4	NS
Triglycerides	9.3 ± 2.5	15.4 ± 1.9	NS
Free fatty acids	9.3 ± 1.1	11.6 ± 2.3	NS
Hydrocarbons	5.2 ± 1.6	7.9 ± 1.4	NS

[a]After Williams *et al.* (1983).
[b]Sum of cholesterol sulfate, free sterols, and sterol esters.
[c]NS, Not significant.

localize to these peripheral membrane preparations (Elias et al., 1984). However, neither steroid sulfatase activity nor cholesterol sulfate localize to lamellar body-enriched subcellular fractions of epidermis (Grayson *et al.*, 1985). Because these organelles are believed to function as secretory vehicles for both stratum corneum intercellular lipids and an array of hydrolytic enzymes (see Elias and Menon, this volume), how both steroid sulfatase and its substrate become concentrated in membrane domains is unclear. Though considered a "microsomal" enzyme, recent ultrastructural immunocytochemical studies also noted localization of the enzyme to plasma membrane, especially in association with coated pits, in both placenta (Dibbelt *et al.*, 1989) and cultured fibroblasts (Willemsen *et al.*, 1988). Cholesterol sulfate, similar to many oxygenated sterols and unlike cholesterol, readily diffuses across the plasma membrane (Ponec and Williams, 1985). In stratum corneum, cholesterol sulfate may simply partition to the hydrophobic intercellular domain, as the cell interior becomes increasingly hydrophilic through dissolution of organellar membranes.

The observation that the proportion of cholesterol sulfate is lower in outer stratum corneum (i.e., squames or scale) than it is in whole stratum corneum sheets (Long *et al.*, 1985; Ranasinghe *et al.*, 1986) supports the concept that steroid sulfatase in stratum corneum continues to be active against its substrate. Ongoing hydrolysis of cholesterol sulfate during stratum corneum transit is postulated to be critical for normal desquamation (Elias *et al.*, 1984; Ranasinghe *et al.*, 1986). It is of interest to note the highly cohesive ungulate hoof is particularly enriched in cholesterol sulfate, wherein it comprises 5–20% of the total lipid in a 1:2 ratio with free cholesterol (Veta *et al.*, 1971; Wertz and Downing, 1984). However, the cholesterol sulfate content of human nail and hair is comparable to the proportion found in epidermis, i.e., 1–3% of total lipid (Serizawa *et al.*, 1990), whereas in RXLI these proportions are significantly increased but no clinical abnormality in either hair or nail is seen in this disorder.

In RXLI epidermis, steroid sulfatase activity, as expected, is absent from all nucleated cell layers and stratum corneum membrane preparations (Elias *et al.*, 1984; Kubilus *et al.*, 1979). The proportion of cholesterol sulfate found in the granular cell layer in RXLI (5.5 ± 0.9%) does not differ from normal epidermis (5.2 ± 1.4%), whereas in RXLI stratum corneum it is significantly increased (3.4 ± 0.5% normal; 11.2 ± 1.1% RXLI) (Elias *et al.*, 1984). The observation that topically applied cholesterol sulfate induces scaling in hairless mice (Maloney *et al.*, 1984; Elias *et al.*, 1985) has provided further credence to the hypothesis that stratum corneum cholesterol sulfate plays a critical role in desquamation. Moreover, in these animals the epidermal labeling indices were normal, indicating a defect in desquamation per se rather than scaling as a consequence of hyperplasia (Elias *et al.*, 1985).

However, it is unclear whether the scaling in RXLI and in this animal model is a function of the excess cholesterol sulfate in stratum corneum or of the correspondingly reduced free sterol content (see Table III). Not only in RXLI scale is there a compensatory decrease in quantity of free cholesterol (Williams and Elias, 1981), but also in the animal model scaling became evident in parallel with a reduction in cholesterol content (Maloney *et al.*, 1984). Similarly, some observers have reported improvement when RXLI patients are treated with topical cholesterol-containing preparations (Lykkesfeld and Hoyer, 1983), although others (Ibsen *et al.*, 1984) have been unable to confirm these results.

To account for the decreased stratum corneum cholesterol content in RXLI and cholesterol sulfate-treated hairless mice (see above), the possibility that cholesterol sulfate may regulate cholesterol synthesis has been considered. Indeed, cholesterol sulfate inhibits cholesterol synthesis in both cultured fibroblasts (Williams *et al.*, 1985) and keratinocytes (Williams *et al.*, 1987b), by inhibition of HMG-CoA reductase at concentrations physiologic to epidermis (i.e., maximal inhibition at 10 μM). Cholesterol sulfate also profoundly inhibits cholesterol synthesis in RXLI fibroblasts, indicating that prior desulfation to cholesterol is not a requirement (Williams *et al.*, 1985). Because extracellular cholesterol sulfate readily equilibrates with intracellular compartments (Ponec and Williams, 1986) and because the outermost nucleated epidermal layer, the stratum granulosum, remains an active site of cholesterol synthesis (Feingold *et al.*, 1983; Monger *et al.*, 1988; Proksch *et al.*, 1991a), stratum corneum cholesterol sulfate could equilibrate with stratum granulosum pools and be in a position to regulate sterologenesis in the upper epidermis. According to this model, steroid sulfatase activity in the stratum corneum may regulate stratum granulosum sterologenesis by regulating stratum corneum cholesterol sulfate content; i.e., hydrolysis of stratum corneum cholesterol sulfate to cholesterol could release the stratum granulosum sterol biosynthetic pathway from feedback inhibition. Hence, cholesterol sulfate may be a molecular regulator of cholesterol synthesis at a site where lipids are specifically generated for barrier function (Williams *et al.*, 1987a). And, it may be this teleological requirement of the stratum granulosum, to attune its sterol biosynthesis to the cholesterol content of the stratum corneum in order to maintain the integrity of the permeability barrier, that accounts for the loss of functional LDL receptors as keratinocytes differentiate (Ponec *et al.*, 1983, 1985; Williams *et al.*, 1987c; Mommaas-Kienhuis *et al.*, 1987) (see Ponec and Feingold, this volume).

Proposed functions for steroid sulfatase and sterol sulfates in other organ systems may also be relevant to the process of normal desquamation and the pathogenesis of scaling in RXLI (for review see Roberts and Liebermann, 1970; Hobkirk, 1985; Roberts, 1987). Regulation of membrane cholesterol sulfate (or desmosterol sulfate, depending on the species) content is postulated to be critical during sperm capacitation (a process of physicochemical change in the acrosomal membrane of spermatozoa prior to membrane fusion with the ovum) (for review see Roberts, 1987). Sterol sulfates are potent inhibitors of acrosin (Burck and Zimmerman, 1980; Bouthillier *et al.*, 1984), a protease involved in capacitation. Sterol sulfates, which are concentrated in acrosomal membranes (Langlais *et al.*, 1981), may be hydrolyzed by sterol sulfatases localized to the female reproductive tract during sperm transit (Lalumiere *et al.*, 1976; Legault *et al.*, 1980). Accordingly, as the sterol sulfates are hydrolyzed, acrosin is released from inhibition. In a similar manner, if cholesterol sulfate also functions as a protease inhibitor in stratum corneum membranes, this could account for the abnormal persistence of numerous desmosomal structures into the outermost layers of RXLI stratum corneum (Mesquita-Guimaraes, 1981; Perrot and Ortonne, 1979; Bazex *et al.*, 1978); desmosomes are normally absent in outer stratum corneum (Allen and Potten, 1975; Arnn and Staehelin, 1981). Protease inactivation could also account for the abnormal persistence of melanosomes (Bazex *et al.*, 1978; Mesquita-Guimaraes, 1981) in RXLI stratum corneum. Though the specific proteases involved in desmosomal degradation have not been delineated, carboxypeptidase and cathepsin B activities are localized to lamellar bodies (Grayson *et al.*, 1985). These and/or other hydrolases may normally degrade desmosomes during stratum corneum transit, and progressive degradation of desmosomes may be an essential prerequisite for normal desquamation (Allen and Potten, 1975; Lundstromm and Egelrud, 1988). Thus, cholesterol sulfate in stratum corneum may normally regulate desmosomal degradation through an effect on protease activity; i.e., cholesterol sulfate would inhibit desmosomal proteolysis, whereas hydrolysis by steroid sulfatase during stratum corneum transit would release these proteases from inhibition, resulting in dissolution of desmosomes. However, to date cholesterol sulfate has been found to be a specific inhibitor of only one serine protease, acrosin (Burck and Zimmerman, 1980).

Finally, Lambeth *et al.* (1987) and Xu and Lambeth (1989) have provided evidence that in adrenal mitochondria, cholesterol sulfate modulates steroid hormogenesis by regulating intramitochondrial cholesterol transport. Though no application to epidermal function is immediately apparent, this observation underscores the evolution in conceptualizing the functional role of sterol sulfates, beyond their traditional assignment as a precursor reservoir for steroid hormone biosynthesis.

5. *Cholesterol Sulfate in Membranes*

The ability of cholesterol sulfate to modify the physical properties of membrane lipids may underlie its proposed function as a stabilizer of membranes. Cholesterol sulfate stabilizes red blood cell membranes, rendering them more resistant to osmotic injury (Bleau *et al.*, 1974, 1975; Lalumiere *et al.*, 1975). Abraham *et al.* (1987) prepared liposomes composed of lipids somewhat similar to the lipid composition of stratum corneum membranes; in their studies, cholesterol sulfate promoted liposomal aggregation in the presence of calcium. Recently, Cheetham *et al.* (1990) have observed that cholesterol sulfate inhibits fusion of Sendai virus envelopes with erythrocyte and liposomal membranes. Moreover, cholesterol sulfate, when added to phospholipid (dielaidoylphosphatidylethanolamine) vesicles, raised the temperature of lipid phase transition. Using differential scanning calorimetry (DSC), Rehfeld*et al.* (1986) demonstrated that cholesterol sulfate, unlike cholesterol, does not form a eutectic mixture when mixed with equimolar concentrations of palmitic acid, which also resulted in an elevated phase transition temperature for cholesterol sulfate-containing mixtures. Subsequently, the lipid phase transitions of normal versus RXLI scale also were examined (Rehfeld *et al.*, 1988). Whereas in normal scale, and in lipid extracts of normal stratum corneum, discrete lipid phase transitions were observed in the physiologic range (Rehfeld and Elias, 1982), these were not apparent in RXLI scale (Rehfeld *et al.*, 1988). Physiologic thermal transitions were evident in both RXLI and normal scale when samples were examined by electron-spin resonance spectroscopy using a highly lipophilic spin probe, perdeuterated di-*t*-butyl nitroxide, however, the lipid environment differed, i.e., the probe partitioned to a more nonpolar environment in RXLI scale (Rehfeld *et al.*, 1988). Furthermore, these differences could be largely reproduced when model mixtures of cholesterol/fatty acid/cholesterol sulfate, in the proportions approximating those found in normal versus RXLI scale, were similarly examined. Hence, Rehfeld *et al.* (1988) proposed that differences in the hydrogen-bonding characteristics of cholesterol-enriched versus cholesterol sulfate-enriched membranes could account for differences in thermal behavior and lipid microenvironment (Fig. 5).

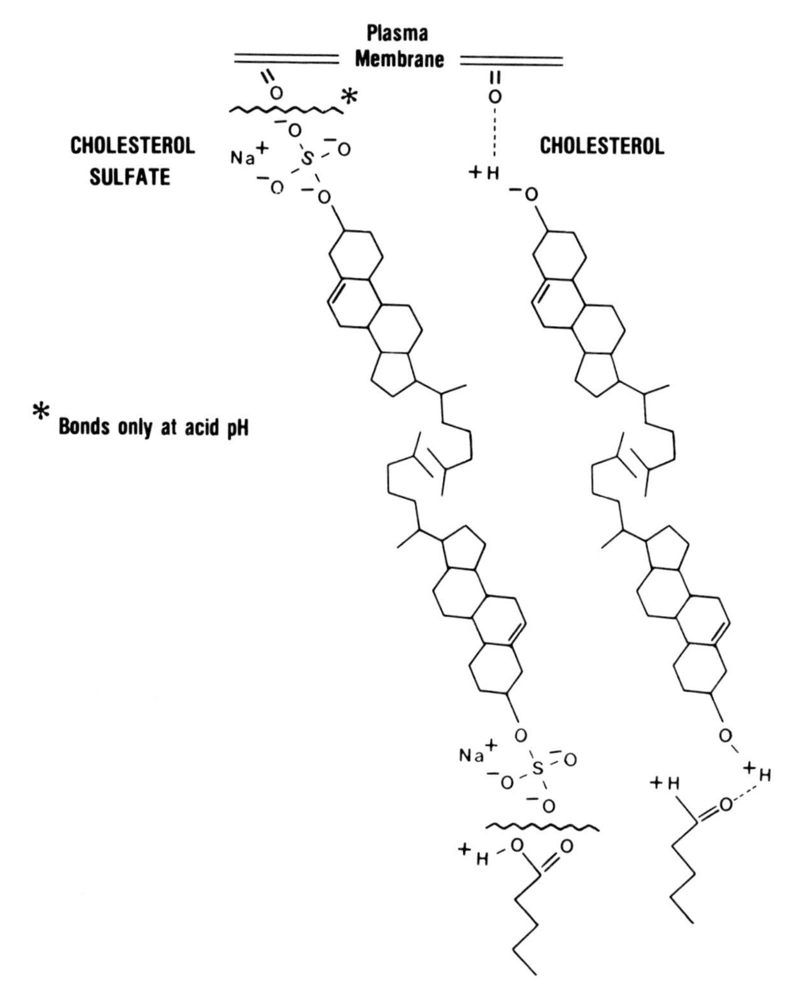

FIG. 5. Model of possible differences in hydrogen bonding between cholesterol sulfate and cholesterol with the carboxyl moieties of fatty acids in stratum corneum membranes. [Reproduced from Rehfeld *et al.* (1988). *J. Invest. Dermatol.* **91,** 499–505, with permission of the Society for Investigative Dermatology.]

6. *Summary*

The identification of steroid sulfatase deficiency as the cause of recessive X-linked ichthyosis has provided a remarkable opportunity to probe the role of an enzyme and its lipid substrates and products in modulating desquamation. Though the full pathogenic sequence remains to be delineated, the observation that the substrate accumulates and product is diminished at the "scene of the crime," i.e., the stratum corneum intercellular membranes, in conjunction with the ability to reproduce the disease in an animal model system, has provided strong evidence directly linking the loss of enzyme function to the disease process. The physical properties of membranes are altered by changes in the cholesterol:cholesterol sulfate ratio. The failure to desquamate in RXLI may then be mediated by the abnormal stability of these cholesterol sulfate-enriched membranes, as has been suggested for other membrane systems, and/or by the persistence of desmosomes into outer stratum corneum, perhaps due to cholesterol sulfate-mediated inhibition of desmosome proteolysis. On the other hand, the similarities between RXLI and the hypocholesterolemic drug-induced scaling disorders discussed previously should be borne in mind, i.e., decreased stratum corneum free sterol content and inhibition of epidermal cholesterol synthesis; cholesterol sulfate is a potent inhibitor of HMG-CoA reductase *in vitro*. In other words, what is missing (sufficient free sterol) may be more critical for normal function and disease pathogenesis than what is present in excess (cholesterol sulfate).

III. Fatty Acids and Acyl Lipids in Desquamation

A. Essential Fatty Acid Deficiency

1. *Introduction*

The dietary requirement for unsaturated fatty acids of the n-3 and n-6 series has been recognized for 60 years (Burr and Burr, 1929). Two species are regarded as the essential fatty acids (EFAs), α-linolenic acid ($C_{18:3}$; n-3) and linoleic acid ($C_{18:2}$; n-6). Through a series of chain elongations and desaturations, arachidonic acid ($C_{20:4}$; n-6) is generated, and subsequently its lipooxygenase and cyclooxygenase metabolites are generated. The skin is a major target of EFA deficiency. Disruption in barrier function develops, indicated by accelerated rates of transepidermal water loss, with scaling and subsequent alopecia. In humans, a similar deficiency syndrome can occur particularly in patients undergoing prolonged lipid-free parenteral nutrition and in infants (for review see Yamanaki *et al.*, 1981). EFA deficiency and the skin have been the subject of several reviews (Prottey, 1976; Elias, 1985; Sherertz, 1986; Hansen, 1986; Ziboh and Chapkin, 1988; Horrobin, 1989).

2. *Experimental EFA Deficiency and the Skin*

That epidermis has a requirement for linoleic acid per se in order to maintain its permeability barrier is well established (for review see Prottey, 1976). Hartop and Prottey (1976) showed that topical linoleic acid and γ-linolenic acid, but not arachidonic acid, dihomo-γ-linolenic acid, or α-linolenic acid could restore barrier function in EFA-deficient rats. Elias *et al.* (1980a) subsequently demonstrated that topical linoleic acid corrects the barrier abnormality in EFA-deficient mice in the presence of pharmacologic blockade of both the cyclooxygenase and lipooxygenase pathways. Furthermore, topical columbinic acid, which undergoes only limited metabolism by either the cyclooxygenase or lipooxygenase pathway, corrects the barrier dysfunction in EFA-deficient rats (Elliott, *et al.*, 1985a,b). Though linoleic acid is found among all epidermal acyl lipids in proportions that reflect serum concentrations (Elias *et al.*, 1979; Lampe *et al.*, 1983b), in three species, the acylglucosylceramide, acylceramide, and acylacid species, linoleic acid is remarkably enriched, where it comprises 50–75% of the ω-esterified fatty acids (Gray *et al.*, 1978; Wertz and Downing, 1982, 1983a,; Bowser *et al.*, 1985). These three linoleate-bearing species share a similar structure in which linoleic acid is esterified to the ω-terminus of a very long-chain (C_{30} to C_{34}) α-hydroxy, monounsaturated fatty acid (see Schurer and Elias, this volume). The hydroxy fatty acid is in amide linkage to the sphingosine base in the acylceramide and acylglucosylceramide.

Whereas the acylacids and acylceramides predominate in the stratum corneum, the acylglucosylceramides are localized to lamellar bodies in the upper nucleated epidermal layers (Gray *et al.*, 1978; Wertz *et al.*, 1984) (see Elias and Menon, this volume). The time course of appearance of radiolabeled linoleic acid is also consistent with the concept that the ceramide and acylacid are derived from the acylglucosylceramide (Bowser *et al.*, 1985; Nugteren *et al.*, 1985). These acylsphingolipids by virtue of their extremely long hydrocarbon chains have been postulated to play a critical role in determining the structure of lamellar granules and the stratum corneum membranes (Wertz and Downing, 1982). In EFA deficiency, oleic acid replaces linoleic acid in these sphingolipid species (Wertz *et al.*, 1983). Elias and Brown (1978) noted distortion of lamellar body architecture in EFA-deficient mice, giving them an empty, "moth-eaten" appearance. Neutral lipid staining of stratum corneum membranes in frozen skin sections was reduced and freeze– fracture replicas demonstrated loss of the broad lamellar sheets found in normal murine stratum corneum. In contrast, Melton *et al.* (1987) observed no ultrastructural alterations in lamellar body or stratum corneum intercellular membranes in EFA-deficient neonatal pig epidermis. However, loss of barrier function was strongly correlated with replacement of linoleic acid by oleic acid in acylsphingolipids (Melton *et al.*, 1987). Menton (1970) also reported normal lamellar body ultrastructure in a study of EFA-deficient mice, but the degree of barrier dysfunction was not stated. The explanation, advanced by Melton *et al.* (1987), that the distorted lamellar granules described by Elias and Brown (1978) were simply tangentially sectioned organelles, would not account for the altered freeze–fracture replicas and loss of neutral lipid staining they also observed.

Whether or not these differences can all be attributed to a more profound deficiency in the neonatal murine model is unclear (Wertz and Downing, 1986).

The introduction of ruthenium tetroxide methodology for the ultrastructural visualization of stratum corneum membrane domains (Madison *et al.*, 1987) has permitted evaluation of the effects of EFA deficiency on the integrity of these membrane structures. Although no abnormalities were described in the membranes of ruthenium-stained sections of porcine EFA-deficient stratum corneum (Melton *et al.*, 1987), Hou *et al.* (1991), using a modification in this technique, found several membrane abnormalities in EFA-deficient neonatal mice. In some regions there was an absence of all intercellular lamellae, except for the lamellae presumed to represent sphingolipid covalently linked to the cornified envelope (Swartzendruber *et al.*, 1987); these regions would correlate with the previous observation of reduced numbers of bilayer leaflets in freeze–fracture replicas of EFA-deficient murine epidermis (Elias and Brown, 1978). In other regions there were increased numbers of lamellae that also displaced alterations in the substructure of individual lamallae (Hou *et al.*, 1991). The freeze–fracture technique might not detect such regions because of its propensity to fracture through the most hydrophobic domain, which in this instance would comprise the remaining covalently bound, most proximal lamella (Rehfeld *et al.*, 1990; Hou *et al.*, 1991). The regions of decreased membrane leaflets may result in the formation of pores through which body water can transpire (Hou *et al.*, 1991). Though the evidence that linoleic acid is required to correct the barrier defect in EFA deficiency is convincing, the hypothesis that substitution of oleic acid ($C_{18:1}$) for linoleic acid ($C_{18:2}$) in the ω-position of the acylceramide constitutes the "molecular defect" in EFA-deficient stratum corneum (Wertz *et al.*, 1983) rests largely on circumstantial evidence. And the multiplicity of ultrastructural changes in murine EFA-deficient stratum corneum is difficult to account for on this basis alone (Hou *et al.*, 1991). An alternate hypothesis has been put forward by Nugteren *et al.* (1985) and Nugteren and Kivits (1987), who propose that linoleic acid is oxygenated by 15-lipooxygenase to 13-hydroxyoctadecadienoic acid, which functions as a signal to regulate epidermal differentiation/proliferation. They note that columbinic acid, a homologue of linoleic acid that undergoes only limited metabolism by either the cyclooxygenase or lipooxygenase pathways (Elliott, *et al.*, 1985a), can also repair the barrier in EFA deficiency and is also susceptible to this lipooxygenase (Nugteren and Kivits, 1987). However, Elias *et al.* (1980a) demonstrated barrier recovery with topical linoleic acid in the presence of lipooxygenase blockade by 5,8,11,14-eicosatetraeonic acid.

The scaling abnormality in EFA deficiency can be corrected by topical prostaglandin E_2 (Ziboh and Hsia, 1972). Similarly, topical eicosatrienoic acid ($C_{20:3}$; n-9), a potent inhibitor of both the lipooxygenase and cyclooxygenase pathways, induces a scaly, hyperproliferative epidermis in EFA-replete hairless mice (Nguyen *et al.*, 1981). Moreover, correction of the scaling abnormality in experimental EFA deficiency appears to be dissociated from correction of the barrier defect. Thus, Prottey (1977) reported that, though topical linoleic acid rapidly corrected the barrier abnormality in EFA-deficient rats, it did not correct the scaling disorder, and, conversely, that topical arachidonic acid rapidly corrected the scaling but not the barrier dysfunction. Houtsmuller (1981) reported that topical arachidonic acid did correct the barrier function; however, Hansen *et al.* (1986) demonstrated that when the barrier defect in EFA-deficient rats was corrected by systemically administered [^{14}C]arachidonic acid, the epidermal acylceramides and acylglucosylceramides were labeled only by [^{14}C]linoleic acid, indicating that retroconversion from arachidonic acid to linoleic acid had occurred. Similarly, with more prolonged applications, topical linoleic acid also corrected the scaling abnormality (Hartop *et al.*, 1978). Elliott *et al.* (1985a,b) observed that the lipooxygenase product of columbinic acid reversed the scaling abnormality in EFA-deficient rats, but did not correct the barrier defect, and its cyclooxygenase product corrected neither.

The scaling abnormality in EFA deficiency has been attributed to a hyperproliferative state (Lowe and DeQuoy, 1978). These authors observed that topical linoleic acid corrected the hyperkeratosis and normalized epidermal DNA synthesis, prior to complete restoration of epidermal prostaglandin content, thus suggesting that alterations in prostaglandin metabolism in EFA deficiency did not account for the epidermal hyperplasia. Recently, Proksch *et al.* (1991b) have shown that it may be the barrier defect in EFA deficiency that drives epidermal hyperplasia; i.e., when EFA-deficient mice are enclosed in a vapor-impermeable wrap, in order to artificially restore their barrier without correcting the underlying EFA-deficient state, epidermal DNA synthesis is normalized. Moreover, in these studies both topical linoleic acid and columbinic acid, but not prostaglandin E_2 normalized DNA synthetic rates.

3. *EFA Deficiency in Dermatologic Disease*

In light of the profound effects of EFA deficiency on epidermal function, it is not surprising that perturbations in EFA metabolism have been proposed in the pathogenesis of a number of dermatologic disorders (for review see Horrobin, 1989). In acne, the initial lesion is hyperkeratosis of the follicular orifice leading to obstruction of the duct. Recently, Downing *et al.* (1986) have proposed that the increased quantity of sebum within the follicle at puberty decreases the effective concentration of linoleic acid (sebum is relatively poor in linoleic acid), resulting in a localized deficiency of linoleic acid in the follicule lumen (see Stewart and Downing, this volume). Such focal EFA deficiency would result in follicular hyperkeratosis, the initial lesion in acne. However, it seems likely that the follicular epithelium, like the epidermis, also obtains EFA from serum lipids, and thus should be protected from "localized" EFA deficiency in this manner.

In psoriasis, a genetically determined skin disorder characterized by focal or generalized, reversible epidermal hyperproliferation, the content of free arachidonic acid in involved epidermis is increased, due to increased phospholipase A_2 activity (Forster *et al.*, 1985), with a relative or absolute shift toward the lipooxygenase rather than cyclooxygenase metabolism (for review see Kragballe and Voorhees, 1985). Perhaps most intriguing is the proposed link, initially advanced by Hansen (1937) and recently refined by Horrobin (1989) and Melnick and Plewig (1989), between atopic dermatitis and abnormal EFA metabolism. A deficiency in Δ^6-desaturase activity, the enzyme that converts linoleic acid to γ-linolenic acid ($C_{18:3}$; n-6), the precursor to leukotrienes and prostaglandins, has been inferred from the pattern of n-6 and n-3 polyunsaturated fatty acids found in sera of atopic patients by several investigators (for review see Horrobin, 1989). Melnik and Plewig (1989) have constructed an elaborate framework linking the proposed Δ^6-desaturase deficiency and consequent effects on prostaglandins and leukotreine metabolism to the immunologic aberrations of the atopic state. Support for this hypothesis has come from clinical studies noting an improvement in atopic dermatitis patients treated systemically with primrose oil, enriched in γ-linolenic acid, which bypasses the Δ^6-desaturase step, but not all clinical studies have observed positive results (for review see Horrobin, 1989). Finally, when considering this hypothesis, it should be borne in mind that a deficiency in Δ^6-desaturase has only been inferred from blood metabolite levels and has not yet been directly demonstrated. It may be relevant here to note that epidermis lacks both the Δ^6-desaturase and the Δ^5-desaturase (Chapkin and Ziboh, 1984; Ziboh and Chapkin, 1988), phenomena that may teleleologically derive from the epidermal requirement to conserve linoleic acid for barrier lipids. As in psoriasis, phospholipase A_2 activity is increased in atopic dermatitis epidermis (Schafer and Kragballe, 1991).

Most recently, the serum fatty acid profile in Darier's disease, an autosomal dominant disorder characterized by discrete hyperkeratotic lesions associated with premature, aberrant cornification, is reported to be deficient in Δ^6-desaturase metabolities, a profile comparable to that associated with atopic dermatitis (Oxholm *et al.*, 1990). However, Grattan *et al.* (1990) have also reported deviations in blood Δ^6 fatty acids in patients with ichthyosis vulgaris, a genetic disorder most probably linked to abnormal filaggrin metabolism (see below) rather than to abnormal lipid metabolism, and similarly abnormal profiles were found in acne and psoriasis. In each disorder the pattern of polyunsaturated metabolities was different, which led the authors to suggest that each might be associated with a unique abnormality in EFA metabolism. However, the actual differences in EFA composition that they observed were modest and may only be distantly related to the underlying defect(s). An instructive analogy here may be the observation of Hernell *et al.* (1982), who reported an altered serum fatty acid profile in patients with Sjögren–Larsson syndrome from which decreased Δ^6-desaturase activity also was inferred. Later, however, Sjögren–Larsson syndrome was shown to be due to fatty alcohol oxidoreductase deficiency (Rizzo *et al.*, 1988, 1989; see below).

4. *Summary*

Experimental EFA deficiency has provided an extremely useful model for exploring the role of stratum corneum lipids in barrier function and desquamation. EFA deficiency leads to alterations in the fatty acid composition of stratum corneum lipids, including substitution of oleic acid substitutes for linoleic acid in certain epidermis-specific acylsphingolipids. EFA deficiency is accompanied by marked disturbances in the ultrastructure of stratum corneum membranes, at least in the murine model. The permeability barrier defect in EFA deficiency is attributed to deficiency of linoleic acid in stratum corneum membrane lipids, rather than to deficiency of a cyclooxygenase or lipooxygenase metabolite, whereas the scaling disorder in EFA deficiency appears to arise through epidermal hyperproliferation, which may be driven through deficiency of a cyclooxygenase or lipooxygenase metabolite of arachidonic acid and/or by defective barrier function. Clinical EFA deficiency is rare, particularly since the introduction of intravenous lipid supplements during prolonged parenteral nutrition. And though abnormal EFA metabolism has been invoked as the cause of a variety of common skin diseases, including acne, atopic dermatitis, and psoriasis, as well as some rare disorders of cornification, such as Dariers disease, each of these attributions awaits definitive proof.

B. Inherited Defects of Acyl-Lipid Metabolism with Abnormal Cornification

1. *Refsum Disease*

Refsum disease is a rare, autosomal recessive disorder characterized by abnormal tissue accumulation of the branched-chain fatty acid, phytanic acid (3,7,11,15-tetramethylhexadecanoic acid) (Refsum, 1946; Klenk and Kahlke, 1963; Richterich *et al.*, 1965; for review see Steinberg, 1989), which is derived from phytol, a component of chlorophyll: The disorder is attributed to deficiency of the enzyme, phytanic acid oxidase (Steinberg *et al.*, 1967a,b; Herndon *et al.*, 1969; Poulos *et al.*, 1984). Refsum disease is characterized by progressive, fluctuating neurologic signs, particularly cerebellar ataxia and peripheral neuropathy, and by retinitis pigmentosa and ichthyosis. The disorder of cornification is not congenital: Indeed, the onset is characteristically delayed until late childhood or thereafter, and it follows rather than precedes neurologic signs. The scaling abnormality improves with dietary restriction of phytanic acid (Sahgal and Olsen, 1975).

The epidermal changes have been studied in only a few patients. In addition to hyperkeratosis, the epidermis shows acanthosis and hypergranulosis (Dykes *et al.*, 1978; Davies *et al.*, 1978) or hypogranulosis (Blanchet-Bardon *et al.*, 1978). Epidermal hyperproliferation was confirmed in one patient by epidermal kinetic studies (Davies *et al.*, 1978). Cytoplasmic lipid vacuoles are seen in the basal layer of the epidermis both by light and electron microscopy (Davies *et al.*, 1978; Blanchet-Bardon *et al.*, 1978). Finally, desmosomal structures reportedly persist into the outer stratum corneum layers (Blanchet-Bardon *et al.*, 1978). As in plasma and other tissues, the epidermal content of phytanic acid was massively increased in one untreated patient, studied just prior to death, accounting for 45, 20, 72, and 35% of long-chain fatty acids in the phospholipid, free fatty acid, triacylglycerol, and sterol ester fractions, respectively (Dykes *et al.*, 1978).

Several hypotheses have been advanced to account for tissue toxicity in Refsum disease (for review see Steinberg, 1989): (1) molecular distortion consequent to substitution of branched-chain for straight-chain fatty acids in tissue lipids; (2) interference with the functions of vitamins E and K due to their structural similarities to phytanic acid; (3) relative deficiency of arachidonic/linoleic acids resulting from substitution by phytanic acid in phospholipics; and (4) a parallel deficiency in α-hydroxylation of straight-chain fatty acids. Because α-hydroxy fatty acids are constituents of the acylceramides and acylglycosylceramides, postulated to be critical for the formation of the stratum corneum intercellular bilayers and lamellar body membranes, respectively (see Elias and Menon, this volume), deficiency in the synthesis of these fatty acids could have profound effects on stratum corneum function. However, Blass *et al.* (1968) found the quantity of skin α-hydroxy fatty acids in two patients to be within the lower limits of normal. Moreover, the delayed onset of ichthyosis in Refsum disease is more consistent with an effect of phytanic acid accumulation.

2. *Sjögren–Larsson Syndrome*

The Sjögren-Larsson syndrome is an autosomal recessive disorder characterized by spasticity, mental retardation, and ichthyosis (Sjögren and Larsson, 1957; Thiele, 1974). The disorder predominates within an inbred population in Northern Sweden, but less frequently cases have been observed in other populations (Jagell *et al.*, 1981). The cutaneous phenotype of Sjögren–Larsson syndrome has been well described, and the scaling appears to result, at least in part, from hyperproliferation (Jagell and Liden, 1982).

Rizzo *et al.* (1987) have described a metabolic cycle in normal fibroblasts whereby fatty alcohols, derived through reduction of the corresponding fatty acyl-CoA derivative by fatty acyl-CoA reductase, are reoxidized to fatty acid through a separate enzymatic reaction catalyzed by fatty alcohol oxidoreductase. Recently, Rizzo *et al.* (1988, 1989) have demonstrated that Sjögren–Larsson syndrome is due to deficiency of fatty alcohol oxidoreductase (fatty alcohol: nicotinamide adenine dinucleotide oxidoreductase) activity. The observed improvement in the cutaneous signs when dietary lipid is restricted to medium-chain triglycerides (Hooft *et al.*, 1967; Guilleminault *et al.*, 1973), in conjunction with the demonstration of deficient fatty alcohol reductase activity in keratinocytes cultured from an affected patient (W. B. Rizzo, 1990 personal communication), provide evidence for a link between the enzyme deficiency and the disorder of cornification. It should be noted, however, that not all patients with this phenotype display deficiency of fatty alcohol oxidoreductase (Koone *et al.*, 1990).

Fatty alcohols are constituents of ether lipids (e.g., plasmalogens) and wax esters. Although cutaneous wax esters are synthesized predominantly by sebaceous glands, this site is unlikely to be the pathogenic locus for the disordered desquamation in the Sjögren–Larsson syndrome, because disease expression begins during infancy and continues during childhood, a time when sebaceous glands are normally quiescent. A more direct result of fatty alcohol oxidoreductase deficiency on epidermal lipid content and metabolism is likely to be the cause, but the substrate–product relationship remains to be defined.

3. *Neutral Lipid Storage Disease*

Chanarin *et al.* (1975) provided the first comprehensive description of neutral lipid storage disease (NLSD), although patients with this syndrome had been described previously (Jordans, 1953; Rozenszajn *et al.*, 1966; Dorfman *et al.*, 1974). This autosomal recessive disorder is characterized by widespread tissue storage of triacylglycerol within non-membrane-bound cytoplasmic droplets. Myopathy and ichthyosis are the most constant clinical signs, with neurosensory deafness, cataracts, and fatty liver, as well as developmental and growth retardation occurring in some patients (for review see Williams *et al.*, 1984).

Although the underlying metabolic defect remains to be defined, studies of NLSD fibroblasts by three groups point to a defect in intracellular triacylglycerol lipolysis (DiDonato *et al.*, 1988; Salvayre *et al.*, 1989; Williams *et al.*, 1991a) (Fig. 6). Because a defective lipase has not been demonstrated directly, this conclusion has been reached through a process of exclusion. The stored lipid is predominantly triacylglycerol, composed of an unremarkable fatty acid profile (Chanarin *et al.*, 1975; Williams *et al.*, 1988c). A primary defect in serum lipoprotein metabolism has been excluded because serum lipoproteins are essentially normal (Chanarin *et al.*, 1975; Williams *et al.*, 1984) and because triacylglycerol storage occurs even when fibroblasts are cultivated in lipid-free media (Williams *et al.*, 1988c). Conflicting data have been reported with regard to fatty acid β-oxidation in NLSD fibroblasts. Angelini *et al.* (1980) noted progressive impairment in conversion of [^{14}C]palmitate to CO_2, but they employed prolonged incubation periods of at least several hours. No defect in fatty acid oxidation is evident when a brief (60 minutes) incubation period is employed (Williams *et al.*, 1988c). Therefore, it is likely that longer pulse periods trap radiolabel progressively in a

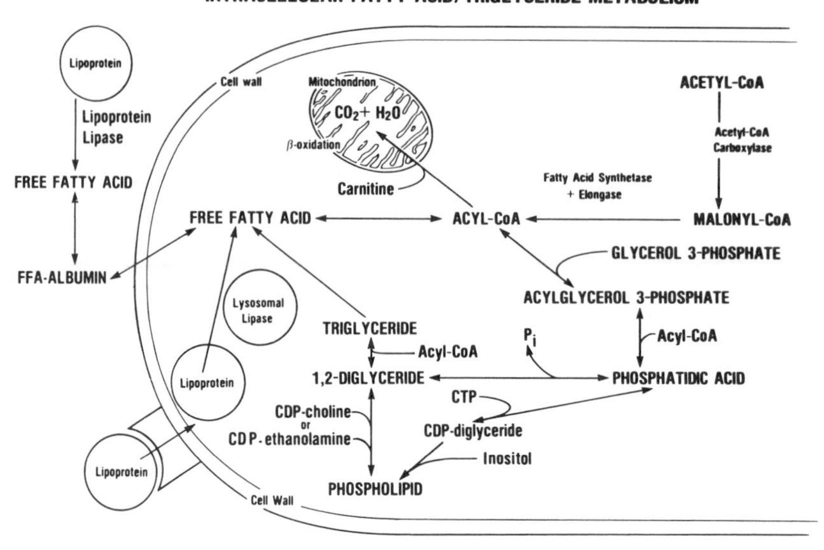

FIG. 6. Diagrammatic representation of the predominant intracellular pathways of fatty acid and triacylglycerol metabolism. Free fatty acids (FFA) may be liberated from lipoprotein triacylglycerol through the action of lipoprotein lipase extracellularly, or by lysosomal acid lipase after uptake of the intact lipoprotein particles. Intracellular triacylglycerol derived from *de novo* synthesis is metabolized by an alternate pathway (see text).

nonmetabolizable triacylglycerol pool (see below). DiDonato *et al.* (1988) also reported normal β-oxidation of long-chain, as well as short- and medium-chain, fatty acids. Moreover, palmitate oxidation in the presence of cyanide is also preserved in NLSD fibroblasts, implying normal peroxisomal fatty acid oxidation as well (DiDonato *et al.*, 1988).

When NLSD fibroblasts are pulsed with either radiolabeled or fluorescently tagged fatty acid or acetate, label progressively accumulates in triacylglycerol (Radom *et al.*, 1987a,b; DiDonato *et al.*, 1988; Salvayre *et al.*, 1989; Williams *et al.*, 1991a). Though these studies point to a primary defect in triacylglycerol catabolism, no impairment in triacylglycerol lipase activity has been demonstrated in NLSD cell homogenates (DiDonato *et al.*, 1988; Salvayre *et al.*, 1989; Williams *et al.*, 1988c, 1991a); hence a functional lipase deficiency within the intact cell, as may arise from deficiency of an activator or transport protein, has been proposed (Williams *et al.*, 1991a). This hypothesis is supported by the exclusion of alternative mechanisms for triacylglycerol storage, such as glycerolipid oversynthesis or alkylglyceride accumulation from a primary defect in ether lipid metabolism (Williams *et al.*, 1991a).

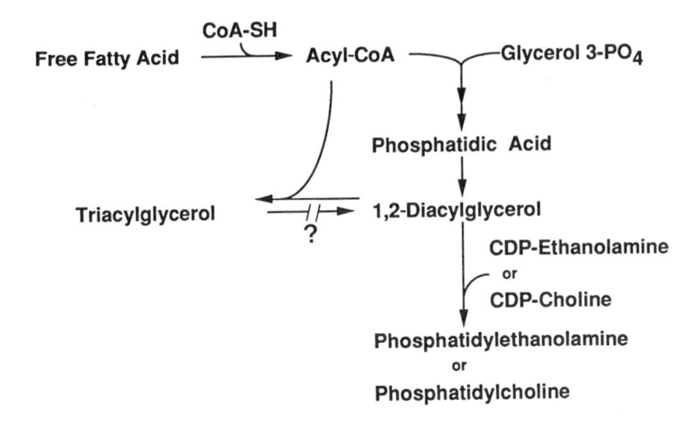

FIG. 7. Proposed site of metabolic block in neutral lipid storage disease. *De novo*-synthesized triacylglycerol normally provides a storage pool of diacylglycerol intended for phosphatidylcholine and phosphatidylethanolamine biosynthesis. In NLSD, *de novo*-synthesized triacylglycerol is formed but cannot be hydrolyzed. Whether this is due to deficiency per se of an intracellular lipase, aberrant enzyme localization, or deficiency of an activator or triacylglycerol apoprotein is unknown. [Reproduced from Williams *et al.* (1991a). *Biochim. Biophys. Acta* (in press), with permission of Elsevier Science Publishers, B.V.]

Though the precise nature of the lipolytic defect remains to be defined, comparisons of the fate of endogenously synthesized versus exogenously supplied triacylglycerol in NLSD versus lysosomal acid lipase-deficient (Wolman's disease) fibroblasts have provided evidence for compartmentalization of these two triacylglycerol metabolic pathways (Radom *et al.*, 1987a,b; Salvayre *et al.*, 1987, 1989; DiDonato *et al.*, 1988) (Fig. 6). In Wolman's disease, exogenously derived (i.e., lipoprotein) triacylglycerol accumulates due to a deficiency of lipase in the lysosomal pathway, whereas in NLSD only endogenously synthesized triacylglycerol accumulates. Although the affected lipase in NLSD has not been defined, the notion of a nonlysosomal, cytosolic lipolytic pathway agrees with prior descriptions of nonlysosomal, acid (Coleman and Haynes, 1983) or neutral (Beaudet *et al.*, 1974; Riddle *et al.*, 1977; Oram *et al.*, 1980; Hoeg *et al.*, 1982; Lengle and Geyer, 1973; Messieh *et al.*, 1983) lipase activities in cultured fibroblasts.

The final step in the predominant biosynthetic pathway for both phosphatidylcholine and phosphatidylethanolanine involves the condensation of cytidine diphosphate-choline (CDP-choline) (or CDP-ethanolamine) with diacylglycerol (for reviews see Bell and Coleman, 1980; Tijburg *et al.*, 1989) (Fig. 7). Hence endogenously synthesized triacylglycerol may function as a reservoir of diacylglycerol intended for phospholipid synthesis; release of diacylglycerol in close proximity to phospholipid biosynthetic enzymes may be essential to prevent activation of protein kinase C in other cell compartments. When NSLD cell lines are pulsed with radiolabeled acetate or oleic acid, the earliest observed abnormality is increased incorporation into phosphatidylcholine and phosphatidylethanolamine (Williams *et al.*, 1991a). Thus, Williams *et al.* (1991a) have postulated that the affected lipase in NLSD is an enzyme that is normally linked to phosphatidylcholine and phosphatidylethanolamine biosynthesis (Fig. 7). This hypothesis is consistent with data from a variety of cultured cell types that have demonstrated a labile pool of newly synthesized triacylglycerol, with radiolabeled fatty acids chasing from triacylglycerol to phospholipid (Elsbach and Farrow, 1969; Cook *et al.*, 1982; Goppelt *et al.*, 1985; Williams *et al.*, 1988d; Schürer *et al.*, 1989). Moreover, Cook and Spence (1985) have demonstrated transfer of the glycerol moiety from triacylglycerol to phospholipid in cultured neuroblastoma cells. One would anticipate such an enzyme to be active in virtually all tissues, an expectation consistent with the widespread organ involvement in NLSD (for review see Elias and Williams, 1985).

As in other tissues, lipid storage is evident in many cell types of the skin, including basal and spinous cell keratinocytes, melanocytes, pilosebaceous and eccrine structures, fibroblasts, mast cells, endothelial cells, and Schwann cells (for review see Elias and Williams, 1985). In addition to cytoplasmic lipid vacuoles in the lower epidermal layers, the number of lamellar bodies appear to be increased in the granular cell layer and their contents are distended by electron-lucent material, which may represent triacylglycerol (Elias and Williams, 1985). After secretion into the intercellular spaces of the stratum corneum, these droplets coalesce, displacing the lamellar material. Vacuoles are also retained within corneocytes. Freeze–fracture replicas similarly demonstrate numerous droplets within lamellar bodies in the stratum granulosum and within the intercellular domains at the stratum granulosum–stratum corneum interface (Elias and Williams, 1985). Although scale lipids were examined in one patient, they did not demonstrate the expected increase in triacylglycerol content. Instead, the proportion of triacylglycerol and free fatty acids was decreased and the proportion of *n*-alkanes increased, a finding similar to that observed in a phenotypically similar disorder of cornification, congenital ichthyosiform erythroderma (Williams and Elias, 1985). However, recent C^{14} dating studies suggest that alkanes in skin are not endogenously derived (Bortz *et al.*, 1989; see below).

4. *Summary*

To date, three genetic disorders of acyl-lipid metabolism that have been described exhibit abnormalities in desquamation. Though these associations provide evidence for the importance of fatty acids and related lipids for epidermal desquamation, little is known regarding the pathogenesis of the disordered cornification for any of these conditions.

IV. Possible Lipid Metabolic Defects in Other Cornification Disorders

The foregoing sections have reviewed those examples of abnormal desquamation linked to discrete abnormalities in epidermal lipid content and/or metabolism. Recent studies also have pointed to potential lipid abnormalities in a number of other genetically transmitted disorders of cornification.

A. Congenital Ichthyosiform Erythroderma and Epidermal Alkanes: Source and Significance

Alkanes are major constituents of plant and insect cuticles, in which they largely mediate barrier function (for reviews see Hadley, 1981, 1989; Lockey, 1985) (see Hadley, this volume). Though long-chain and very long-chain aliphatic hydrocarbons (*n*-alkanes) have been identified in both the stratum corneum and skin surface lipids of mammals (Haati, 1961; Lampe *et al.*, 1983b), because of the accessibility of the skin surface to environmental contamination and because of the absence of known enzymatic pathways in mammals for the generation of alkanes, they have been attributed to surface petroleum-product contamination (Gloor *et al.*, 1974; for reviews see Downing, 1976; Lester, 1979). With the discovery by Williams and Elias (1982, 1984) that scales from patients with certain disorders of cornification contain large quantities of *n*-alkanes (comprising up to 45% of scale lipids), the issue of an endogenous source of these hydrocarbons in skin was reopened. Moreover, the observations that pathological scale with a high alkane content also displayed a correspondingly reduced triacylglycerol and free fatty acid content but a normal total lipid weight percentage suggested a product–precursor relationship (Williams and Elias, 1982). Subsequently, the observation that a distinctive lipid profile distinguished biochemically between two phenotypically distinct patient groups, congenital ichthyosiform erythroderma (CIE; possessing increased scale alkanes) and lamellar ichthyosis (possessing normal scale alkanes), provided further impetus to the notion that these alkanes might have an endogenous origin (Williams and Elias, 1984, 1985a). However, Bortz *et al.* (1989) were unable to identify any alkanes of modern age in pooled skin surface lipids from normal volunteers. Though these studies left open the possibility that alkane-containing emollients used by one subject [topical emollients are commonly petroleum derived and thus contain predominantly very long-chain alkanes (Williams and Elias, 1985a)] could have obscured the presence of small quantities of endogenously generated alkanes (Elias and Williams, 1990), we have subsequently confirmed their findings in the alkane fractions from two CIE patients using a highly sensitive method for ^{14}C dating (Turteltaub *et al.*, 1990; M. L. Williams *et al.*, 1990 unpublished observations). Thus, it may be concluded that scale alkanes from both normal and pathological skin have an exogenous origin.

Yet, several questions remain regarding the presence and significance of alkanes in pathological scale. Large quantities of alkanes (i.e., approximately 25% of scale lipids) were obtained (1) from sites that had been left untreated with topical emollients for 4 weeks prior to scale collection, (2) from sites that were continuously covered with a plastic wrap, and (3) from sites treated with a non-alkane-based emollient (Williams and Elias, 1985a). Because CIE is a hyperproliferative epidermal disorder (Frost *et al.*, 1966; Hazell and Marks, 1985), prolonged retention of scales contaminated by alkane-based emollients seems an unlikely explanation for these observations. Yet, topically applied medicaments may be identified on body sites far removed from their original application (Johnson *et al.*, 1983). Ghadially *et al.* (1991) have recently observed that alkanes (Vaseline), when topically applied to hairless mice, penetrate through stratum corneum intercellular domains, displacing the membrane bilayers. And in CIE, significant quantities of *n*-alkanes (i.e., ~10% of total lipid) were recovered from the stratum granulosum cell layers (Williams and Elias, 1984). Because long-chain alkanes are potent stimulators of epidermal hyperplasia (Peters and White, 1980; Harris *et al.*, 1980; Elias *et al.*, 1985), these exogenous lipids may play a role in disease pathogenesis. Finally, the apparent preferential retention of exogenous alkanes by scale in some disorders of cornification, but not in others, remains an intriguing, unexplained association.

Though a metabolic defect resulting in alkane accumulation can no longer be considered the cause of CIE, several ultrastructural studies have suggested an underlying disorder of lipid metabolism (Anton-Lamprecht, 1972; Arnold *et al.*, 1988; Kanerva *et al.*, 1983; Niemi and Kanerva, 1989; Ghadially *et al.*, 1990). Anton-Lamprecht and others (for review see Arnold *et al.*, 1988) have described ultrastructural criteria that define three subsets of patients: (1) numerous intercellular lipid droplets in stratum corneum; (2) crystalloid inclusions, which they postulate represent cholesterol crystals; and (3) unusual membranous inclusions. Recently, Ghadially *et al.* (1990), using ruthenium tetroxide to visualize stratum corneum membrane domains ultrastructurally, described a variety of alterations in CIE, including phase separation, disorganization of lipid lamellae, and, in some patients, unusual alterations in the substructure of individual lamellae. Most patients also exhibited lipid inclusions within corneocytes. Some also displayed crystalloid inclusions, similar to those described by Anton-Lamprecht in ichthyosis congenita (type II), and finally, some patients displayed atypical, distorted lamellar bodies. Until sufficient longitudinal studies have established the stability of these ultrastructural changes over time, and pedigree studies have confirmed the uniformity of these features within kindreds, it may be premature to assert that all of these ultrastructural differences reflect genetic heterogeneity. Nor can one assume that these changes indicate an underlying lipid metabolic defect. It is possible that lipid vacuoles simply reflect epidermal hyperproliferation, because similar vacuoles also are present in psoriatic stratum corneum (Brody,

1962) and in "differentiated" keratinocyte cell cultures (Williams *et al.*, 1988a; Bodde *et al.*, 1990) (see Ponec, this volume). Alternatively, Menon *et al.* (1991b) has proposed that both the secretion of lamellar bodies and their complement of hydrolytic enzymes may be deficient in CIE. In this regard, Bergers *et al.* (1990) have described abnormalities in the profile of hydrolytic enzymes in scale from two subsets of these patients, one group exhibiting decreased β-glucosidase and phospholipase activities and the other exhibiting decreased short-chain carboxyesterase activity. Abnormal remodeling of lamellar body-derived lipids could result in the ultrastructural abnormalities in stratum corneum membranes observed by Ghadially *et al.* (1990), whereas a secretory failure could give rise to the retained lipids within corneocytes described by Anton-Lamprecht and others (for review see Arnold *et al.*, 1988). Again, however, it seems equally plausible that alterations in lamellar body content and/or secretion might arise secondarily as a consequence of epidermal hyperproliferation and incomplete differentiation.

B. Harlequin Ichthyosis and Lamellar Body Deficiency

Harlequin ichthyosis is an autosomal recessive disorder characterized at birth by massive platelike scales, which distort facial features and interfere with respiratory movements and feeding. As a consequence, the disorder is usually fatal in the perinatal period. In the few recently described, long-term survivors, however, the phenotype has been observed to change dramatically; after the massive, platelike scales of the perinatal period are shed, a severe exfoliative erythroderma ensues (Lawlor and Peiris, 1985; Williams and Elias, 1987; Rogers and Scarf, 1989; Roberts, 1989). This phenotypic change may occur in response to environmental factors, because the phenotype reverts back to thick, constrictive hyperkeratotic plates when the skin is occluded postnatally (Williams and Elias, 1987).

Several investigations have noted lipid vacuoles, unusual membranous structures, and an absence of normal lamellar bodies ultrastructurally (Buxman *et al.*, 1979; Fleck *et al.*, 1989; Dale *et al.*, 1990b). In the case reported by Fleck *et al.* (1989), large concentric lamellated structures were present in the apical region of upper nucleated cell layers, which were secreted into the intercellular spaces of the stratum corneum, implying that they represented abnormal lamellar bodies. Yet, freeze–fracture replicas displayed little secreted abnormal lamellar material. Instead, most regions of the plasma membrane were occupied by desmosomal structures and structures resembling tight junctions, which were not present in normal stratum corneum, and showed retention of abnormal lamellar structures within the corneocyte. In contrast, Buxman *et al.* (1979) and Dale *et al.* (1990b) did not observe atypical lamellar bodies; instead, they described numerous lipid vacuoles with an absence of lamellar bodies in nucleated cell layers and an absence of intercellular lamellae within the stratum corneum.

Abnormalities in structural proteins have been sought by several investigators. Craig *et al.* (1970) reported an altered X-ray diffraction pattern of stratum corneum, in which the normal α-helical keratin pattern was completely replaced by a cross-β pattern. Though subsequent patients have not shown this diffraction pattern (Baden *et al.*, 1982; Buxman *et al.*, 1979), some express keratin peptides that are associated with epidermal hyperproliferation (Baden *et al.*, 1982; Dale *et al.*, 1990b). The most comprehensive study is that of Dale *et al.* (1990b), who examined in 10 patients the expression of the epidermal keratins and profilaggrin/filaggrin. Whereas six expressed "hyperproliferative" keratins, in the remainder keratin expression was normal. In seven patients, profilaggrin (the major protein of keratohyalin granules) was present in the epidermis but was not processed normally within the stratum corneum to filaggrin, whereas in the remainder little or no profilaggrin could be detected. Whether these marked differences in structural protein expression reflect genetic heterogeneity, as Dale *et al.* (1990b) suggest, or whether they correspond to the dramatic changes in cutaneous phenotype that take place in postnatal survivors (*vide supra*) is unclear.

One may speculate that a deficiency in stratum corneum lipid lamellae results in markedly increased rates of transepidermal water loss upon exposure to the xeric stress of postnatal life. In experimental animals, Proksch *et al.* (1991b) have shown that loss of barrier integrity stimulates epidermal DNA synthesis. Thus, increased transepidermal water loss may trigger epidermal hyperplasia in these infants, which in turn would alter keratin and perhaps profilaggrin expression.

Buxman *et al.* (1979) reported a massive (i.e., 10- to 30-fold) increase in sterol and triacylglycerol content of stratum corneum from one patient studied 9 months postnatally. These data, however, are difficult to interpret because the values for normal stratum corneum sterol and triacylglycerol content that they report are surprisingly low, but the proportion of sterol esters is excessively high. For example, Buxman *et al.* (1979) reported the normal range of stratum corneum from sterols to be 1.7–2.5 mg/g dry weight, of which 85% is esterified. However, stratum corneum is normally 5–10% lipid by weight, of which 12–20% is free sterol (Lampe *et al.*, 1983b; Williams and Elias, 1981); this corresponds to a normal range of 9–17 mg/g dry weight free sterol, depending upon the source (scale versus whole stratum corneum sheet) and body site. Moreover, sterol esters comprise less than 30% of total sterols. Similarly, the normal range for scale or stratum corneum triacylglycerols reported by others (Lampe *et al.*, 1983b; Williams and Elias, 1982) is higher (8–16 mg/g) than Buxman *et al.* (1979) report (0.75–2.0 mg/g). Because they do not report the source of their laboratory normals, it is impossible to account for these discrepancies. However, if their values for sterols and triacylglycerols in their harlequin ichthyosis patient (total sterols, 19.8 mg/g, 8.2% esterified; triacylglycerols, 32 mg/g) are compared to these normal values from the literature, the sterol content may be within the normal range, whereas the triacylglycerol content may be increased twofold, rather than 30-fold. These data may be compatible with an underlying lipid metabolic defect, as the authors suggest, but they could also reflect epidermal hyperplasia, because triacylglycerol accumulation is prominent in the hyperplastic phenotype of cultured keratinocytes (Williams, *et al.*, 1988a,b; Ponec *et al.*, 1989; Bodde *et al.*, 1990) (see Ponec, this volume). Triacylglycerol synthesis is also increased in psoriatic skin (Summerly *et al.*, 1978), and lipid vacuoles are prominent ultrastructurally (Brody, 1962).

In summary, harlequin ichthyosis appears to represent a severe defect in lamellar body organellogenesis, which results in deficient intercellular lamellae in the stratum corneum. Whether this arises as part of a more general abnormality of epidermal differentiation or whether some of the biochemical abnormalities described (e.g., altered profilaggrin processing and expression, altered keratin expression) may occur as part of the pathological response to a defective barrier is unknown.

C. Peroxisomal Deficiency in Chondrodysplasia Punctata/Ichthyosiform Disorders

Peroxisomes are ubiquitous intracellular organelles involved in a variety of metabolic processes, including oxidations, transaminations, and lipid biosynthesis and catabolism (for review see Lazarow, 1987). Recently, a great deal of attention has been directed toward genetic diseases characterized either by deficiency of a single peroxisomal enzyme or by a more global deficiency in peroxisomal functions (for reviews see Schutgrens *et al.*, 1986; Singh *et al.*, 1988). Ichthyosis characteristically occurs in two peroxisomal disorders, the autosomal recessive, rhizomelic chondrodysplasia punctata (Heymans *et al.*, 1986; Hoefler *et al.*, 1988; Poulos *et al.*, 1988; Heikoop *et al.*, 1990) and the X-linked dominant, chondrodysplasia punctata (Conradi–Hünermann syndrome) (Holmes *et al.*, 1987; Kalter *et al.*, 1989). As in the global peroxisomal deficiency disorder, Zellweger syndrome, more than one peroxisomal function is deficient in rhizomelic chondrodysplasia punctata. However, unlike Zellweger syndrome, normal-appearing peroxisomes are present in rhizomelic, chondrodysplasia punctata (Heymans *et al.*, 1986; Hoefler *et al.*, 1988). Plasmalogen synthesis is reduced and phytanic acid oxidation is impaired (Heymans *et al.*, 1986; Poulos *et al.*, 1988; Hoefler *et al.*, 1988); moreover, 3-oxoacyl-CoA thiolase is present in a precursor form (Heikoop *et al.*, 1990) and is aberrantly localized (Balfe *et al.*, 1990). Yet, certain other peroxisomal functions are preserved, including peroxisomal fatty acid oxidation, and as a result, plasma levels of very long-chain fatty acids, pipecolic acid, and bile acid intermediates are not increased (Poulos *et al.*, 1988; Hoefler *et al.*, 1988). Kalter *et al.* (1989) reported deficiency of dihydroxyacetone phosphate acyltransferase (DHAP-AT), a peroxisomal enzyme that catalyzes the first step in plasmalogen synthesis, in a patient with Conradi–Hünermann syndrome, although plasma levels of phytanic acid, as well as very long-chain fatty acids and bile acid intermediates, were normal.

CHILD syndrome (congenital hemilateral ichthyosiform erythroderma with limb defects) is an X-linked dominant trait that shares a number of phenotypic features, including chondrodysplasia punctata, with the Conradi–Hünermann syndrome (Happle *et al.*, 1980). Recently, Williams *et al.* (1991b) have demonstrated global peroxisomal deficiency in fibroblasts taken from the involved skin in CHILD syndrome, whereas fibroblasts from the contralateral, uninvolved skin exhibited normal peroxisomal enzyme functions. Involved skin fibroblasts were deficient in catalase activity and in DHAP-AT. Moreover, the number of peroxisomes in involved fibroblasts was markedly decreased when examined by diaminobenzidine (DAB) ultracytochemistry, though rare DAB^+ structures were identified.

Skin fibroblasts from the involved side also synthesize and secrete large quantities of prostaglandin E_2 into the media (Goldyne and Williams, 1989). Moreover, the growth of involved skin fibroblasts is retarded, but it can be normalized by inhibition of prostaglandin synthesis with indomethacin. Prostaglandin E_2 inhibits the growth of cultured fibroblasts (Goldyne and Williams, 1989), but it stimulates the growth of keratinocytes (Pentland and Needleman, 1986). Thus, these studies raise the possibility that the ichthyosiform erythroderma in involved skin represents epidermal hyperplasia that has been induced by dermal fibroblast prostaglandin E_2 production (Goldyne and Williams, 1989), a hypothesis that may be supported by the recent observation that keratinocytes cultured from involved skin are phenotypically normal (Dale *et al.*, 1990b). Because peroxisomes are involved in the oxidation of both arachidonic acid and prostaglandins, peroxisomal deficiency may be the proximate cause of these perturbations in prostaglandin metabolism.

V. Conclusion: Lipid and Nonlipid Effectors of Desquamation

From the foregoing it is apparent that the circumstantial evidence linking altered epidermal lipid metabolism and abnormal stratum corneum lipid composition to pathological desquamation is strong. Yet, in no instance has the pathogenic sequence been fully delineated. Molecular substitution of one lipid for another may alter the structure and functional properties, such as fluidity and water permeability, of stratum corneum membranes, and may also impact on the function of nonlipid constituents. Desmosomes, which represent sites of cell anchorage within viable epithelial cells, persist as intact transmembrane structures in the inner to mid-stratum corneum on most body sites (for review see Arnn and Staehelin, 1981; Chapman and Walsh, 1990). Desmosomes are particularly numerous on the thick stratum corneum of palms and soles, sites that are also lipid depleted

(Lampe *et al.*, 1983b). Thus, desmosomes may provide for the tighter cohesiveness of the inner stratum corneum layers (King *et al.*, 1979). Conversely, proteolytic degradation of these structures may be a prerequisite for normal desquamation (Allen and Potten, 1975; Lundstromm and Egelrud, 1988; Chapman and Walsh, 1990), which may be mediated in its final stages through the development of clefts and rifts between the remaining lipid bilayers. Abnormal persistence of desmosomes is seen ultrastructurally in several disorders of cornification, including RXLI, CIE, and Refsum disease (Bazex *et al.*, 1978; Anton-Lamprecht and Kahlke, 1974; Menon *et al.*, 1991b). Because proteases are copackaged with lipids and other hydrolytic enzymes within lamellar bodies (Grayson *et al.*, 1985; Elias *et al.*, 1988), failure to form and/or secrete sufficient numbers of "mature" lamellar bodies could alter stratum corneum membrane organization and result in desmosomal persistence, as observed in CIE stratum corneum ultrastructurally (Menon *et al.*, 1989b, 1991b; Ghadially *et al.*, 1991a). Alternatively, stratum corneum lipids may regulate desmosomal degradation through protease inhibition/activation. Such a mechanism might account for the abnormal persistence of desmosomes in RXLI (see Section II,C,1).

Mechanical factors may be important in the final stages of normal desquamation. Intercellular lipids and nonlipid constituents, such as glycoproteins, also, postulated to play a role in stratum corneum cohesion (Brysk *et al.*, 1989), may be extracted from the outer stratum corneum layers during bathing. At the same time, corneocytes take up water and swell, exerting mechanical stresses upon the weakened, intercellular cohesive forces. The ability of corneocytes to take up water is due to their rich endowment in free amino acids, derived in part through proteolysis of the structural protein, filaggrin (Scott, *et al.*, 1982; Scott, 1986; Scott and Harding, 1986). Filaggrin is also postulated to play a role in keratin filament assembly in the inner stratum corneum and is derived from a large precursor protein, profilaggrin, the major constituent of the epidermal keratohyalin granule (for review see Dale *et al.*, 1990c). Ichthyosis vulgaris appears to arise from a genetic defect in the regulation of profilaggrin synthesis (Sybert *et al.*, 1985; Fleckman *et al.*, 1987). In ichthyosis vulgaris, which represents a retention type of hyperkeratosis, absent or reduced filaggrin content may result in insufficient quantities of osmotically active materials in outer corneocytes to exert the mechanical forces required to destabilize intercellular membrane bonds. Whether or not this mechanism accounts for the abnormal desquamation in ichthyosis vulgaris, this example serves as a reminder that not all disorders of cornification can be attributed to abnormalities in stratum corneum lipids.

Whatever the role of lipids in corneocyte cohesion, it is well established that stratum corneum lipids provide the permeability barrier to water loss. Disturbances in stratum corneum lipid content and membrane organization may result in impaired barrier function (e.g., essential fatty acid deficiency; see Section III,A). Because of the critical importance of an intact barrier to survival in a terrestrial environment, a variety of epidermal homeostatic responses to barrier dysfunction would be anticipated. Recently, it has been shown that acute perturbations in barrier function lead to a burst in DNA synthesis, whereas a more prolonged disruption (e.g., EFA deficiency) may result in epidermal hyperplasia (Proksch *et al.*, 1991b). Epidermal hyperplasia in turn results in abnormal desquamation, possibly through flooding of the cornified compartment with incompletely formed, functionally incomplete units. For example, immature lamellar bodies may not all be secreted, whereas those secreted may contain an insufficient complement of lipids and/or hydrolytic enzymes. Thus, the manner in which alterations in stratum corneum lipids induce abnormal desquamation may be complex and indirect.

References

Abraham, W., Wertz, P. W., Landmann, L., and Downing, D. T. (1987). *J. Invest. Dermatol.* **88,** 212–214.

Ahrens, R. A., Dupont, J., and Thompson, M. J. (1965). *Proc. Soc. Exp. Biol. Med.* **118,** 436–440.

Alberts, A. W. (1988). *Am. J. Cardiol.* **62,** 125–155.

Allen, T. D., and Potten, C. S. (1975). *J. Ultrastruct. Res.* **51,** 94–105.

Anderson, P. C., and Martt, J. M. (1965). *Arch. Dermatol.* **92,** 181–183.

Angelini, C., Philippart, M., Borrone, C., Bresolin, N., Cantini, M., and Lucke, S. (1980). *Ann. Neurol.* **7,** 5–10.

Anton-Lamprecht, I. (1972). *Arch. Derm. Forsch.* **243,** 88–100.

Anton-Lamprecht, I., and Kahlke, W. (1974). *Arch. Dermatol. Forsch.* **250,** 185–206.

Arnn, J., and Staehelin, L. A. (1981). *Int. J. Dermatol.* **5,** 330–339.

Arnold, M.-L., Anton-Lamprecht, I., Melz-Rotonfuss, B., and Hartschuh, W. (1988). *Arch. Dermatol. Res.* **280,** 268–278.

Avigan, J., Steinberg, D., Vroman, H. F., Thompson, M. J., and Mosettif, E. (1960). *J. Biol. Chem.* **235,** 3123–3126.

Baden, H. P., Kubilus, J., Rosenbaum, K., and Fletcher, A. (1982). *Arch. Dermatol.* **118,** 14–18.

Balfe, S., Hoefler, G., Chen, W. W., and Watkins, P. A. (1990). *Pediatr. Res.* **27,** 304–310.

Ballabio, A., Parenti, G., Carrozzo, R., Sebastio, G., Andria, G., Buckle, V., Fraser, N., Craig, I., Rocchi, M., Romeo, G., Jobsis, A. C., and Perico, M. G. (1987). *Proc. Natl. Acad. Sci. U.S.A.* **84,** 4519–4523.

Ballabio, A., Carrozzo, R., Parenti, G., Gil, A., Zollo, M., Persico, G., Gillard, E., Affara, N., Yates, J., Ferguson-Smith, M. A., Frants, R. R., Eriksson, A. W., and Andria, G. (1989). *Genomics* **4,** 36–40.

Bazex, A., Bazex, J., Gauthier, Y., and Surleve-Bazeille, J.-E. (1978). *Ann. Dermatol. Venereol.* **105,** 753–756.

Beaudet, A. L., Lipson, M. H., Ferry, G. D., and Nichols, B. L., Jr. (1974). *J. Lab. Clin. Med.* **84,** 54–62.

Bedin, M., Fournier, T., Mouhadjer, and Pointis, G. (1988). *J. Steroid Biochem.* **30,** 439–441.

Bell, R. M., and Coleman, R. A. (1980). *Annu. Rev. Biochem.* **49,** 459–487.

Bergers, M., Traupe, H., Dünnwald, S. C., Mier, P., von Dooren-Greebe, R., Steijlen, P., and Happle, R. (1990). *J. Invest. Dermatol.* **94,** 407–412.

Bergner, A., and Shapiro, L. J. (1981). *J. Clin. Endocrinol. Metab.* **53,** 221–223.

Bergner, A., and Shapiro, L. J. (1988). *J. Inherited Metab. Dis.* **11,** 403–415.

Bick, D., Curry, C. J. R., McGill, J. R., Schorderet, D. F., Bux, R. C., and Moore, C. M. (1989). *Am. J. Med. Genet.* **33,** 100–107.

Bilheimer, D. W., Grundy, S. M., Brown, M. S., and Goldstein, J. L. (1983). *Proc. Natl. Acad. Sci. U.S.A.* **80,** 4124–4128.

Blanchet-Bardon, C. L., Anton-Lamprecht, I., Puissant, A., and Schnyder, U. W. (1978). *In* "The Ichthyosis" (R. Marks and P. J. Davies, eds.) pp. 65–69.

Blass, J. P., Avigan, J., and Steinberg, D. (1968). *Biochim. Biophys. Acta* **187,** 36–41.

Bleau, G., Bodley, F., Longpre, J., Chapdelaine, A., and Roberts, K. D. (1974). *Biochim. Biophys. Acta* **352,** 1–9.

Bleau, G., Lalumiere, G., Chapdelaine, A., and Roberts, K. D. (1975). *Biochim. Biophys. Acta* **375,** 220–223.

Bodde, H. E., Holman, B., Spies, F., Weerhem, A., Kempenar, J., Mommaas M., and Ponec, M. (1990). *J. Invest. Dermatol.* **95,** 108–116.

Bonifas, J. M., Morley, B. J., Oakey, R. E., Kan, Y. W., and Epstein, E. H., Jr. (1987). *Proc. Natl. Acad. Sci. U.S.A.* **84,** 9248–9251.

Bortz, J. T., Wertz, P. W., and Downing, D. T. (1989). *J. Invest. Dermatol.* **93,** 723–727.

Bouthillier, M., Bleau, G., Chapdelaine, A., and Roberts, K. D. (1984). *Biol. Reprod.* **31,** 936–941.

Bowser, P. A., Nugteren, D. H., White, R. J., Houtsmuller, U. M. T., and Prottey, C. (1985). *Biochim. Biophys. Acta* **834,** 419–428.

Brod, J., Berrebi, C., Fiat, F., Noblet, J. P., and Prunieras, M. (1987). *Pharmacol. Skin* **1,** 223–230.

Brody, I. (1962). *J. Ultrastruct. Res.* **6,** 341–353.

Brysk, M. M., Rajaramar, S., Penn, P., Barlow, E., and Bell, T. (1989). *Exp. Cell Biol.* **57,** 60–66.

Burck, P. J., and Zimmerman, R. E. (1980). *J. Reprod. Fertil.* **58,** 121–125.

Burr, G. O., and Burr, M. M. (1929). *J. Biol. Chem.* **82,** 345–367.

Buxman, M. M., Goodkin, P. E., Fahrenbach, W. H., and Dimond, R. L. (1979). *Arch. Dermatol.* **115,** 189–193.

Chanarin, I., Patel, A., Slavin, G., Wills, E. J., Andrews, T. M., and Stewart, C. (1975). *Br. Med. J.* **1,** 553–555.

Chang, P. L., Varey, P. A., Rosa, N. E., Ameen, M., and Davidson, R. G. (1986). *J. Biol. Chem.* **261,** 14443–14447.

Chang, P. L., Mueller, O. T., Lafrenie, R. M., Varey, P. A., Rosa, N. E., Davidson, R. G., Henry, W. M., and Shows, T. B. (1990). *Am. J. Hum. Genet.* **46,** 729–737.

Chapkin, R. S., and Ziboh, V. A. (1984). *Biochem. Boiphys. Res. Commun.* **124,** 784–792.

Chapman, S. J., and Walsh, A. (1990). *Arch. Dermatol. Res.* **262,** 304–310.

Chatterjee, S., Clarke, K. S., and Kwiterovich, P. O. (1986). *J. Biol. Chem.* **261,** 13474–13479.

Cheetham, J. J., Epand, R. M., Andrews, M., and Flanagan, T. D. (1990). *J. Biol. Chem.* **265,** 12404–12409.

Choi, H. Y., and Hobkirk, R. (1986). *J. Steroid Biochem.* **25,** 985–989.

Clayton, R. B., Nelson, A. N., and Frantz, I. D., Jr. (1963). *J. Lipid Res.* **4,** 166–178.

Coleman, R. A., and Haynes, E. B. (1983). *Biochim. Biophys. Acta* **751,** 230–240.

Conary, J. T., Nauerth, A., Burns, G., Hasilik, A., and Figura, K. (1986). *J. Biochem.* **158,** 71–76.

Conary, J. T., Lorkowski, G., Schmidt, B., Pohlmann, R., Nagel, G., Meye, H. E., Krentler, C., Cully, J., Hasilik, A., and Figura, K. (1987). *Biochem. Biophys. Res. Commun.* **144,** 1010–1017.

Cook, H. W., and Spence, M. W. (1985). *Can. J. Biochem. Cell Biol.* **63,** 919–926.

Cook, H. W., Clarke, J. T. R., and Spence, M. W. (1982). *J. Lipid Res.* **23,** 1292–1300.

Cooke, A., Gillard, E. F., Yates, J. R. W., Mitchell, M. J., Aitken, D. A., Weir, D. M., Affara, N. A., and Ferguson-Smith, M. A. (1988). *Hum. Genet.* **79,** 49–52.

Craig, J. M., Goldsmith, L. A., and Baden, H. P. (1970). Pediatr. **46,** 437–440.

Crawford, M. D. A. (1982). *J. Inherited Metab. Dis.* **5,** 153–163.

Curry, C. J. R., Magenis, R. E., Brown, M., Lanman, J. T., Tsai, J., O'Lague, P., Goodfellow, P., Mohandas, T., Bergner, E. A., and Shapiro, L. J. (1984). *N. Engl. J. Med.* **311,** 1010–1014.
Dale, B. A., Holbrook, K. A., Fleckman, P., Kimball, J. R., Brumbaugh, S., and Sybert, V. P. (1990a). *J. Invest. Dermatol.* **94,** 6–18.
Dale, B. A., Kimball, J. R., Fleckman, P., Holbrook, K. A., and Herbert, A. (1990b). *Clin. Res.* **38,** 652A.
Dale, B. A., Resing, K. A., and Haydock, P. V. (1990c). *In* "Cellular and Molecular Biology of Intermediate Filaments" (R. D. Goldman and P. M. Steinart, eds.), pp. 343–412. Plenum, New York.
Daniel W. L. (1985). *Isozymes: Curr. Top. Biol. Med. Res.* **12,** 189–228.
Davies, M. G., Reynolds, D. J., Marks, R., and Dykes, P. J. (1978). *In* "The Ichthyosis" (R. Marks and P. J. Davies, eds.), pp. 51–64.
Dibbelt, L., and Kuss, E. (1986). *Biol. Chem. Hoppe-Seyler* **367,** 1223–1229.
Dibbelt, L., Herzog, V., and Kuss, E. (1989). *Biol. Chem. Hoppe-Seyler* **370,** 1093–1102.
DiDonato, S., Garavaglia, B., Striscinglio, P., Borrone, C., and Andria, G. (1988). *Neurology* **38,** 1107–1110.
Dijkstra, A. C., Vermeesch-Markslag, A. M. G., Vromans, E. W. M., Happle, R., van de Kerkhof, P. C. M., Zwanenburg, B., Vos, F., and Vermorken, A. J. M. (1987). *Acta Derm.- Venereol.* **67,** 369–376.
Dorfman, M. L., Hershko, C., Eisenberg, S., and Sagher, F. (1974). *Arch. Dermatol.* **110,** 261–266.
Downing, D. T. (1976). *In* "Chemistry and Biochemistry of Natural Waxes" (P. E. Kolattukdy, ed.), pp. 17–48. Elsevier, New York.
Downing, D. T., Stewart, M. E., Wertz, P. W., and Strauss, J. S. (1986). *J. Am. Acad. Dermatol.* **14,** 211–215.
Downing, D. T., Stewart, M. E., and Strauss, J. S. (1987). *In* "Dermatology in General Medicine" (T. B. Fitzpatrick, A. Z. Eisen, K. Wolff, I. M. Freedberg, and K. F. Austen, eds.), 3rd Ed., pp. 185–190. McGraw-Hill, New York.
Dykes, P. J., Marks, R., Davies, M. G., and Reynolds, D. J. (1978). *J. Invest. Dermatol.* **70,** 126–129.
Edwards, P. A., Lan S.-F., and Fogelman, A. M. (1983). *J. Biol. Chem.* **258,** 10219–10222.
Elias, P. M. (1983). *J. Invest. Dermatol.* **80,** 44–49.
Elias, P. M. (1985). *In* "Dermatopharmacology and Dermatotoxicology" (H. I. Maibach and N. J. Lowe, eds.), pp. 272–289. Karger, Basel.
Elias, P. M., and Brown, B. E. (1978a). *Lab. Invest.* **39,** 574–583.
Elias, P. M., and Williams, M. L. (1985). *Arch. Dermatol.* **121,** 1000–1008.
Elias, P. M., and Williams, M. L. (1990). *J. Invest. Dermatol.* **94,** 730–731.
Elias, P. M., Brown, B. E., Fritsch, P. O., Goerke, R. J., Gray, G. M., and White, R. J. (1979). *J. Invest. Dermatol.* **73,** 339–348.
Elias, P. M., Brown, B., and Ziboh, V. A. (1980a). *J. Invest. Dermatol.* **74,** 230–233.
Elias, P. M., Menon, G. K., Grayson, S., and Brown, B. E. (1980b). *J. Invest. Dermatol.* **91,** 3–10.
Elias, P. M., Lampe, M. A., Chung, J.-C., and Williams, M. L. (1983a). *Lab. Invest.* **80,** 44–49.
Elias, P. M., Bonar, L., Grayson, S., and Baden, H. P. (1983b). *J. Invest. Dermatol.* **80,** 213–214.
Elias, P. M., Williams, M. L., Maloney, M. E., Bonifas, J. A., Brown, B. E., Grayson, S., and Epstein, E. H., Jr. (1984). *J. Clin. Invest.* **74,** 1414–1421.
Elias, P. M., Williams, M. L., Maloney, M. E., Fritsch, P. O., and Chung, J.-C. (1985). *In* "Models in Dermatology" (H. Maibach and N. Lowe eds.), Vol. 1, pp. 105–126. Karger, Basel.
Elias, P. M., Menon, G. K., Grayson, S., and Brown, B. E. (1988). *J. Invest. Dermatol.* **91,** 3–10.
Elliott, W. J., Morrison, A. R., Sprecher, H. W., and Neeldeman, P. (1985a). *J. Biol. Chem.* **260,** 987–992.
Elliott, W. J., Sprecher, H., and Needleman, P. (1985b). *Biochim. Biophys. Acta* **835,** 158–160.
Elsbach, P., and Farrow, S. (1969). *Biochim. Biophys. Acta* **176,** 438–441.
Emmons, G. T., Rosenblum, E. R., Malloy, J. M., McManus, K. R., and Campbell, I. M. (1980). *Biochem. Biophys. Res. Commun.* **96,** 34–38.

Endo, A. (1985). *In* "Regulation of HMG-CoA Reductase" (B. Preiss, ed.), pp. 49–78. Academic Press, New York.
Epstein, E. H., Jr., and Leventhal, M. E. (1981). *J. Clin. Invest.* **67,** 1257–1262.
Epstein, E. H., Jr., Krauss, R. M., and Shackelton, C. H. L. (1981). *Science* **214,** 659–660.
Epstein, E. H., Jr., Bonifas, J. M., Baarber, T. C., and Haynes, M. (1984a). *J. Invest. Dermatol.* **83,** 332–335
Epstein, E. H., Jr., Williams, M. L., and Elias, P. M. (1984b). *J. Am. Acad. Dermatol.* **10,** 866–868.
Epstein, E. H., Jr., Williams, M. L., and Elias, P. M. (1987) *In* "Current Problems in Dermatology" (H. Honigsmann, ed.), pp. 32–44. Karger, Basel.
Famagalli, R., and Niemiro, R. (1964). *Life Sci.* **3,** 555.
Feingold, K. R., Brown, B. E., Lear, S. R., Moser, A. H., and Elias, P. M. (1983). *J. Invest. Dermatol.* **81,** 365–369.
Feingold, K. R., Mao-Qiang, M., Menon, G. K., Cho, S. S., Brown, B. E., and Elias, P. M. (1990). *J. Clin. Invest.* **86,** 1738–1745.
Feingold, K. R., Mao-Qiang, M., Proksch, E., Menon, G. K., Brown, B., and Elias, P. M. (1991). *J. Invest. Dermatol.* **96,** 201–209.
Fleck, R. M., Barnades, M., Schulz, W. W., Roberts, L. J., and Freeman, R. G. (1989). *J. Am. Acad. Dermatol.* **21,** 999–1000.
Fleckman, P., Holbrook, K. A., Dale, B. A., and Sybert, V. P. (1987). *J. Invest. Dermatol.* **88,** 640–645.
Forster, S., Ilderton, E., Norris, J. F. G., Summerly, R., and Yardley, H. J. (1985). *Br. J. Dermatol.* **112,** 135–147.
France, J. T., and Liggins, G. C. (1969). *J. Clin. Endocrinol. Metab.* **29,** 138–143.
Freinkel, R. K., and Aso, K. (1969). *J. Invest. Dermatol.* **52,** 148–154.
Freinkel, R. K., and Fiedler-Weiss, V. (1974). *J. Invest. Dermatol.* **62,** 458–462.
Frost, P. (1973). *J. Invest. Dermatol.* **60,** 541–552.
Frost, P., and Van Scott, E. J. (1966). *Arch. Dermatol.* **94,** 113–126.
Frost, P., Weinstein, G. D., and Van Scott, E. J. (1966). *J. Invest. Dermatol.* **47,** 561–567.
Frost, P., Weinstein, G. D., Bothwell, J. W., and Wildnauer, R. (1968). *Arch. Dermatol.* **98,** 230–233.
Gara, A., Estrada, E., Rothman, S., and Lorenz, A. L. (1964). *J. Invest. Dermatol.* **43,** 559–564.
Geiger, J.-M., and Hartmann, H.-R. (1986). *Arch. Dermatol. Res.* **278,** 426–428.
Ghadially, R., Menon, G. K., Williams, M. L., and Elias, P. M. (1990). *Clin. Res.* **38,** 223A.
Ghadially, R., Feingold, K. R., and Elias, P. M. (1991). *Clin. Res.* **39,** 538A.
Gillard, J. F., Affara, N. A., Yates, W. R. W., Goudie, D. R., Lambert, J., Aitken, D. A., and Ferguson-Smith, M. A. (1987). *Nucleic Acids Res.* **15,** 3977–3985.
Gloor, M., Josephs, H., and Friedrich, H. C. (1974). *Arch. Dermatol. Forsch.* **250,** 277–284.
Gniot-Szulzycka, J., and Januszewska, B. (1986). *Acta Biochem. Blon.* **33,** 203–215.
Goldyne, M. E., and Williams, M. L. (1989). *J. Clin. Invest.* **84,** 357–360.
Goppelt, M., Kohler, L., and Resch, K. (1985). *Biochim. Biophys. Acta* **833,** 463–472.
Grattan, C., Burton, J. L., Manku, M., Stewart, C., and Horrobin, D. F. (1990). *Clin. Exp. Dermatol.* **15,** 174–176.
Gray, G. M., and Yardley, J. (1975). *J. Lipid Res.* **16,** 441–447.
Gray, G. M., White, R. J., and Majer, J. R. (1978). *Biochim. Biophys. Acta* **528,** 127–137.
Grayson, S., and Elias, P. M. (1982). *J. Invest. Dermatol.* **78,** 128–135.
Grayson, S., Johnson-Winegar, A. G., Wintroub, B. U., Epstein, E. H., Jr., and Elias, P. M. (1985). *J. Invest. Dermatol.* **85,** 289–295.
Grubauer, G., Feingold, K. R., and Elias, P. M. (1989). *J. Lipid. Res.* **30,** 89–96.
Guilleminault, C., Harpey, J. P., and Lafourcade, J. (1973). *Neurology* **23,** 367–373.
Haati, E. (1961). *Scand. J. Clin. Lab. Invest.* **13,** 1–108 (Suppl. 59).
Hadley, N. F. (1981). *Biol. Rev. Cambridge Philos. Soc.* **56,** 23–47.
Hadley, N. F. (1989). *Prog. Lipid Res.* **28,** 1–33.
Hansen, A. E. (1937). *Am. J. Dis. Child.* **53,** 933–946.

Hansen, H. S. (1986). *Trends Biochem. Sci. (Pers. Ed.)* **11,** 263–265.
Hansen, H. S., Jensen, B., and von Wettstein-Knowles, P. (1986). *Biochim. Biophys. Acta* **878,** 284–287.
Happle, R., Koch, H., and Lenz, W. (1980). *Eur. J. Pediatr.* **134,** 27–33.
Harris, R. R., MacKenzie, I. C., and Williams, R. A. P. (1980). *J. Invest. Dermatol.* **74,** 402–426.
Hartop, P. J., and Prottey, C. (1976). *Br. J. Dermatol.* **95,** 255–264.
Hartop, P. J., Allenby, C. F., and Prottey, C. (1978). *Clin. Exp. Dermatol.* **3,** 259–267.
Hazell, R., and Marks, R. (1985). *Arch. Dermatol.* **121,** 489–493.
Heikoop, J. C., Van Roermund, C. W. T., Just, W. W., Ofman, R., Schutgens, R. B. H., Heymans, H. S. A., Wanders, R. J. A., and Tager, J. N. (1990). *J. Clin. Invest.* **86,** 126–130.
Herndon, J. H., Jr., Steinberg, D., and Uhlendorf, B. W. (1969). *N. Engl. J. Med.* **282,** 1034–1038.
Hernell, O., Holmgren, G., Jagell, S. F., Johnson, S. B., and Holman, R. T. (1982). *Pediatr. Res.* **16,** 45–49.
Herrmann, F. H., Grimm, U., and Hadlich, J. (1987). *J. Inherited Metab. Dis.* **10,** 89–94.
Heymans, H. S. A., Oorthays, J. W. E., Nelck, G., Wanders, R. J. A., Dingemans, K. P., and Schutgens, R. B. H. (1986). *J. Inherited Metab. Dis.* **2,** 329–331.
Hobkirk, R. (1985). *Can. J. Biochem. Cell Biol.* **63,** 1127–1144.
Hobkirk, R., Renard, R., and Raeside, J. I. (1989). *J. Steroid Biochem.* **32,** 387–392.
Hoefler, G., Hoefler, S., Watkins, P. A., Chen, W. W., Moser, A., Baldwin, V., McGillivary, B., Charrow, J., Friedman, J. M., Rutledge, L., Hashiomoto, T., and Moser, H. W. (1988). *J. Pediatr.* **112,** 726–733.
Hoeg, J. M., Demosky, S. J., Jr., and Brewer, H. B., Jr. (1982). *Biochim. Biophys. Acta* **711,** 59–65.
Holmes, R. D., Wilson, G. N., and Hajra, A. K. (1987). *N. Eng. J. Med.* **316,** 1608.
Hooft, C., Kriekemans, J., van Acker, K., Devos, E., Traen, S., and Verdonk, G. (1967). *Helv. Paediatr. Acta* **31,** 447–458.
Horlick, L., and Avigan, J. (1963). *J. Lipid Res.* **4,** 160–165.
Horrobin, D. F. (1989). *J. Am. Acad. Dermatol.* **20,** 1045–53.
Hotz, W. (1983). *Adv. Lipid Res.* **20,** 195–216.
Hou, S. Y. E., White, S. H., Menon, G. K., Grayson, S., Ghadially, R., and Elias, P. M. (1991). *J. Invest. Dermatol.* **96,** 215–223.
Houtsmuller, V. M. T. (1981). *Prog. Lipid Res.* **20,** 889–897.
Hoyer, H., Lykkesfeldt, G., Ibsen, H. H., and Brandrup, F. (1986). *Dermatologica* **172,** 184–190.
Ibsen, H. H., Brandrup, F., and Secher, B. (1984). *Lancet* **ii,** 645.
Iwamori, M., Moser, H. W., and Kishimoto, Y. (1976). *Arch. Biochem. Biophys.* **174,** 199–208.
Jagell, S., and Liden, S. (1982). *Clin. Genet.* **21,** 243–252.
Jagell, S., Gustavson, K.-H., and Holmgren, G. (1981). *Clin. Genet.* **19,** 233–256.
Jetten, A. M., George, M. A., Pettit, G. R., Herald, C. L., and Rearick, J. I. (1989a). *J. Invest. Dermatol.* **93,** 108–115.
Jetten, A. M., George, M. A., Nervi, C., Boone, L. R., and Rearick, J. I. (1989b). *J. Invest. Dermatol.* **92,** 203–209.
Jobsis, A. C., de Groot, W. P., Meijer, A. E. F. H., and Van Der Loos, C. M. (1983). *Br. J. Dermatol.* **108,** 567–572.
Johnson, R., Nusbaum, B. P., Horwitz, S. N., and Frost, P. (1983). *Arch. Dermatol.* **119,** 660–663.
Jordans, G. H. W. (1953). *Acta Med. Scand.* **145,** 419–423.
Kalter, D. C., Atherton, D. J., and Clayton, P. T. (1989). *J. Am. Acad. Dermatol.* **21,** 248–256.
Kanerva, L., Niemi, K.-M., Lawharanta, J., and Lassus, A. (1983). *Am. J. Dermatopathol.* **5,** 555–567.
Karkainen, J., Nikkari, T., Ruponen, S., and Hautz, E. (1965). *J. Invest. Dermatol.* **44,** 333–338.
Kawano, J., Kotani, T., Ohtaki, S., Minamino, N., Matsuo, H., Oinuma, T., and Aikawa, E. (1989). *Biochim. Biophys. Acta* **997,** 199–205.
Kellum, R. E. (1967). *Arch. Dermatol.* **95,** 218–220.

King, C. S., Barton, S. P., Nocholls, S., and Marks, R. (1979). *Br. J. Dermatol.* **100,** 165–172.
Klenk, E., and Kahlke, W. (1963). *Hoppe-Seyler's Z. Physiol. Chem.* **333,** 133–139.
Koone, M. D., Rizzo, W. B., Elias, P. M., Williams, M. L., Lightner, V., and Pinnell, S. (1990). *Arch. Dermatol.* **126,** 1485–1490.
Koppe, J. G., Marinkovic-Ibsen, A., Rijken, Y., DeGroot, W. P., and Jobsis, A. C. (1978). *Arch. Dis. Child.* **53,** 803–806.
Kragballe, K., and Voorhees, J. J. (1985). *Curr. Probl. Dermatol.* **119,** 548–552.
Kubilus, J., Tarascio, J., and Baden, H. P. (1979). *Am. J. Hum. Genet.* **31,** 50–53.
Lalumiere, G., Longpre, J., Trudel, J., Chapdelaine, A., and Roberts, K. D. (1975). *Biochim. Biophys. Acta* **394,** 120–128.
Lalumiere, G., Bleau, G., Chapdelaine, A., and Roberts, K. D. (1976). *Steroids* **27,** 247–259.
Lambeth, J. D., Xu, X. X., and Glover, M. (1987). *J. Biol. Chem.* **262,** 9181–9188.
Lampe, M. A., Burlingame, A. L., Whitney, J., Williams, M. L., Brown, B. E., Roitman, E., and Elias, P. M. (1983a). *J. Lipid Res.* **24,** 120–130.
Lampe, M. A., Williams, M. L., and Elias, P. M. (1983b). *J. Lipid Res.* **24,** 131–140.
Langlais, J., Zollinger, M., Plante, L., Chapdelaine, A., Bleau, G., and Roberts, K. D. (1981). *Proc. Natl. Acad. Sci. U.S.A.* **78,** 7266–7270.
Lawlor, F., and Peiris, S. (1985). *Br. J. Dermatol.* **112,** 585–590.
Lazarow, P. B. (1987). *J. Inherited Metab. Dis.* **1,** 11–22.
Legault, Y., Belau, G., Chapdelaine, A., and Roberts, K. D. (1980). *Biol. Reprod.* **23,** 720–725.
Lengle, E., and Geyer, R. P. (1973). *Biochim. Biophys. Acta* **296,** 411–425.
Lester, D. E. (1979). *Prog. Food Nutr. Sci.* **3,** 1–66.
Li, X. M., Yen, P., Mohandras, T., and Shapiro, L. J. (1990). *Nucleic Acids Res.* **18,** 2783–2788.
Lockey, K. H. (1985). *Comp. Biochem. Physiol. B* **81B,** 263–273.
Long, S. A., Wertz, P. W., Strauss, J. S., and Downing, D. T. (1985). *Arch. Dermatol. Res.* **277,** 284–287.
Lowe, N. J., and DeQuoy, P. (1978). *J. Invest. Dermatol.* **70,** 200–203.
Lundstrom, A., and Egelrud, T. (1988). *J. Invest. Dermatol.* **91,** 340–343.
Lykkesfeldt, G., Nielsen, M. D., and Lykkesfeldt, A. E. (1984). *Obstet. Gynecol.* **64,** 49–54.
Lykkesfeldt, G., Bennett, P., Lykkesfeldt, A. E., Micic, S., Moller, S., and Svenstrup, B. (1985a). *Clin. Endocrinol. (Oxford)* **23,** 385–393.
Lykkesfeldt, G., Hoyer, H., Ibsen, H. H., and Brandrup, F. (1985b). *Clin. Genet.* **28,** 231–237.
Ma, P. T., Gil, G., Sudhof, T. C., Bilheimer, D. W., Goldstein, J. L., and Brown, M. S. (1986). *Proc. Natl. Acad. Sci. U.S.A.* **83,** 8370–8374.
MacIndoe, J. H. (1988). *Endocrinology (Baltimore)* **123,** 1281–1287.
Madison, K. C., Swartzendruber, D. C., Wertz, P. W., and Downing, D. T. (1987). *J. Invest. Dermatol.* **88,** 714–718.
Maloney, M. E., Williams, M. L., Epstein, E. H., Jr., Law, M. Y. L., Fritsch, P. O., and Elias, P. M. (1984). *J. Invest. Dermatol.* **83,** 253–256.
McGuire, J. (1982). *In* "Dermatology Update" (S. L. Moschelle, ed.), pp. 197–215. Elsevier, New York.
Melnik, B. (1989). *In* "The Ichthyoses" (H. Traupe, ed.), pp. 15–42. Springer-Verlag, Berlin.
Melnik, B. C., and Plewig, G. (1989). *J. Am. Acad. Dermatol.* **21,** 557–563.
Melton, J. L., Wertz, P. W., Swartzendruber, D. C., and Downing, D. T. (1987). *Biochim. Biophys. Acta* **921,** 191–197.
Menon, G. K., Hou, E. S. Y., Grayson, S., and Elias, P. M. (1989a). *Clin. Res.* **37,** 233A.
Menon, G. K., Placzek, D., Hincenbergs, M., and Williams, M. L. (1989b). *J. Invest. Dermatol.* **92,** 480A.
Menon, G. K., Placzek, D., Hincenbergs, M., and Williams, M. L. (1989c). *Clin. Res.* **37,** 620A.
Menon, G. K., Feingold, K. R., Man, M.-Q., and Elias, P. M. (1991a). In press.

Menon, G. K., Williams, M. L., Ghadially, R., and Elias, P. M. (1991b). *Arch. Dermatol.* (in press).
Menton, D. N. (1970). *J. Morphol.* **132,** 181–206.
Merrill, A. H., Jr. (1988). *Biochim. Biophys. Acta* **754,** 284–291.
Mesquita-Guimaraes, J. (1981). *Dermatologica* **162,** 157–166.
Messieh, S., Clarke, J., Jr., Cooke, R., and Spence, M. W. (1983). *Pediatr. Res.* **17,** 770–774.
Messmer, T. O., Wang, E., Stevens, V. L., and Merrill, A. H., Jr. (1989). *J. Nutr.* **119,** 534–538.
Meyer, J. C., and Grundmann, H. P. (1980). *Arch. Dermatol. Res.* **269,** 213–215.
Meyer, J. C., Grundmann, H. P., and Schnyder, U. (1979). *Arch. Dermatol. Res.* **266,** 95–97.
Meyer, J. C., Groh, V., Giger, V., Weiss, H., Varbelow, H., and Schnyer, U. W. (1982). *Dermatologica* **164,** 249–257.
Meyer, J. C., Grundmann, H., and Weiss, H. (1984). *Dermatologica* **169,** 305–310.
Migeon, B. R., Shapiro, L. J., Norum, R. A., Mohandas, T., Axelman, J., and Dabora, R. L. (1982). *Nature (London)* **299,** 838–840.
Milewich, L., Sontheimer, R. D., and Herndon, J. H., Jr. (1990). *Arch. Dermatol.* **126,** 1312–1314.
Mohandas, T., Shapiro, L. J., Sparkes, R. S., and Sparkes, M. C. (1979). *Proc. Natl. Acad. Sci. U.S.A.* **76,** 5779–5783.
Mommaas-Kienhuis, A.-M., Grayson, S., Wijsman, M. C., Vermeer, B. J., and Elias, P. M. (1987). *J. Invest. Dermatol.* **89,** 513–517.
Monger, D. J., Williams, M. L., Feingold, K. R., Brown, B. E., and Elias, P. M. (1988). *J. Lipid Res.* **29,** 603–612.
Moriyasu, M., and Ito, A. (1982). *J. Biochem. (Tokyo)* **92,** 1197–1204.
Moriyasu, M., Ito, A., and Omura, T. (1982). *J. Biochem. (Tokyo)* **92,** 1189–1195.
Munroe, D. G., and Chang, P. L. (1987). *Am. J. Hum. Genet.* **40,** 102–114.
Nakamura, T., Matsuzawa, Y., Okano, M., Kitano, Y., Funahashi, T., Yamashita, S., and Tarui, S. (1988). *Atherosclerosis* **70,** 43–52.
Nelson, K., Keinanen, B. M., and Daniel, W. L. (1983). *Experientia* **39,** 740–742.
Nguyen, T. T., Ziboh, V. A., Uematsu, S., McCullough, J. L., and Weinstein, G. (1981). *J. Invest. Dermatol.* **76,** 3834–387.
Nicolaides, N. (1963). *In* "Advances in Biology of the Skin" (W. Montagna, R. A. Ellis, and A. F. Silver, eds.), Vol. 4, pp. 167–187. Macmillan, New York.
Nicolaides, N., Fu, H. C., Ansari, M. N. A., and Rice, G. R. (1972). *Lipids* **7,** 506–517.
Niemi, K.-M., and Kanerva, L. (1989). *Am. J. Dermatopathol.* **11,** 149–156.
Nikkari, T. (1974). *J. Invest. Dermatol.* **62,** 257–267.
Noel, H., Beauregard, G., Potier, M., Bleau, G., Chapdelaine, A., and Roberts, K. D. (1983). *Biochim. Biophys. Acta* **758,** 88–90.
Nugteren, D. H., and Kivits, G. A. A. (1987). *Biochim. Biophys. Acta* **921,** 135–141.
Nugteren, D. H., Christ-Hazelhof, E., van der Beck, A., and Houtsmuller, U. M. T. (1985). *Biochim. Biophys. Acta* **834,** 429–436.
Oram, J. F., Shafir, E., and Bierman, E. L. (1980). *Biochim. Biophys. Acta* **619,** 214–227.
Oxholm, A., Oxholm, P., Bang, F. C., and Horrobin, D. F. (1990). *Arch. Dermatol.* **126,** 1308–1311.
Parsons, W. B., Jr., and Flinn, J. H. (1959). *Arch. Intern. Med.* **103,** 783–790.
Pathak, R. K., Lasky, K. L., and Anderson, R. G. W. (1986). *J. Cell Biol.* **102,** 2158–2168.
Pentchev, P. G., Comly, M. E., Kruth, H. S., Vanier, M. T., Wenger, D. A., Patel, S., and Brady, R. O. (1985). *Proc. Natl. Acad. Sci. U.S.A.* **82,** 8247–8251.
Pentchev, P. G., Kruth, H. S., Comly, M. E., Butler, J. D., Venier, M. T., Wenger, D. A., and Patel, S. (1986). *J. Biol. Chem.* **261,** 16775–16780.
Pentland, A. P., and Needleman, P. (1986). *J. Clin. Invest.* **77,** 246–251.
Perrot, H., and Ortonne, J. P. (1979). *Arch. Dermatol. Res.* **265,** 123–131.
Peters, R. F., and White, A. M. (1980). *Br. J. Dermatol.* **98,** 301–314.
Ponec, M., and Williams, M. L. (1986). *Arch. Dermatol. Res.* **279,** 32–36.

Ponec, M., Havekes, L., Kempenaar, J., and Vermeer, B. J. (1983). *J. Invest. Dermatol.* **81,** 125–130.
Ponec, M., Havekes, L., Kempenaar, J., Lavrijsen, S., Wijsman, M., Boonstra, J., and Vermeer, B. J. (1985). *J. Cell. Physiol.* **125,** 98–106.
Ponec, M., Kempenaar, J., Weerheim, W., and Boonstra, J. (1987). *J. Cell. Physiol.* **133,** 358–364.
Ponec, M., Weerheim, A., Kempenaar, J., Elias, P., and Williams, M. (1989). *In Vitro* **25,** 689–696.
Poulos, A., Pollard, A. C., Mitchell, J. D., Wise, G., and Mortimer, G. (1984). *Arch. Dis. Child.* **59,** 222–229.
Poulos, A., Sheffield, L., Sharp, P., Sherwood, G., Johnson, D., Beckman, K., Fellenberg, A. J., Wraith, J. E., Chow, C. W., Usher, S., and Singh, H. (1988). *J. Pediatr.* **113,** 685–690.
Proksch, E., Elias, P. M., and Feingold, K. R. (1991a). *Biochim. Biophys. Acta* **1083,** 71–79.
Proksch, E., Feingold, K. R., Mao-Qiang, M., and Elias, P. M. (1991b). *J. Clin. Invest.* **87,** 1668–1673.
Prottey, C. (1976). *Br. J. Dermatol.* **94,** 579–585.
Prottey, C. (1977). *Br. J. Dermatol.* **97,** 29–38.
Radom, J., Salvayre, R., Maret, A., Negre, A., and Douste-Blazy, L. (1987a). *Biochim. Biophys. Acta* **920,** 131–139.
Radom, J., Salvayre, R., Negre, A., Maret, A., and Douste-Blazy, L. (1987b). *Eur. J. Biochem.* **164,** 703–708.
Ranasinghe, A. W., Wertz, P. W., Downing, D. T., and MacKenzie, I. C. (1986). *J. Invest. Dermatol.* **86,** 187–190.
Ranney, R. E., and Daskalakis, E. G. (1964). *Proc. Soc. Exp. Biol. Med.* **116,** 999.
Rearick, J. I., and Jetten, A. M. (1986). *J. Biol. Chem.* **261,** 13898–13904.
Refsum, S. (1946). *Acta Psychiatr. Scand., Suppl.* No. 38, 9–303.
Rehfeld, S. J., and Elias, P. M. (1982). *J. Invest. Dermatol.* **79,** 1–3.
Rehfeld, S. J., Williams, M. L., and Elias, P. M. (1986). *Arch. Dermatol. Res.* **278,** 259–263.
Rehfeld, S. J., Plachy, W. Z., Williams, M. L., and Elias, P. M. (1988). *J. Invest. Dermatol.* **91,** 499–505.
Rehfeld, S. J., Plachy, W. Z., Hou, S. Y., and Elias, P. M. (1990). *J. Invest. Dermatol.* **95,** 217–223.
Reihner, E., Rudling, M., Stahlberg, D., Berglund, L., Ewerth, S., Bjorkhem, I., Einarsson, K., and Angelin, B. (1990). *N. Engl. J. Med.* **323,** 224–228.
Richterich, R., Moser, M., and Rossi, E. (1965). *Humangenetik* **1,** 322–332.
Riddle, M. C., Fujimoto, W., and Ross, R. (1977). *Biochim. Biophys. Acta* **488,** 359–369.
Rizzo, W. B., Craft, D. A., Dammann, A. L., and Phillips, M. W. (1987). *J. Biol. Chem.* **36,** 17412–17419.
Rizzo, W. B., Dammann, A. L., and Craft, D. A. (1988). *J. Clin. Invest.* **81,** 738–744.
Rizzo, W. B., Dammann, A. L., Craft, D. A., Black, S. H., Tilton, A. H., Africk, D., Chaves-Carballo, E., Holmgren, C., and Jagell, S. (1989). *J. Pediatr.* **115,** 228–234.
Roberts, K. D. (1987). *J. Steroid Biochem.* **27,** 337–341.
Roberts, K. D., and Lieberman, S. (1970). *In* "Chemical and Biological Aspects of Steroid Conjugation" (S. Bernstein and S. Solomon, eds.), pp. 219–290. Springer-Verlag, New York.
Roberts, L. J. (1989). *J. Am. Acad. Dermatol.* **21,** 335–339.
Robertson, D. A., Freeman, C., Nelson, P. V., Morris, C. P., and Hopwood, J. J. (1988). *Biochem. Biophys. Res. Commun.* **157,** 218–224.
Rogers, M., and Scarf, C. (1989). *Pediatr. Dermatol.* **6,** 216–221.
Rose, F. A. (1982). *J. Inherited Metab. Dis.* **5,** 145–152.
Rothman, S. (1950). *Arch. Dermatol.* **62,** 814–819.
Rothman, S. (1964). *In* "The Epidermis" (W. Montagna and W. C. Lobitz, Jr., eds.), pp. 35–40. Academic Press, New York.
Rozenszajn, L., Klajman, A., Yaffe, D., and Efrati, P. (1966). *Blood* **28,** 258–265.
Ruiter, M., and Meyler, L. (1960). *Dermatologica* **120,** 139–144.
Ruokonen, A. (1978). *J. Steroid Biochem.* **9,** 939–946.

Ruokonen, A., Oikarinen, A., Palatsi, R., and Huhtaniemi, I. (1980). *Br. J. Dermatol.* **102,** 245–248.

Ruokonen, A., Oikarinen, A., and Vihko, R. (1986). *J. Steroid Biochem.* **25,** 113–119.

Sahgal, V., and Olsen, W. O. (1975). *Arch. Intern. Med.* **135,** 585–587.

Salvayre, R., Negre, A., Maret, A., Radom, J., and Douste-Blazy, L. (1987). *Eur. J. Biochem.* **170,** 453–458.

Salvayre, R., Negre, A., Radom, J., and Douste-Blazy, L. (1989). *FEBS Lett.* **250,** 35–39.

Schafer, L., and Kragbelle, K. (1991). *J. Invest. Dermatol.* **96,** 10–15.

Schmickel, R. D. (1986). *J. Pediatr.* **109,** 231–241.

Schnur, R. E., Trask, B. J., Van den Engh, G., Punnett, H. H., Kistenmacher, M., Tomeo, M. A., Naids, R. E., and Nussbaum, R. L. (1989). *Am. J. Hum. Genet.* **45,** 706–720.

Schuchman, E. H., Jackson, C. E., and Desnick, R. J. (1990). *Genomics* **6,** 149–158.

Schürer, N. Y., Monger, D. J., Hincenbergs, M., and Williams, M. L. (1989). *J. Invest. Dermatol.* **92,** 196–202.

Schutgens, R. B. H., Heymans, H. S. A., Wanders, R. J. A., van der Bosch, H., and Tager, J. M. (1986). *Eur. J. Pediatr.* **144,** 430–440.

Scott, I. R. (1986). *J. Invest. Dermatol.* **87,** 460–465.

Scott, I. R., and Harding, C. R. (1986). *Dev. Biol.* **115,** 84–92.

Scott, I. R., Harding, C. R., and Barrett, J. G. (1982). *Biochim. Biophys. Acta* **719,** 110–117.

Serizawa, S., Nagai, T., Ito, M., and Sato, Y. (1990). *Clin. Exp. Dermatol.* **15,** 13–15.

Shapiro, L. J. (1985). *Adv. Hum. Genet.* **16,** 331–381.

Shapiro, L. J., and Weiss, R. (1978). *Lancet* **i,** 70–72.

Shapiro, L. J., Weiss, R., Buxman, M. M., Vidgoff, J., and Dimond, R. L. (1978). *Lancet* **ii,** 756.

Shapiro, L. J., Yen, P., Pomerantz, D., Martin, E., Rolewic, L., and Mohandas, T. (1989). *Proc. Natl. Acad. Sci. U.S.A.* **86,** 8477–8481.

Sherertz, E. F. (1986). *In* "Nutrition and the Skin" (D. A. Roe, ed.), pp. 117–130. Alan R. Liss, New York.

Simpson, G. M., Blair, J. H., and Cransvick, E. H. (1964). *Clin. Pharmacol. Ther.* **5,** 310–321.

Singer, I. I., Scoh., S., Kazazis, D. M., and Haff, J. W. (1988). *Proc. Natl. Acad. Sci. U.S.A.* **85,** 5264–5268.

Singh, I., Johnson, G. H., and Brown, F. R., III (1988). *Am. J. Dis. Child.* **142,** 1297–1301.

Sjögren, T., and Larsson, T. (1957). *Acta Psychiatr. Neurol. Scand.* **32,** 1–113.

Sokol, J., Blanchette-Mackie, E. J., Kruth, H. S., Dwyer, N. K., Amende, L. M., Butler, J. D., Robinson, E., Patel, S., Brady, R. O., Comly, M. E., Vanier, M. T., and Pentchev, P. G. (1988). *J. Biol. Chem.* **263,** 3411–3417.

Spence, M. W., and Callahan, J. W. (1989). *In* "The Metabolic Basis of Inherited Disease" (C. R. Scriver, A. L. Beaudet, W. S. Sly, and D. Valle, eds.), pp. 1655–1676. McGraw-Hill, New York.

Stein, C., Hille, A., Seidel, J., Rijnbout, S., Waheed, A., Schmidt, B., Geuze, H., and von Figura, K. (1989). *J. Biol. Chem.* **264,** 13865–13872.

Steinberg, D. (1989). *In* "The Metabolic Basis of Inherited Disease" (C. R. Scriver, A. L. Beaudet, W. S. Sly, and D. Valle, eds.), pp. 1533–1550. McGraw-Hill, New York.

Steinberg, D., and Avigan, J. (1960). *J. Biol. Chem.* **235,** 3127–3130.

Steinberg, D., Herndon, J. H., Jr., Uhlendorf, B. W., Mize, C. E., and Milne, G. W. A. (1967a). *Science* **156,** 1740–1742.

Steinberg, D., Mize, C. E., Avigan, J., Fales, H. M., Eldjarnn, K., Stokke, O., and Refsum, S. (1967b). *J. Clin. Invest.* **46,** 313–322.

Stewart, M. E., Downing, D. T., Pochi, P. E., and Strauss, J. S. (1978). *Biochim. Biophys. Acta* **529,** 380–386.

Stewart, M. E., Benoit, A. M., Downing, D. T., and Strauss, J. S. (1984). *J. Invest. Dermatol.* **82,** 74–78.

Stewart, M. E., Steele, W. A., and Downing, D. T. (1989). *J. Invest. Dermatol.* **92,** 371–378.
Summerly, R., Ilderton, E., and Gray, G. M. (1978). *Br. J. Dermatol.* **99,** 279–288.
Swanbeck, G., and Thyresson, N. (1962). *Acta Derm.-Venereol.* **42,** 445–457.
Swartzendruber, D. C., Wertz, P. W., Madison, K. C., and Downing, D. T. (1987). *J. Invest. Dermatol.* **88,** 709–713.
Sybert, V. P., Dale, B. A., and Holbrook, K. A. (1985). *J. Invest. Dermatol.* **84,** 191–194.
Taylor, N. F. (1982). *J. Inherited Metab. Dis.* **5,** 164–176.
Thiele, V. (1974). *Hum. Genet.* **22,** 91–118.
Thomas, G. H., Tuck-Muller, C. M., Miller, C. S., and Reynolds, L. W. (1989). *J. Inherited Metab. Dis.* **12,** 139–151.
Tiepolo, L., Zuffardi, O., and Rodewald, A. (1977). *Hum. Genet.* **39,** 277–281.
Tiepolo, L., Zuffardi, O., Fraccaro, M., di Natale, D., Gargaantini, L., Muller, C. R., and Ropers, H. H. (1980). *Hum. Genet.* **54,** 205–206.
Tijburg, L. B. M., Geelen, M. J. H., and van Golde, L. M. G. (1989). *Biochim. Biophys. Acta* **1004,** 1–19.
Turteltaub, K. W., Felton, J. S., Gledhill, B. L., Vogel, J. S., Southon, J. R., Caffee, M. W., Finkel, R. C., Nelson, D. E., Proctor, I. D., and Davis, J. C. (1990). *Proc. Natl. Acad. Sci. U.S.A.* **87,** 1–5.
Vaccaro, A. M., Salvioli, R., Muscillo, M., and Renola, L. (1987). *Enzyme* **37,** 115–126.
van Diggelen, O. P., Konstantinidou, A. E., Bousema, M. T., Boer, M., Bakx, T., and Jobsis, A. C. (1989). *J. Inherited Metab. Dis.* **12,** 273–280.
Verdery, R. B., and Theolis, R., Jr. (1982). *J. Biol. Chem.* **257,** 1412–1417.
Veta, N., Kawamura, S., Kanagawa, I., and Yamakawa, T. (1971). *J. Biochem. (Tokyo)* **70,** 881–883.
Wertz, P. W., and Downing, D. T. (1982). *Science* **217,** 1261–1262.
Wertz, P. W., and Downing, D. T. (1983a). *J. Lipid Res.* **24,** 753–758.
Wertz, P. W., and Downing, D. T. (1983b). *J. Lipid Res.* **24,** 759–765.
Wertz, P. W., and Downing, D. T. (1984). *J. Lipid Res.* **25,** 1320–1323.
Wertz, P. W., Cho, E. S., and Downing, D. T. (1983). *Biochim. Biophys. Acta* **753,** 350–355.
Wertz, P. W., Downing, D. T., Freinkel, R. K., and Traczyk, T. N. (1984). *J. Invest. Dermatol.* **93,** 193–195.
Wilkinson, D. I. (1969). *J. Invest. Dermatol.* **52,** 339–343.
Wilkinson, D. I., and Farber, E. M. (1967). *J. Invest. Dermatol.* **49,** 526–530.
Willems, P. J., de Bruijn, W. A., Groenhuis, A., Mooyaart, B. R., and Berger, R. (1986). *J. Inherited Metab. Dis.* **9,** 156–162.
Willemsen, R., Kroos, M., Hoogeveen, A. T., van Dongen, J. M., Parenti, G., van der Loos, C. M., and Reuser, J. J. (1988). *Histochem. J.* **20,** 41–51.
Williams, M. L. (1983). *Pediatr. Dermatol.* **1,** 1–24.
Williams, M. L. (1986). *Pediatr. Dermatol.* **3,** 476–486.
Williams, M. L., and Elias, P. M. (1981). *J. Clin. Invest.* **68,** 1404–1410.
Williams, M. L., and Elias, P. M. (1982). *Biochem. Biophys. Res. Commun.* **107,** 322–328.
Williams, M. L., and Elias, P. M. (1984). *J. Clin. Invest.* **74,** 296–300.
Williams, M. L., and Elias, P. M. (1985a). *Arch. Dermatol.* **121,** 477–488.
Williams, M. L., and Elias, P. M. (1985b). *In* "Pathogenesis of Skin Disease" (B. Thiers and R. L. Dobson, eds.), pp. 519–551. Churchill-Livingstone, New York.
Williams, M. L., and Elias, P. M. (1987). *In* "Dermatologic Clinics" (J. C. Alper, ed.), pp. 155–178. Saunders, Philadelphia, Pennsylvania.
Williams, M. L., Grayson, S., Bonifas, J. M., Epstein, E. H., Jr., and Elias, P. M. (1983). *In* "The Stratum Corneum" (R. Marks and G. Plewig, eds.), pp. 79–85. Springer-Verlag, New York.
Williams, M. L., Koch, T. K., McDonnell, J. J., Frost, P., Epstein, L. B., Grizzard, W. S., and Epstein, C. H. (1984). *Am. J. Med. Genet.* **20,** 711–726.

Williams, M. L., Wiley, M., and Elias, P. M. (1985). *Biochim. Biophys. Acta* **945,** 349–357.

Williams, M. L., Feingold, K. R., Grubauer, G., and Elias, P. M. (1987a). *Arch. Dermatol.* **123,** 1535–1538.

Williams, M. L., Rutherford, S. A., and Feingold, K. R. (1987b). *J. Lipid Res.* **28,** 955–967.

Williams, M. L., Rutherford, S. L., Mommaas-Kienhuis, A.-M., Grayson, S., Vermeer, B. J., and Elias, P. M. (1987c). *J. Cell. Physiol.* **132,** 428–440.

Williams, M. L., Brown, B. E., Monger, D. J., Grayson, S., and Elias, P. M. (1988a). *J. Cell. Physiol.* **136,** 103–110.

Williams, M. L., Hincenbergs, M., and Holbrook, K. A. (1988b). *J. Invest. Dermatol.* **91,** 263–268.

Williams, M. L., Monger, D. L., Hincenbergs, M., Rehfeld, S. J., and Grunfeld, C. (1988c). *J. Inherited Metab. Dis.* **11,** 131–143.

Williams, M. L., Rutherford, S. L., Ponec, M., Hincenbergs, M., Placzek, D., and Elias, P. M. (1988d). *J. Invest. Dermatol.* **91,** 86–91.

Williams, M. L., Coleman, R. A., Placzek, D., and Grunfeld, C. (1991a). *Biochim. Biophys. Acta* **1096,** 162–167.

Williams, M. L., Emami, S., Hanley, K. P., Menon, G., and Goldyne, M. (1991b). *Clin. Res.* **39,** 377A.

Williams, M. L., Menon, G., and Hanley, K. P. (1991c). In press.

Wilson, J. D. (1963). *In* "Advances in Biology of the Skin" (W. Montagna, R. A. Ellis, and A. F. Silver, eds.), Vol. 4, pp. 148–166. Macmillan, New York.

Winkelmann, R. K., Perry, H. O., and Achor, R. W. P. (1963). *Arch. Dermatol.* **87,** 372–377.

Xu, X. X., and Lambeth, J. D. (1989). *J. Biol. Chem.* **264,** 7222–7227.

Yamanaki, W. K., Clemans, G. W., and Hutchinson, M. L. (1981). *Prog. Lipid Res.* **19,** 187–215.

Yardley, H. J. (1969). *Br. J. Dermatol.* **81,** 29–38.

Yen, P. H., Allen, E., Marsh, B., Mohandras, T., Wang, N., Taggart, R. T., and Shapiro, L. J. (1987). *Cell* **49,** 443–454.

Yoshike, T., Matsui, T., and Ogawa, H. (1985). *Br. J. Dermatol.* **112,** 431–433.

Ziboh, V. A., and Chapkin, R. S. (1988). *Prog. Lipid Res.* **27,** 81–105.

Ziboh, V. A., and Hsia, S. L. (1972). *J. Lipid Res.* **13,** 458–467.

Zuckerman, N. G., and Hagerman, D. D. (1966). *Arch. Biochem. Biophys.* **135,** 410–415.

Chemistry and Function of Mammalian Sebaceous Lipids

MARY ELLEN STEWART AND DONALD T. DOWNING

The Marshall Dermatology Research Laboratories
University of Iowa College of Medicine
Iowa City, Iowa 52242

I. Introduction

Sebaceous glands are small organs of mammalian skin that secrete a lipid mixture, known as sebum, onto the skin surface. Anatomically, sebaceous glands are appendages of the epidermis and share some of its characteristics. For example, both types of tissues are made up of cells that have a short lifetime and have to be constantly replaced, and both are active sites of lipid synthesis. However, the types of lipids made are completely different.

The lipid composition of sebum has been analyzed in detail for about 20 species. Each species produces sebum of a unique composition, although there are often similarities in composition within genera. Intermediates in the biosynthetic pathway to cholesterol, which do not accumulate in other tissues, are fairly common sebum constituents. Even more common are a variety of mono- and diesters, which typically contain unusual fatty acids. Moreover, the number of different fatty acids present in the sebaceous esters is often very large. Because of the unusual structures and complex mixtures present in sebum, its analysis presents special problems.

Sebaceous glands secrete sebum by a holocrine process. That is to say, the sebaceous cells retain the sebum that they synthesize until it is released by degradation of the entire cell. The degradation appears to be an orderly process during which cell constituents such as protein, DNA, and phospholipid are recycled in some way so that they do not appear in sebum. The holocrine nature of sebum secretion means that, at any given time, different cells within the same gland are engaged in quite different metabolic activities, depending on their stage of differentiation.

The activity of sebaceous glands is under hormonal control. Androgenic hormones cause an increase in gland size by stimulating both the rate of cell division and the rate of lipid accumulation. The increase in androgen levels at puberty causes a large increase in the rate of sebum secretion. In humans, the increase is associated with the appearance of adolescent acne. For this reason, the effects of hormones on sebaceous glands have been studied extensively.

The biochemistry of sebaceous glands has proved difficult to study because techniques for obtaining lipogenesis in homogenates or in cell culture are not available. However, some information has been obtained by *in vitro* incubation of intact glands and by *in vivo* studies.

II. Anatomy and Distribution

Mammalian skin consists of a thin outer layer of epidermis and a thicker inner layer of dermis. Although sebaceous glands are located in the dermis, they are actually appendages of the epidermis. Ducts connect sebaceous glands to the hair follicles (which are also epidermal appendages). Sebaceous glands are also found under the oral epithelium and other smooth and hairless areas, in which locations the ducts lead directly to the surface. The number of sebaceous glands per unit area is greatest on the head, (Benfenati and Brillanti, 1939). The largest glands (Yamada, 1932) and the greatest variability in size (Kligman and Shelley, 1958) also occur in these areas. The palms and soles have no sebaceous glands.

Specialized organs with sebaceous-type structures also are found in mammals. Glands at the borders of the eyelids, called meibomian glands, appear histologically very similar to skin sebaceous glands, but secrete a different lipid mixture. Meibomian gland function has been reviewed recently in this series (Tiffany, 1987). Other specialized glands include the hamster flank organ, the gerbil ventral gland, and the rat and mouse preputial glands. These are thought to be scent-marking organs. The chemical composition of their secretions has not been well studied.

Microscopically, sebaceous glands consist of an outer layer of germinative cells and inner layers of cells that become increasingly large and lipid-filled as they progress toward the sebaceous duct. The germinative cells resemble undifferentiated epidermal cells in structure, having large nuclei, mitochondria, tonofilaments, and other intracellular inclusions. After differentiation, the cells become filled with lipid droplets and the nuclei appear smaller and distorted in shape. Electron microscopy (Fig. 1) reveals that near the sebaceous duct, the cells abruptly become more densely stainable by heavy metals (Ito *et al.*, 1984), apparently due to disruption of the nuclei and the release of chromatin into the cytoplasm. Degradation of chromatin and other nonlipid cell constituents apparently occurs soon after this stage because the final product is a clear oil (Kligman and Shelley, 1958; Stewart *et al.*, 1983). The protein may be recycled into lipid, but the fate of the phosphorus and nitrogenous end products is not known, since these elements are not found in the excreted sebum.

III. Comparative Chemistry of Cutaneous Lipids

A. Variation in Skin Lipid Class Composition among Vertebrates

1. *Mammals*

Profound variation in sebum composition among mammals is of considerable biological interest and has provided some fascinating chemical problems. Surveys of surface lipid composition of large numbers of species, using thin-layer chromatography (TLC), have emphasized this variability (Nicolaides *et al.*, 1968; Lindholm *et al.*, 1981). In addition, the skin surface lipids from a few mammalian species have been examined in considerable detail and the structures and chain-length compositions of individual lipids have been determined (Downing, 1976; Wheatley, 1986).

2. *Birds*

Extensive studies have also been carried out on the composition of the predominantly nonpolar lipids produced by the uropygial glands possessed by most birds (Jacob, 1976; Downing, 1986). These glands have been regarded as the avian analog of the mammalian sebaceous gland, but are limited to one pair of large structures opening onto the dorsum of the tail. As in mammals, the composition of the lipids from the uropygial glands differs widely in composition between species, both in the classes of lipids present and in the structures of the aliphatic chains (Jacob, 1976; Downing, 1986). In birds, the absence of lipid-producing glands from most of the body surface is compensated by the copious production of nonpolar lipids by the epidermis, which, as a result, has been called a "planar

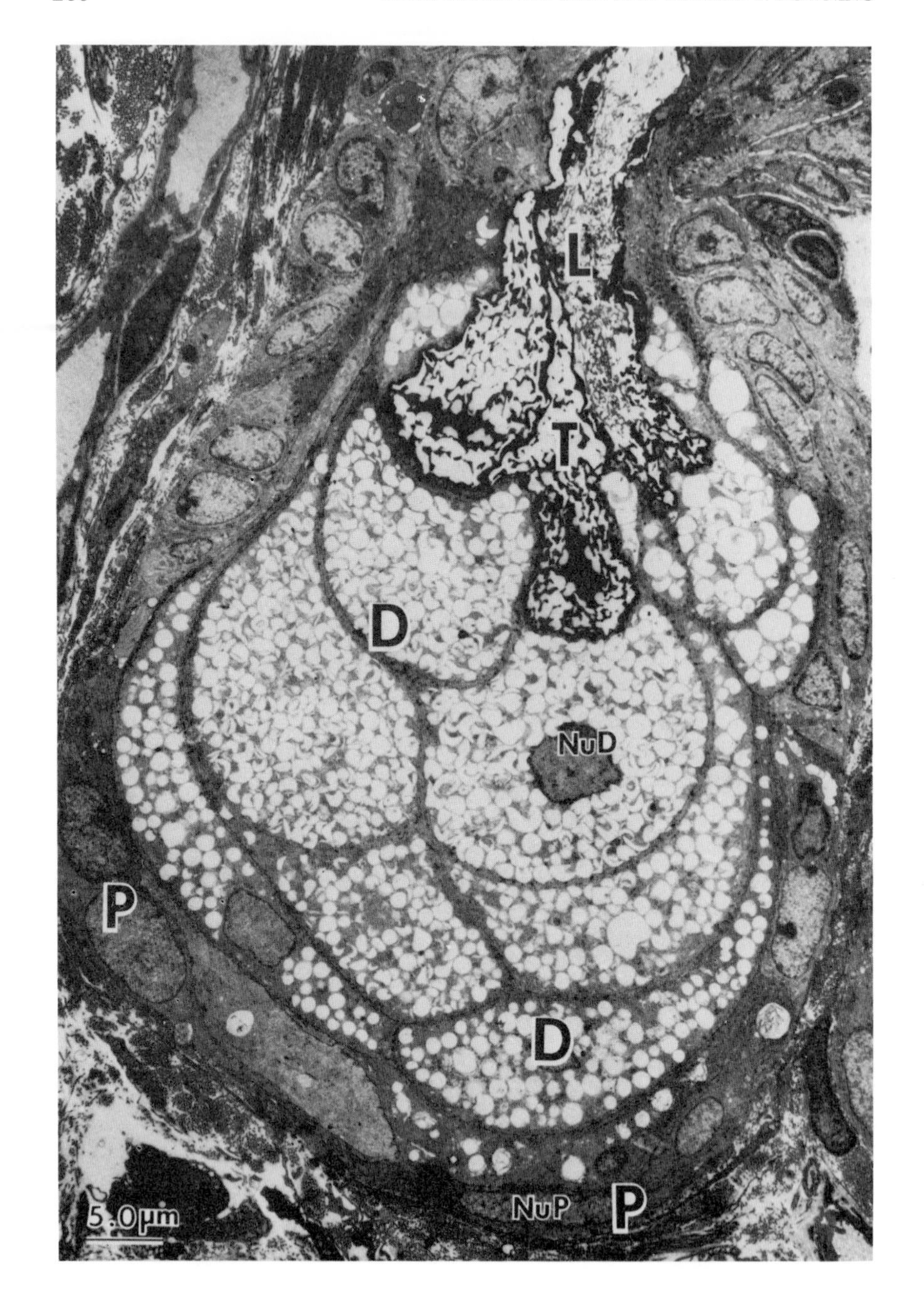
L
T
D
NuD
P
D
5.0μm
NuP
P

sebaceous gland" (Lucas, 1972). The lipid exudes from the epidermal cells as they cornify (Menon *et al.*, 1981), and in the domestic chicken amounts to almost half the weight of the stratum corneum (D. T. Downing *et al.*, unpublished observations). The nonpolar epidermal lipid of the chicken is slightly different in composition from the lipid produced by the chicken uropygial glands (Wertz *et al.*, 1986). It seems likely that the nonpolar epidermal lipids of birds will be found to differ in composition between species as do the uropygial lipids, and to reflect the composition of the glandular lipids in each species.

3. *Reptiles*

The epidermis of reptiles also produces significant amounts of nonpolar lipids. Cholesterol and free fatty acids were major components of the lipids from cast skins of 24 species examined, but the less polar lipids differed widely between the species (Burken *et al.*, 1985). Skin lipids from the Florida indigo snake contain a complex mixture of saturated and monounsaturated C_{29} to C_{33} methyl ketones, the corresponding secondary alcohols, and a series of primary alcohols having two fewer carbon atoms (Ahern and Downing, 1974). Similar series of methyl ketones in garter snakes may serve a pheromonal function (Mason *et al.*, 1989), but such lipids may be widespread among snakes (Burken *et al.*, 1985). In addition, reptiles possess specialized lipid-producing cutaneous glands (Quay, 1986), but these appear to function as the source of pheromones rather than for producing significant amounts of lipid for anointing the epidermis.

B. Variation in Sebum Lipid Class Composition among Mammals

Nicolaides *et al.* (1968) compared the skin surface lipids of 14 mammals and the uropygial gland lipids of four birds. They concluded that only humans produced sebum having a significant amount of triglycerides or squalene. In all the other species, including chimpanzee and baboon, the surface lipids contained instead a variety of mono- and diesters. In a survey of 46 mammals, Lindholm *et al.* (1981) also found profound differences in surface lipid composition between species, but showed that closely related species may have similar sebum composition.

In addition to the chromatographic surveys of sebum composition, comprehensive analyses have been made of the surface lipids from a limited number of species, many of which are included in Table I.

FIG. 1. Electron micrograph of a rabbit sebaceous gland. Uranyl acetate–lead citrate staining. The peripheral germinative cells (P) have no lipid droplets; these accumulate after differentiation and eventually fill the differentiating cells (D). Nuclei of the peripheral germinative cells (NuP) are ellipsoidal, whereas those of the differentiating cells (NuD) are irregular in shape. Toward the lumen (L) of the gland, terminally differentiated cells (T) have no intact nuclei and the cytoplasm stains darkly. Finally, lipids are released through the sebaceous duct into the hair canal. [Reproduced from Ito, M., *et al.* (1984). *J. Invest. Dermatol.* **82,** 381–385. Copyright by Williams & Wilkins, 1984.]

Table I

CLASS COMPOSITION (wt%) OF SOME MAMMALIAN SKIN SURFACE LIPIDS[a]

Lipid	Human[1]	Sheep[2]	Rabbit[3]	Rat[3]	Mouse[4]	Guinea pig[5]	Gerbil[6]	Dog[7]	Cat[3]	Cow[8]	Mink[9]	Mole[10]	Horse[11]	Donkey[12]	Mule[12]	Siberian horse[12]	Onager[12]	Zebra[12]	Otter[13]	Beaver[13]	Kinkajou[13]
Squalene	12	—	—	—	—	—	—	—	—	—	—	70	—	—	—	—	—	—	44	80	94
Sterol esters	3	46	—	27	10	33	10	42	12	3	5	5	38	24	40	44	29	35	—	—	—
Wax esters	25	10	4	17	5	—	—	—	—	—	92	15	—	—	—	—	—	—	—	—	—
Hydroxyacid diesters	—	5	66	12	—	—	—	—	66	38	—	—	—	—	—	—	—	—	—	—	—
Diol diesters	—	4	—	14	66	24	36	32	—	8	—	—	1	9		1	2	4	—	—	—
ω-Lactones	—	—	—	—	—	—	—	—	—	—	—	—	47	56	54	52	67	57	—	—	—
Glyceryl ethers	—	—	—	—	—	28	—	—	—	—	—	—	—	—	—	—	—	—	—	—	—
Wax triesters	—	—	—	—	—	—	—	—	—	30	—	—	—	—	—	—	—	—	—	—	—
Triglycerides	41	—	—	8	6	—	26	—	—	4	—	1	—	—	—	—	—	—	—	—	—
Sterol diesters	—	12	—	—	—	—	—	—	—	—	—	—	—	—	—	—	—	—	—	—	—
Free fatty acids	16	—	—	1	1	—	5	—	—	2	1	1	—	—	—	—	—	—	—	—	—
Free fatty alcohols	—	—	—	—	—	6	9	—	—	1	1	1	—	—	—	—	—	—	—	—	—
Free sterols	1	12	4	6	13	9	8	9	3	4	1	3	14	10	6	3	2	3	—	—	—
Unidentified	2	11	26	15	—	—	5	12	19	10	—	4	—	—	—	—	—	—	56	20	6

[a]Superscript numbers for each species refer to the following references: (1) Downing *et al.* (1969); (2) Downing *et al.* (1975); (3) Wheatley (1986); (4) Wilkinson and Karasek (1966); (5) Downing and Sharaf (1976); (6) Yeung *et al.* (1981); (7) Sharaf *et al.* (1977); (8) Downing and Lindholm (1982); (9) Colton *et al.* (1986); (10) Downing and Stewart (1987); (11) Downing and Colton (1980); (12) Colton and Downing (1983); (13) Lindholm and Downing (1980).

The occurrence of the various skin surface lipid classes is described below, with the classes listed in order of increasing polarity.

1. Squalene

Human sebum was thought to be the only cutaneous lipid to contain a significant amount of squalene, until a TLC survey indicated a large proportion of this constituent in surface lipids from the otter, the beaver, and the kinkajou (Lindholm and Downing, 1980; Lindholm *et al.*, 1981). A species of mole has also been found to produce sebum having a major proportion of squalene (Downing and Stewart, 1987). It is interesting that, aside from humans, all of these species inhabit damp environments.

2. Sterols and Sterol Esters

Esters of cholesterol and other sterols are among the most common constituents of surface lipids (Table I), sometimes representing a major proportion of the lipid. In some species, the esterified sterol consists exclusively of cholesterol, whereas in other species, other sterols predominate (Table II). The esterified fatty acids often include unusually long-chained species (Wheatley, 1986). In the human, very-long-chained steryl esters are found in vernix caseosa (fetal sebum) and in sebum from prepubertal children, but not in adult sebum (Nicolaides *et al.*, 1972; Stewart and Downing, 1990). Free sterols are ubiquitous in mammalian surface lipids, but usually in low concentration and consisting predominantly of cholesterol, some or all of which may originate in the epidermis.

3. Wax Esters

Waxes consisting of long-chain fatty acids esterified with long-chain alcohols have been thought of as typical constituents of sebum, but actually are present in less than half of the species that have been examined, and only in one species do these esters exceed 25% of the surface lipid (Table I).

4. Wax Diesters

Two types of wax diesters have been identified in sebum. In Type I, each molecule consists of a hydroxyacid having the carboxyl group esterified with a fatty alcohol and the hydroxyl group esterified with a fatty acid. In Type II, each molecule consists of a diol bearing two esterified fatty acids. Both types have been identified as major constituents in a few species, and have also been found together in some species (Table I). The diol diesters present in surface lipids of the domestic dog (Sharaf *et al.*, 1977) are unusual in containing one long-chain fatty acid and one isovaleric acid moiety in each molecule. Thin-layer chromatograms indicate that several canines (blue fox, red fox, coyote, and wolf), belonging to three genera, each produce similar diesters in their sebum (Lindholm *et al.*, 1981), but the detailed composition of these esters has yet to be examined.

Table II
STEROLS IN THE SURFACE LIPIDS OF SOME MAMMALIAN SPECIES

	Species[b]								
Sterol[a]	Rabbit[(1)]	Rat[(2)]	Mouse[(2)]	Guinea pig[(1)]	Gerbil[(3)]	Sheep[(4)]	Cow[(5)]	Horse[(6)]	Mink[(7)]
Cholesterol	91	23	36	91	100	50	100	100	100
Desmosterol	—	—	—	—	—	5	—	—	—
Lanosterol	—	3	—	—	—	34	—	—	—
Dihydrolanosterol	—	—	—	—	—	11	—	—	—
Agnosterol	—	8	—	—	—	2	—	—	—
Lathosterol	3	39	64	9	—	—	—	—	—
Methylsterols	—	27	—	—	—	—	—	—	—

[a]Data are expressed as a percentage of total sterols.

[b]Superscript numbers for each species refer to the following references: (1) Wheatley and James (1957); (2) Nikkari (1965); (3) Yeung *et al.* (1981); (4) Downing *et al.* (1975); (5) Downing and Lindholm (1982); (6) Downing and Colton (1980); (7) Colton *et al.* (1986).

5. *Lactones*

After the surface lipids of the horse were found to contain a high proportion of C_{30} to C_{34} ω-lactones (Downing and Colton, 1980), thin-layer chromatograms indicated that other members of the genus also produced these unusual compounds (Lindholm *et al.*, 1981). However, chemical examination showed that the lactones (equolides) produced by various equine species differ significantly in chain structures, as described later. No other mammals have been found to produce ω-lactones in their sebum, but hydrolyzed sheep sebum does contain a small proportion of ω-hydroxyacids that might be derived from lactones.

6. *Glyceryl Ethers*

The preputial gland lipids of the rat and mouse contain 1-*O*-alkyl-2,3-diacyl-*sn*-glycerols (Wheatley, 1986), but only the guinea pig has been found to have such compounds in its surface lipids (Downing and Sharaf, 1976). Subsequent biosynthetic studies with the guinea pig showed that the glyceryl ether diesters are not synthesized in the skin (Gaul *et al.*, 1985).

7. *Wax Triesters*

Triesters formed from one alkane-1,2-diol, one α-hydroxyacid, and two fatty acids have been found in surface lipids only of the cow, along with both Type I and Type II diesters, and partial hydrolysis products of each of the polyesters (Downing and Lindholm, 1982).

8. *Triglycerides*

Triglycerides and free fatty acids together amount to 57% of human sebum. The free fatty acids are not formed in the gland, but are released from triglycerides by bacteria in the hair canal and on the skin surface. Several other species have been reported to have triglycerides in their surface lipids, but in each case the constituent fatty acids resembled dietary lipids and the triglycerides probably were contaminants. There is no such doubt about the triglycerides of human surface lipids, because most of the fatty acids have unique positions of unsaturation (Nicolaides *et al.*, 1964) and/or branched chains.

9. *Free Fatty Acids*

Free fatty acids are a major component of skin surface lipids only in humans, where they are derived almost entirely from bacterial hydrolysis of sebaceous triglycerides. The proportion of free fatty acids in human skin surface lipids differs widely between subjects, depending on the degree of bacterial hydrolysis (Downing *et al.*, 1969). In other species, which all lack triglycerides, free fatty acids sometimes are minor constituents of the surface lipid and could be derived from the stratum corneum, where free fatty acids compose part of the lipid mixture (Kooyman, 1932; Reinertson and Wheatley, 1959; Gray *et al.*, 1982; Wertz *et al.*, 1987; Hedberg *et al.*, 1988). Wax mono- and diesters are regarded as being resistant to bacterial hydrolysis. However, the occurrence of small proportions of partially hydrolyzed diesters and triesters in surface lipids of the cow (Downing and Lindholm, 1982) indicates that some nonglyceridic esters might be susceptible to bacterial hydrolysis on the skin surface.

10. *Free Fatty Alcohols*

Fatty alcohols rarely occur uncombined in skin surface lipids. It was this observation that led to the conclusion that wax esters are resistant to hydrolysis by cutaneous bacteria. Nevertheless, free alcohols do occur in the surface lipids of the guinea pig and the gerbil (Table I), where they may be secreted as such by the sebaceous glands.

11. *Free Sterols*

Where sterols occur unesterified in the skin surface lipids, they usually consist predominantly or exclusively of cholesterol, part or all of which probably originates in the epidermis, where cholesterol forms a major part of the lipid mixture (Kooyman, 1932; Reinertson and Wheatley, 1959; Gray *et al.*, 1982; Wertz *et al.*, 1987).

C. Differences in Aliphatic Chain Structures between Species

1. *Branched Chains*

Although branched-chain fatty acids and alcohols are viewed as characteristic of sebaceous lipids, there is wide variation between species in the proportions and

Table III

PERCENTAGE OF THE ALIPHATIC COMPONENTS OF SHEEP WOOL LIPIDS[a]

Carbon number	Component[b]: Fatty acids n	i	a	ω-Hydroxyacids n	i	a	α-Hydroxyacids n	i	a	Fatty alcohols n	i	a	α,β-Diols n	i	a
12	0.1	—	0.3	—	—	—	—	—	—	—	—	—	—	—	—
13	0.2	—	1.1	—	—	—	—	—	—	—	—	—	—	—	—
14	1.9	4.1	—	—	—	—	2.3	0.5	—	—	—	—	0.1	—	—
15	1.2	—	7.7	—	—	—	0.9	—	1.1	—	—	—	—	—	0.1
16	2.5	6.7	—	—	—	—	39.7	1.4	—	0.1	0.1	—	2.2	0.3	—
17	0.2	—	4.1	—	—	—	0.7	—	0.4	—	—	0.1	—	—	0.6
18	1.5	4.9	—	—	—	—	2.1	24.3	—	0.5	0.4	—	0.9	17.0	—
19	—	—	6.3	—	—	—	0.1	—	3.1	—	—	0.3	0.2	—	5.8
20	0.9	5.2	—	—	—	—	0.3	3.2	—	2.9	7.0	—	1.0	11.8	—
21	—	—	5.8	—	—	—	0.1	—	2.4	0.1	—	3.8	0.4	—	12.3
22	0.8	2.4	—	—	—	—	0.3	3.6	—	2.6	4.0	—	0.8	16.2	—
23	—	—	3.2	—	—	—	—	—	5.2	0.1	—	1.8	0.4	—	15.0
24	3.0	3.2	—	—	—	—	0.2	5.8	—	6.6	5.0	—	0.4	10.2	—
25	—	—	6.2	—	—	—	—	—	2.7	—	—	8.4	—	—	4.1
26	2.2	5.2	—	0.9	—	—	—	—	—	4.0	14.3	—	—	0.1	—
27	—	—	5.5	—	—	0.5	—	—	—	—	—	18.1	—	—	—
28	1.1	2.2	—	7.7	—	—	—	—	—	1.4	5.0	—	—	—	—
29	—	—	3.9	—	—	4.6	—	—	—	—	—	6.3	—	—	—
30	0.5	1.3	—	37.8	3.4	—	—	—	—	0.6	—	—	—	—	—
31	—	—	2.6	—	—	12.3	—	—	—	—	—	2.3	—	—	—
32	—	—	—	19.0	4.6	—	—	—	—	—	—	—	—	—	—
33	—	—	—	—	—	7.7	—	—	—	—	—	1.5	—	—	—
34	—	—	—	1.5	—	—	—	—	—	—	—	—	—	—	—
Totals	16.0	35.1	46.4	66.9	8.0	25.1	46.7	38.8	14.9	18.9	35.8	42.6	6.4	55.6	37.9

[a]From Downing *et al.* (1960).
[b]Abbreviations: n, normal; i, iso; a, anteiso.

structures of the branched compounds. Some species, such as the cow (Downing and Lindholm, 1982) and the mink (Colton *et al.*, 1986), produce exclusively straight-chain fatty acids. In contrast, branched-chain fatty acids of the iso and anteiso structures predominate in sheep sebum (Table III), from which they were first isolated by fractional distillation (Weitkamp, 1945). Subsequent studies showed that the fatty alcohols, fatty diols, and hydroxyacids of sheep sebum all contain a high proportion of branched-chain components (Downing *et al.*, 1960). Detailed chronicles have been published of the numerous investigations that led to the present knowledge of sheep sebum fatty acids (Motiuk, 1979a) and fatty alcohols (Motiuk, 1979b).

In human sebum, straight-chain acids and alcohols predominate, but branched-chain compounds were readily apparent when analyses were first performed by

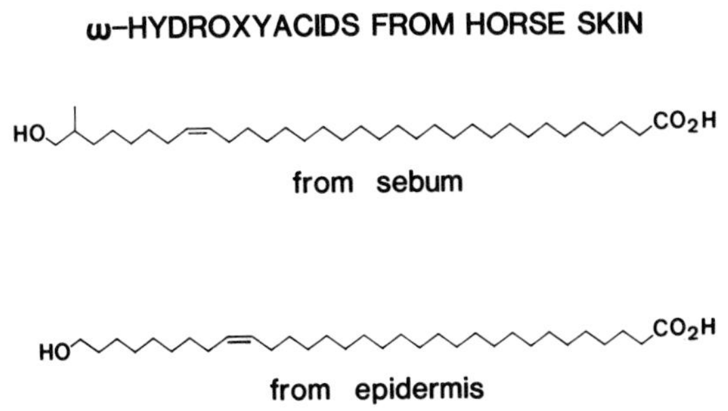

ω-HYDROXYACIDS FROM HORSE SKIN

HO ... CO_2H

from sebum

HO ... CO_2H

from epidermis

FIG. 2. Representative ω-hydroxyacids from horse sebaceous lactones and horse epidermal acylglucosylceramides. The lactones contain virtually 100% branched-chain acids, whereas the acylglucosylceramides contain almost 100% straight-chain acids (Wertz *et al.*, 1983).

gas chromatography (James and Wheatley, 1956). In addition to the iso and anteiso chain structures, the saturated fatty acids of human sebum were found to contain members having methyl branches on one or more of the even-numbered carbon atoms throughout the chain (Nicolaides and Apon, 1977).

Species differences in sebaceous branched-chain fatty acids appear to be genetically controlled. This was clearly demonstrated in a study of surface lipids from members of the genus *Equus* (Colton and Downing, 1983). In each species, the predominant lipids are the equolides, consisting of the giant-ring lactones formed by cyclizing C_{30} to C_{34} ω-hydroxyacids. In the domestic horse, the hydroxyacid chains are almost entirely iso-branched, whereas in the donkey the chains are entirely straight-chained. In the hybrid of these species, the mule, the equolide chains consist of equal parts of branched-chain and straight-chain structures.

The observations with the domestic horse were extended to comparing the chain structures of the sebaceous equolides with the ω-hydroxyacids of similar chain length present in the acylglucosylceramides of horse epidermis (Wertz *et al.*, 1983). The epidermal hydroxyacids contained only 0.2% of branched-chain compounds in contrast to 99.7% branched-chain components in the equolides (Fig. 2). This result illustrates the metabolic autonomy of differentiated sebaceous cells.

2. *Unsaturated Chains*

In human sebum, about half of the fatty acid and fatty alcohol chains are monounsaturated (Table IV), and the species apparently is unique in producing these series by Δ^6 desaturation (Weitkamp *et al.*, 1947; Nicolaides, 1967; Nicolaides *et al.*, 1964, 1972; Downing and Greene, 1968b). Methylene-interrupted desaturation of monoenes produced by Δ^6 desaturation also occurs, resulting in a series of unique dienoic acids (Nicolaides and Ansari, 1969). The unique positions of unsaturation provide evidence that the lipid classes, including the triglycerides, are syn-

Table IV

POSITIONAL ISOMERS OF FATTY ACID MONOENES OF SEBACEOUS WAX ESTERS

Monoene carbon skeleton[a]	Mole% of total monoenes	Mole% at Δ positions indicated							
		Δ^5	Δ^6	Δ^7	Δ^8	Δ^9	Δ^{10}	Δ^{11}	Δ^{12}
n-14	7.57	—	100	—	—	—	—	—	—
n-16	47.17	—	95	—	4	1	—	—	—
n-18	7.81	—	38	2	52	5	3	—	—
n-20	0.44	—	—	11	9	3	61	6	10
n-22 to 28	0.12	—	—	—	—	—	—	—	—
n-13	0.05	—	—	—	—	—	—	—	—
n-15	5.36	—	100	—	—	—	—	—	—
n-17	1.97	—	75	7	17	1	—	—	—
n-19	0.18	—	55	13	38	8	34	2	—
n-21 to 27	0.03	—	—	—	—	—	—	—	—
i-16	21.63	—	100	—	—	—	—	—	—
i-18	1.35	2	39	2	57	—	—	—	—
i-20	0.08	6	7	1	24	4	59	—	—
i-22 to 28	0.08								
ai-15	2.36	1	97	1	1	—	—	—	—
ai-17	3.66	—	100	—	—	—	—	—	—
ai-19	0.12	—	47	9	44	—	—	—	—
ai-21 to 27	0.02	—	—	—	—	—	—	—	—

[a]Abbreviations: n, normal, i, iso, ai, anteiso. Trace constituents are omitted. Adapted from Nicolaides *et al.* (1972).

thesized *de novo* in the sebaceous glands and are not derived from dietary or circulating lipids. Similar assurance is provided in several other species by the presence in their sebum of high concentrations of monounsaturated chains that appear to have been formed by chain extension following Δ^9 desaturation of myristic and palmitic acids, instead of stearic acid, as is predominant in other tissues.

In contrast to the methylene-interrupted dienoic fatty acids of human sebum, dienoic acids of quite a different type have been found among the equolides of horse sebum (Frost *et al.*, 1984). The positions of unsaturation in these chains indicate an initial Δ^9 desaturation of the iso-C_{18} saturated fatty acid, followed by chain extension to C_{24}–C_{32}, a second Δ^9 desaturation, and further chain extension to 34, 36, or 38 carbons (Figs. 3 and 4).

D. DIFFERENCES IN STEROL STRUCTURES BETWEEN SPECIES

Although some species have only cholesterol in the sterol esters or the free sterols of their surface lipids, others appear to have a complete block in the

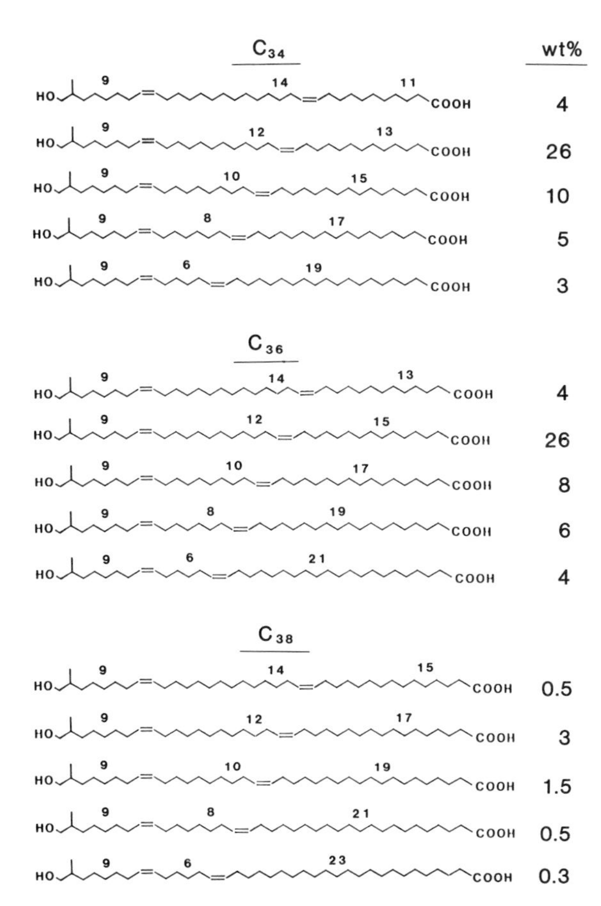

FIG. 3. Composition of the dienoic ω-hydroxyacids from horse sebaceous lactones. The double bonds appear to have been introduced by two successive Δ^9 desaturations, separated and followed by variable degrees of chain extension (Frost *et al.*, 1984).

biosynthesis of cholesterol in their sebaceous glands (Table III). This may be the case with each of the five species that produce squalene as a major sebum constituent (Table I); the small amount of cholesterol in their surface lipids may be present in the sebaceous cells prior to their differentiation and/or may be a product of the epidermal cells. In other species, notably the rat and the sheep, several sterols are each present in significant amounts and presumably result from several partial interruptions in the biosynthetic pathway (Fig. 5). Alternatively, complete blockage at different points in the pathway at different stages in sebocyte differentiation could produce a similar result. A time-course study of sebum synthesis *in vivo* might distinguish between these possibilities.

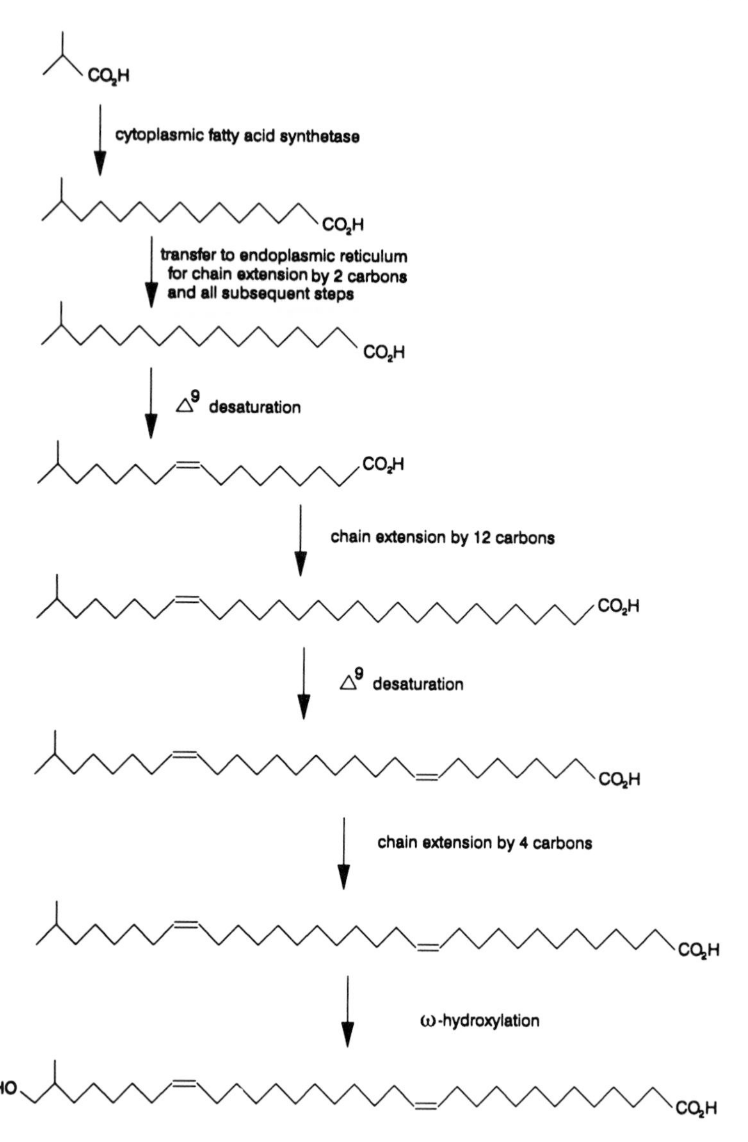

FIG. 4. Inferred pathway for biosynthesis of the dienoic hydroxyacids found as lactones (equolides) in the sebaceous glands of the horse (Frost *et al.*, 1984).

In addition to the quantitatively significant sterols in the surface lipids, small amounts of 7-dehydrocholesterol, and possibly 7-hydroxycholesterol, are biologically significant constituents as precursors of vitamin D. In humans, the vitamin D is produced in the epidermis (Wheatley and Reinertson, 1958; Holick 1986),

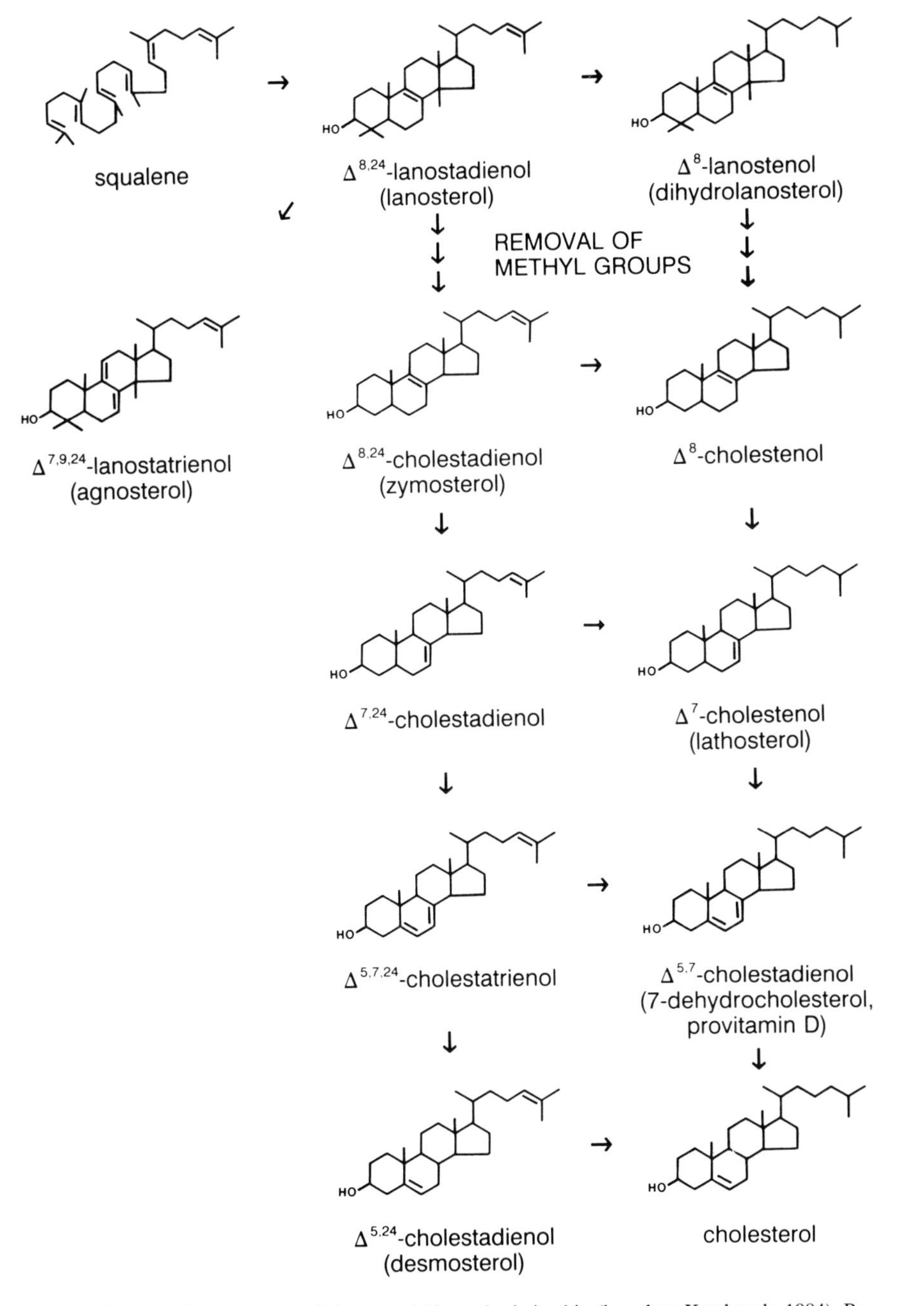

FIG. 5. Possible pathways of cholesterol biosynthesis in skin (based on Kandutsch, 1984). By comparison with Tables I and II, it can be postulated where the points of inhibition are in various species. Thus, there appears to be inhibition of squalene cyclization in the human, beaver, otter, kinkajou, and mole, of demethylation in the sheep and rat, of double-bond isomerization in the rat and mouse, and of side-chain reduction in the sheep.

but in furred animals it may be produced on the pelage from precursors excreted in sebum, and then either absorbed through the epidermis or ingested during grooming.

IV. Analysis of Sebum

Because the sebum of most mammals is a complex mixture of lipid classes, each of which may contain homologous series of several chain structures, the analysis of sebum is not a trivial pursuit. Many of the methods that have been developed for this purpose have been described in considerably greater detail than is possible here (Downing and Stewart, 1985). The present treatment is limited to outlining the procedures that are most frequently required and to describing the principles for their use.

A. Isolation and Analysis of Lipid Classes

1. *Thin-Layer Chromatography*

Most of the lipid classes that are present in sebum can be resolved adequately enough by TLC on silica gel for quantitative analysis and for preparative separation. Using commercially prepared plates (Alltech Associates; Deerfield, Illinois; cat. # 16330), and successive development with hexane, then toluene, then hexane/diethyl ether/acetic acid (70:30:1), almost all sebaceous lipids can be resolved. With 10–20 μg of total lipid applied to 6-mm-wide lanes scored in the adsorbent, and visualization of the developed chromatograms by charring with sulfuric acid, photodensitometry can provide quantitative analyses of constituents amounting to as little as 1–2% of the sebum (Downing, 1968; Downing and Stranieri, 1980).

In studies involving species whose sebum has not been rigorously characterized, TLC mobility alone cannot be relied upon as a criterion of identity. For this purpose, preparative TLC can be used for the isolation of lipid classes for further identification. As much as 50 mg of total lipid can be applied across the width of a TLC plate and resolved with the multiple development system. Visualization can be achieved by spraying the plate with a fluorescent agent such as 2′, 7′-dichlorofluorescein or 8-hydroxy-1,3,6-pyrenetrisulfonic acid (both Eastman Kodak Co.; Rochester, New York) and viewing under UV light. Alternatively, lipid bands on the chromatogram can be located by scanning the chromatogram with a photodensitometer at a wavelength of 200 nm. The located bands can then be scraped off and eluted with ether to recover each of the resolved lipids.

Two lipid classes that are often difficult to resolve from each other by TLC, depending on the chain lengths of the constituent fatty acids. are the wax esters and cholesterol esters. These can be separated by column chromatography on magnesium hydroxide (Stewart and Downing, 1981), because of the preferential retention of cholesterol esters by this adsorbent.

2. *Spectrometry*

Infrared and nuclear magnetic resonance (NMR) spectrometry have been useful in the initial characterization of lipid classes from sebum, but usually some form of chemical evidence is necessary for positive identification. In addition, since most aliphatic constituents consist of homologous series, gas chromatography must be employed for resolution, followed by mass spectrometry for identification of individual homologues.

In some instances, NMR may be sufficient to establish the structures of individual lipid classes, even if they consist of one or more homologous series. Variations in chain length and chain structure will usually affect only the methyl and methylene signals in the spectrum, leaving clear downfield signals for diagnosis of protons adjacent to functional groups and unsaturation. Examples of this use of NMR are provided by the spectra obtained of giant-ring lactones of branched-chain or straight-chain ω-hydroxyacids, and of natural mixtures of the two, from the sebum of various equine species (Fig. 6).

3. *Alkaline Hydrolysis*

Since most sebaceous lipids contain ester linkages, it is usual to first apply alkaline hydrolysis to liberate carboxylic and hydroxylic moieties for detailed analysis. In early studies, before the ability to resolve individual lipid classes was developed, saponification was always the first step in analysis, so that only the total acids and alcohols were produced, and no information could be obtained regarding the individual lipid classes.

Early studies also produced the impression that wax esters were highly resistant to saponification, but it is now known that under mild, homogeneous reaction conditions the hydrolysis can be achieved rapidly. In ethanol or methanol solution containing 5% water and 1 M NaOH, wax esters can be completely hydrolyzed in 30 minutes at 45°C. Esters of secondary alcohols may require more vigorous conditions (70°C for 2 hours). The liberated fatty alcohols or sterols can then be extracted with hexane after addition of water. However, this solvent partition does not provide a clear-cut or efficient separation of the acids and alcohols. A much cleaner separation is obtained by precipitating the calcium salts of the acids by addition of calcium chloride and extraction of the unsaponifiables with acetone (Downing et al., 1960). Recently, the practice has been adopted of treating the initial saponification mixture with excess BCl_3/methanol and separating the resulting mixture of alcohols and fatty acid methyl esters by TLC (Downing and Stewart, 1985).

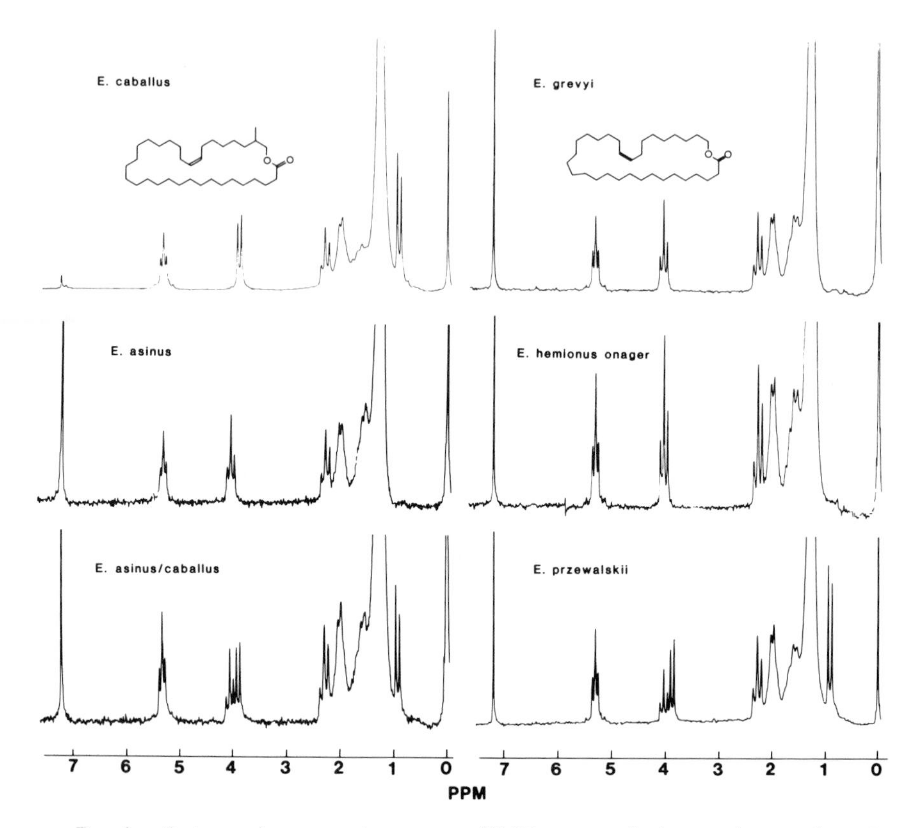

FIG. 6. Proton nuclear magnetic resonance (NMR) spectra of sebaceous lactones from six species of the genus *Equus*. Methyl group signals (0.92 ppm) are absent in the NMR spectra of lactones from the donkey (*E. asinus*), zebra (*E. grevyi*), and Asiatic ass (*E. hemionus onager*), indicating straight, ω-substituted chains. In the horse (*E. caballus*), the 0.92-ppm signal is a doublet, indicating a methyl side chain; the ω-methylene protons adjacent to the alkyl oxygen also produce a doublet, at 4.06 ppm, indicating that the methyl group is attached to the ω-1 carbon. In the mule (*E. asinus/caballus*) and Siberian horse (*E. przewalskii*), the ω-methylene protons produced a mixture of doublets and triplets, indicating mixtures of methyl-branched and straight chains. The spectra were obtained at 90 MHz in deuterochloroform (Colton and Downing, 1983).

B. STRUCTURE DETERMINATION OF ALIPHATIC SERIES

1. *Fatty Acids*

In extensive studies by Nicolaides (1971) and Nicolaides and Apon (1976, 1977), capillary gas chromatography was used to resolve the methyl esters of saturated methyl-branched fatty acids from human sebum, and mass spectrometry was used for locating the position of branching in each isomer. A pattern of equivalent chain lengths was established, from which individual fatty acid methyl esters now can be recognized from their gas chromatographic relative retention times. Also, retention times can be predicted for acids that possess more than one methyl branch. The homologous series of branched acids identified by Nicolaides and Apon cover virtually all of the variations of chain branching that have been discovered in mammalian sebum, but, because of the vast differences in sebum composition between species, caution should still be exercised in assigning identity when examining new species.

2. *Hydroxyacids*

Although hydroxyacids are widespread in mammalian sebum, only α- and ω-hydroxyl locations have been noted so far. It is fortuitous that in these positions, the location of the hydroxyl function is readily established. Reduction of hydroxyacid methyl esters with lithium aluminum hydride produces diols. The diols produced from α- or β-hydroxyacids form an isopropylidene adduct that is useful in chromatographic isolation. The α,β-diols will react with periodate, distinguishing them from α, γ-diols, which also will form an adduct with acetone. The ω-hydroxyacid methyl esters are easily recognized by oxidation of the hydroxyl to a carboxyl with chromic acid in acetone solution (Carey and Sundberg, 1983). In all other positions the hydroxyl group will be oxidized to a ketone, the position of which can be located chemically or by mass spectrometry.

3. *Fatty Alcohols*

In the mammalian species that have been studied, the sebaceous fatty alcohols have chain branching and unsaturation similar to those found in the fatty acids of the same species. The logistics of analysis may therefore be simplified by oxidizing the alcohols to fatty acids with chromic acid in acetone, and then using the same procedures that have been developed for analysis of fatty acids. Likewise, where diols have been found, these have been either α,β- or α,ω-isomers, which can be handled similarly to the corresponding materials produced by $LiAlH_4$ reduction of hydroxyacids.

4. *Sterols*

Preliminary examinations of sebum are greatly assisted by the characteristic positions of migration of sterols and sterol esters on TLC, and by the specific colors that many sterols produce when chromatograms are slowly heated after spraying with sulfuric acid. The colors produced by each sterol are similar to those obtained in the Liebermann–Burchard reaction, and can be seen with less than 1 μg of sterol on the chromatogram.

After isolation of the sterols, the acetates or silyl derivatives can be subjected to gas chromatography/mass spectrometry for more positive identification. This procedure is necessary in view of the wide range of sterols that have been found in sebum from various species, and the frequent occurrence of numerous intermediates of cholesterol biosynthesis that differ only by the number or position of double bonds or methyl groups. However, it is usual to find most of the intermediates in the form of esters, whereas cholesterol occurs as both free sterol and sterol esters.

5. *Positions of Unsaturation*

In a number of species, unsaturated fatty acids from sebum have been shown to have double-bond positions that are unlike those in other tissues. For this reason, positions of unsaturation need to be rigorously established for each new species that is examined. With the use of capillary gas chromatography, most double-bond-positional isomers can be resolved and the unsaturation located by mass spectrometry. Otherwise, location of positions of unsaturation by chemical means requires, first, the separation of the fatty acid methyl esters according to degree of unsaturation (Downing and Stewart, 1985), and then isolation of each chain length by gas chromatography or high-performance liquid chromatography (HPLC). Since each monounsaturated chain will usually contain more than one positional isomer, mass spectrometry alone is not useful for the location or quantitative analysis of the distribution of isomers even in the pure chain-length fractions.

A variety of chemical procedures have been developed that involve splitting the unsaturated acids at the double bond(s) and analysis of the products by gas chromatography. The most commonly used procedure is ozonization and reduction (Nicolaides *et al.*, 1972). Periodate/permanganate oxidation has also been employed for scission at positions of unsaturation; loss of the volatile short-chain fragments can be avoided by handling them as tetramethylammonium salts that are pyrolyzed to methyl esters in the gas chromatograph (Downing and Greene, 1968a). These methods have shown that, in some instances, each pure chain-length fraction of monounsaturated acids may contain up to 10 positional isomers (Downing and Greene, 1968b), which would be difficult to analyze by gas–liquid chromatography (GLC)/mass spectrometry.

C. Distinguishing Sebum from Other Skin Surface Lipids

Sebum is remarkable for containing classes of lipids that are not found in other tissues and that can be used as markers for the presence and amount of sebum on the skin surface. However, it should be remembered that the class composition of sebum differs profoundly between species. Furthermore, unusual lipids from other endogenous and exogenous sources may be found on the skin. Specialized lipid-secreting cutaneous glands are found in different species in a variety of locations, such as the hamster flank organ, the gerbil ventral gland, and the preputial glands of many rodents. In addition, many mammals possess Harderian glands behind the eye, which can produce copious amounts of lipid that could be transferred to the fur. In all species, the epidermis produces polar and nonpolar lipids, predominantly ceramides, cholesterol, and free fatty acids, with small proportions of cholesterol esters and diesters and triglycerides. The surface lipids of laboratory animals may also become contaminated with lipids from a variety of sources, including food, feces, bedding, and atmospheric pollution. In human subjects, cosmetics and emollients may be sources of contamination. Instructions should be given to discontinue the use of such products before collection of lipids.

In spite of all of these potential sources of contamination, significant interference with the determination of sebum composition is infrequent, because the quantity of sebum produced usually far outweighs most potential sources of contamination. In young adult humans, the amount of sebum produced on the head, chest, and back (averaging about 1 mg/cm^2/day) is sufficient to overcome any significant effect of epidermal lipids on sebum composition (Greene *et al.*, 1970), and normal frequency of washing seems adequate to minimize most environmental contamination. However, in areas of low sebum production and in children and elderly subjects, epidermal and environmental lipids may make a significant contribution. In animals, a steady flow of sebum from the base to the tip of the hairs tends to sweep contaminating lipids to the surface of the pelage, from which they are transferred to environmental surfaces.

Although several sources of potential contamination of sebum are recognized, definition of these is not an easy task. One method that appears to have the potential to decide this question is isotopic labeling. When [^{14}C]acetate is injected intradermally, the label is incorporated into both sebum and epidermal lipid. Lipids that are not synthesized at the site will be unlabeled, and therefore recognizable as contaminants. In this way, a major constituent of guinea pig surface lipids, the glyceryl ether diesters, was found to be exogenous (Gaul *et al.*, 1985), possibly originating in the Harderian glands of the eye (Yamazaki *et al.*, 1981). Likewise, the paraffin hydrocarbons that are almost ubiquitous in surface lipids were unlabeled after intradermal injection of [^{14}C]acetate in pigs (Hedberg *et al.*, 1988). It has since been found that the paraffin hydrocarbons of human surface lipids are devoid of even the natural background level of ^{14}C, and therefore must be contaminants of fossil origin (Bortz *et al.*, 1989).

In certain circumstances, the fatty acid composition of a particular surface lipid component will provide evidence of its origin in the sebaceous glands or in one of the specialized lipid-producing cutaneous organs. Conversely, the presence of high concentrations of linoleic acid in the skin surface triglycerides of the gerbil (Yeung *et al.*, 1981) and the cow (Noble *et al.*, 1974) suggests that in these instances the triglycerides were not produced by the sebaceous glands. In fact, triglycerides have frequently been reported in surface lipids but probably are contaminants in all species except humans, where the branched chains and unique positions of unsaturation indicate their sebaceous origin. In the case of animals that are given a very fatty diet, fecal excretion of undigested triglyceride or direct contamination from the food may be the source of the triglycerides. When this occurs, it should be possible to detect the fact by comparison of the compositions of the surface triglycerides with those of the food.

V. Physiology of Sebum Secretion

As detailed above, the lipids synthesized *de novo* by the sebaceous glands are different from circulating lipids and often are unique to the species. In addition, sebaceous glands have at least two other peculiarities. One is the holocrine mechanism of secretion, which contrasts with the merocrine mechanism of many secretory glands. Another somewhat unusual feature is that the activity of the glands is profoundly influenced by the sex hormone status of the animal. These facts have important consequences for the design of experiments and the interpretation of experimental observations.

A. Holocrine Secretion of Sebum

Holocrine secretion involves continuous cell division in the germinative epithelium of the gland, gradual accumulation of the product in each cell as it grows and differentiates, and eventual complete disruption of the cell as it releases its product. Undifferentiated germinative sebocytes on the periphery of the sebaceous gland contain little or no sebum, but presumably have a normal complement of cholesterol and phospholipids derived from the lipids in the circulation. When sebocytes differentiate, they appear to lose the ability to take up preformed lipids from the circulation, since few of the lipid classes or chain structures in sebum have counterparts in the circulating lipids. Even new phospholipids appear to be made from sebaceous-type fatty acids (Stewart *et al.*, 1978; Colton and Downing, 1985b).

At the end of the cell's life, all of the subcellular organelles are digested, including the membrane phospholipids, fatty acids from which are incorporated into additional sebum. Membrane cholesterol may be esterified with membrane fatty acids or with fatty acids synthesized *de novo*. As the result of this reprocessing of the cell's initial endowment of lipid, sebum probably always contains at least a small amount of exogenous cholesterol and exogenous fatty acids in addition to the lipid synthesized *de novo*. However, this exogenous lipid may be highly diluted, since during sebum accumulation, a sebocyte may increase in volume by up to 150-fold (Tosti, 1974), most of which represents newly synthesized lipid.

Because lipid is not released from a sebaceous cell until it has completed its life cycle, it is important to know the average life span or turnover time of sebaceous cells. The turnover time of human sebaceous cells has been determined by incorporation, migration, and loss of tritiated thymidine, and was reported to be about 14 days (Plewig and Christophers, 1974). On the other hand, by intradermal injection of [^{14}C]acetate, the average time between synthesis and excretion of lipid was shown to be about 8 days in humans, 6 days in sheep (Downing *et al.*, 1975), 5 days in rats (Colton and Downing, 1985c) and guinea pigs (Gaul *et al.*, 1985), and as long as 21 days in horses (Colton and Downing, 1985a).

Because of the time that elapses between lipid synthesis and its delivery onto the skin surface, there will be a delay before any change in the composition of the sebum synthesized becomes apparent on the skin surface. The effect of this delay was seen in experiments involving fasting human subjects (Pochi *et al.*, 1970; Downing *et al.*, 1972). In these subjects, triglyceride and wax ester secretion was reduced about 50%, whereas squalene biosynthesis was unaffected. As a result, the concentration of squalene in the sebum doubled during fasting, but the full effect was not seen in surface sebum until 10 days after fasting began. After normal diet was resumed, 10 days elapsed before sebum returned to its prefast composition.

The rate of secretion in a holocrine system depends not only on turnover time, but also on the number of cells making sebum and on the amount of product generated by each cell. Since the musculature of the pilosebaceous apparatus is inadequate for expression of product (Kligman and Shelley, 1958), material is delivered at a constant rate by the force of continuing cell division and product accumulation. Changes in sebum excretion occur only as a result of changes in the rate of cell division or in the amount of sebum produced per cell. An implication of these observations is that changes in sebum excretion occur on a time scale of days rather than of minutes or hours. However, it has been repeatedly observed that, after the skin surface is defatted by washing or solvent extraction, the surface film of sebum is replaced much more rapidly than can be explained by the normal rate of sebum biosynthesis. Kligman and Shelley (1958) recognized that the sebaceous glands themselves do not respond to defatting, but continue to produce sebum at a steady rate. They proposed that the rapid restoration of the surface lipid was provided by a reservoir of sebum in the hair canal, from which sebum tends to be withdrawn by capillary action when the skin surface (which is criss-crossed by tiny crevices) is defatted.

B. Measurement of Sebum Secretion Rate

The existence of a follicular reservoir of sebum tends to inflate measurements of sebum secretion. However, the effect of the reservoir can be minimized by a technique involving continuous absorption of sebum from the surface for 14 hours, following by continuous collection at the same site for a subsequent interval (Downing *et al.*, 1982; Collison *et al.*, 1987). A number of studies have used this technique for measurement of sebum secretion in children (Stewart and Downing, 1985a), in adults over the entire human life span (Jacobsen *et al.*, 1985), and in acne patients and normal controls (Harris *et al.*, 1983).

A measure of sebum secretion rate can also be obtained by determining the sebum content of full-thickness skin biopsies (Downing *et al.*, 1981). Since the average delay between sebum synthesis and its excretion onto the skin surface in man is about 8 days (Downing *et al.*, 1975), it can be inferred that at any given time the skin will contain the product of 8 days of lipogenesis by the glands. Although this method has been used for human subjects, it clearly is not suitable for routine use. The method should, however, be especially useful in animal studies, wherein the reservoir depletion and surface collection of lipids are usually made difficult by the pelage and by the habits of the animal.

Knowledge that the skin contains 8 days' production of sebum can also be used to calculate the rate of flow of sebum through the follicular canal (Downing and Strauss, 1982). The number of follicles per unit area and the average length and diameter of the follicular lumen were measured in unfixed sections of human scalp skin. All of the parameters were used to calculate the rate of flow in the follicular canals. The time required to traverse the measured length of the canals was then estimated to be about 12 hours, close to the time required to deplete the follicular reservoir.

C. Hormonal Influences on Sebum Composition

The most important hormones affecting sebaceous glands are the sex steroids. Androgens stimulate and estrogens suppress sebaceous gland activity. Rony and Zakon (1943) first showed that injections of testosterone in prepubertal boys stimulated increases in sebaceous gland size and in the number of sebaceous glands per follicle. Suppression of sebaceous gland activity by a synthetic estrogen was shown by Jarrett (1955). These effects were quantified by Strauss and Pochi (1963) using a gravimetric assay of sebum production (Strauss and Pochi, 1961). Methyltestosterone was found to stimulate sebum production not only in prepubertal children, but also in elderly women and in men in whom sebum production had been suppressed with ethynyl estradiol. Estrogen suppressed sebum production not only in young men, but also in young women and in male castrates. Adrenal androgens also stimulate sebum production (Drucker *et al.*, 1972) and presumably are responsible for maintaining sebaceous gland activity in women and eunuchs.

Extensive research by Ebling (1974) and colleagues on rats suggests that androgens stimulate sebum secretion by stimulating lipid synthesis as well as by stimulating mitosis. That the two processes are controlled independently was indicated by the observation that estradiol suppresses lipid secretion without affecting mitosis. Similar evidence has not been obtained for humans, but histologically it can be observed that the sebocytes of prepubertal children are often devoid of visible lipid droplets (Strauss and Pochi, 1963), whereas the differentiated cells of adults are greatly distended with lipid (Tosti, 1974).

As discussed above, the lipid that accumulates in differentiated cells is synthesized *de novo*. However, since secretion is holocrine, this newly synthesized lipid eventually becomes mixed with lipid acquired before differentiation. The amount of lipid acquired by a cell before differentiation would be relatively constant, whereas the amount of lipid synthesized after differentiation would depend on the hormonal state of the animal. If the membrane lipid is of the usual type (i.e., cholesterol and phospholipids composed of circulating fatty acids), but the newly synthesized lipid is of distinctive sebaceous types, then it can be seen that hormonal stimulation would change the composition of the sebum finally secreted by changing the ratio between the two types of lipid. Thus, the composition of animal sebum may be different in castrates compared to intact animals, and may also change with seasonal variations in gonadal activity. In humans, it would be expected that prepubertal children (who have levels of circulating androgens near zero) would have sebum of a composition different from that of adults, and this has been demonstrated.

The fact that the class composition of children's skin surface lipid is different from that of adults has been known since Eckstein (1927) reported that the hair fat of young children has a high cholesterol content compared to that of adults. Ramasastry *et al.* (1970) investigated the class composition of skin surface lipid from birth to puberty. Peak concentration of total cholesterol (free cholesterol plus cholesterol esters) occurred at about age 6, which was also the age when wax ester concentration was the lowest. At the time, the interpretation of the changes in lipid class composition was that cholesterol and cholesterol esters are of epidermal origin and compose a large proportion of skin surface lipid in children because there is little sebaceous wax esters and triglycerides to dilute them. However, further investigation suggested that cholesterol esters are secreted mainly by the sebaceous glands (Stewart *et al.*, 1984), and that the composition of sebum changes during puberty.

The percentage of cholesterol and its esters in sebum can provide a direct expression of sebaceous gland activity when incorporated into the following ratio: wax esters:(cholesterol + cholesterol esters) [WE:(CH + CE)]. This ratio has been found to be directly proportional to the absolute rate of wax ester secretion in young children (Stewart and Downing, 1985a). This observation indicates that, as wax ester secretion increases with increased biosynthetic activity by each sebaceous cell, cholesterol secretion remains constant or increases much more slowly, providing a sort of internal standard. Pochi *et al.* (1977) used the WE:(CH + CE) ratio to demonstrate a positive correlation between sebaceous gland activity and the urinary excretion of adrenal androgens in children aged 5–10. In teenagers and adults, a correlation between the WE:(CH + CE) ratio and wax ester secretion has not been demonstrated. This may be because the WE:(CH + CE) ratio is a measure of sebum synthesis per cell, whereas variations in adult sebum secretion may depend both on sebum synthesis per cell and on the number of sebaceous glands per unit area of skin. Also, the percentage of cholesterol esters is difficult to measure accurately in adult sebum because the amount is small compared to the wax esters, with which cholesterol esters overlap chromatographically.

During puberty, changes in fatty acid composition of sebum also occur (Nazzaro-Porro *et al.*, 1979; Sansone-Bazzano *et al.*, 1980). Like the changes in lipid class composition, the changes in fatty acid composition appear to be caused by increasing lipid synthesis per cell. For example, fatty acids derived from the circulation are more dilute in the sebum from more active glands compared to less active glands. The clearest example involves linoleate, which, since it is available only from dietary lipid, must be acquired through the circulation. Morello *et al.* (1976) found a lower percentage of linoleate in the skin surface lipid of acne patients than in control subjects, who normally have a lower rate of sebum secretion (although secretion rate was not measured). In a group of young children, Stewart *et al.* (1986b) found that the proportion of linoleate in wax esters was inversely proportional to the WE:(CH + CE) ratio. Similar results were found (Stewart *et al.*, 1989) in a group of older boys for linoleate in each of the sebaceous lipid classes that contain fatty acids (wax esters, cholesterol esters, triglycerides, and free fatty acids). The latter study also showed that other apparently exogenous fatty acids ($C_{18:0}$, $C_{18:1\Delta}{}^{9}$, $C_{18:1\Delta}{}^{11}$) also decreased in concentration with increasing sebaceous gland activity. Conversely, endogenously synthesized fatty acids, particularly $C_{16:1\Delta}{}^{6}$, increased in concentration in all lipid classes with increasing sebum secretion. The correlation between the WE:(CH + CE) ratio and the fatty acid composition of sebaceous lipids indicates that both aspects of sebum composition are controlled by a single mechanism. This mechanism seems to be the mixing, just before sebum is secreted, of a hormonally controlled amount of endogenous lipid with a relatively constant amount of lipid derived from the circulation.

An important practical implication of hormonal control of sebum production and composition relates to the understanding and treatment of acne vulgaris. It is known that individuals with acne tend to have high rates of sebum secretion compared to individuals without acne. Also, acne does not appear in children until adrenal androgen secretion starts to increase in early puberty. Furthermore, oral medications that reduce sebum secretion, such as estrogens, antiandrogens, and 13-*cis*-retinoic acid, can alleviate acne. The reason why high sebum secretion rates are associated with acne is not known. However, the dilutional effect of increasing sebaceous gland activity on sebum linoleate suggests that the sebum of active glands may be acnegenic because it induces essential fatty acid deficiency in the epithelium of the follicle. This theory of acne is expounded in detail elsewhere (Downing *et al.*, 1986).

VI. Biochemistry of Sebum

Knowledge of the biochemistry of sebum synthesis is in a rather primitive stage compared to what is known about the formation of lipids in other organs. The classical biochemical techniques of homogenizing tissue, isolating subcellular fractions, and purifying enzymes for characterization have not been applied successfully in mammalian sebaceous glands. Rather, information has come from *in vivo* studies, from incubations of whole skin or isolated sebaceous glands, from inferences from the composition of sebum, and from related processes in other organs.

A. *In Vivo* Studies

Probably the most important information obtained from *in vivo* studies concerns the timing of events in sebum biosynthesis. The time between synthesis and surface excretion has been measured for several species (see above). In these experiments, a labeled lipid precursor was injected intradermally and the time required for excreted lipid to reach peak specific activity was observed. In most species, all the sebaceous lipid classes reached peak specific activity at the same time after labeling.

The results in the horse (Colton and Downing, 1985a) were more complicated. Peak specific activities for the two major lipid classes, cholesterol esters and equolides, occurred at different times, namely, 10–16 days for cholesterol esters and 15–21 days for equolides. It appeared that the cholesterol esters were synthesized later in the life of the sebaceous cell (i.e., closer to the time of excretion) than were the equolides. A further complication was that the equolides apparently are synthesized in two steps, separated in time. This was shown by isolating the monounsaturated equolides, cleaving the molecules at the double bond, and then hydrolyzing the ester bond. The peak specific activity for the carboxyl end of the molecules appeared in lipid excreted several days earlier than detection of the peak specific activity for the hydroxyl end. It was hypothesized that fatty acids synthesized early in the life of the cell, and residing for a while in cell membranes as phospholipids, are later elongated, hydroxylated, and cyclized to equolides.

In another type of *in vivo* experiment, which has been done in humans (Downing *et al.*, 1977) and the horse (Colton and Downing, 1985b), separate sites were injected with [^{14}C]acetate over a period of several weeks and then biopsied at one time. (In humans, sites on the scalp that were to be removed in the course of a hair transplant were used.) In the human study it was found that squalene and

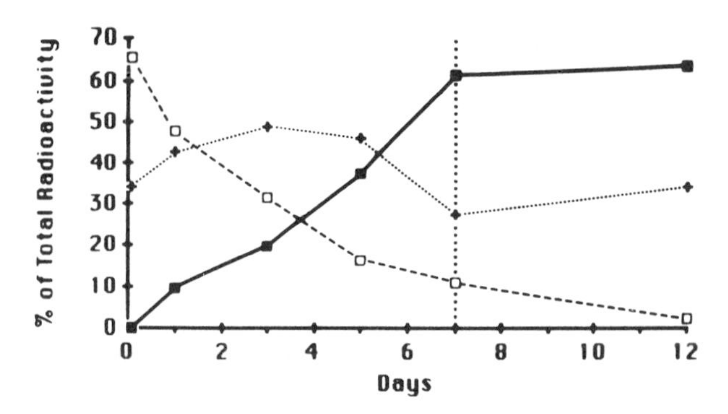

FIG. 7. Distribution of radioactivity between phospholipids (□), equolides (■), and cholesteryl esters (+) in biopsies taken from horse skin at different times after intradermal injection of [1-^{14}C] acetate. The labeling in equolides continued to increase, at the expense of the phospholipids, until labeled lipid began to be lost to the skin surface after 7 days (Colton and Downing, 1985b).

triglycerides were labeled immediately after injection and the total radioactivity in these classes remained constant until labeled lipid began to be lost to the surface, at about day 5. Wax esters were also labeled immediately, but their radioactivity increased during the days following injection. The interpretation was that wax esters are synthesized, in part, from fatty acids recycled from some other lipid class, which had become highly labeled shortly after injection of acetate. This lipid class was thought to be phospholipids. Evidence that sebaceous phospholipids do serve as precursors of sebum was obtained in the experiment with the horse. Radioactivity was initially high in phospholipids and low in equolides, but during 12 days, the proportion of radioactivity in phospholipids fell, while that in equolides rose, until equolides accounted for most of the total radioactivity (Fig. 7).

The natural substrate for lipid biosynthesis in sebaceous glands is a matter of some disagreement. Acetate is generally used as the labeled precursor in biochemical experiments, but has been regarded simply as a convenient way to label a pool of acetyl-CoA that is mostly derived from glucose (Wheatley, 1974). However [^{14}C]glucose, injected subcutaneously in either horse or rat, produced very little labeling of sebaceous lipids (Colton and Downing, 1985c), casting some doubt on its role as the physiological carbon source. The identity of the physiological substrate has also been sought in *in vitro* experiments, which are discussed below.

B. *In Vitro* STUDIES

For several reasons the interpretation of *in vitro* studies of sebaceous lipogenesis is difficult. As outlined above, the synthesis of sebum may be a multistep process in which fatty acids synthesized for membrane phospholipids are later incorporated into sebaceous lipids. The entire sequence may take several days and, for it to be observed, differentiation *in vitro* would have to occur. In fact, the usual *in vitro* experiment runs for only a few hours and, even if continued longer, there is no evidence that differentiation occurs. There is also the problem of separating sebaceous glands from other tissue, particularly adipose tissue.

1. Human Skin

The first *in vitro* study of lipogenesis in human skin (Nicolaides *et al.*, 1955) was aimed at clarifying the (then controversial) role of squalene in cholesterol biosynthesis. Human skin seemed an appropriate tissue because it produces large amounts of squalene, which could be purified fairly easily and assayed for radioactivity. The study found that squalene had a specific radioactivity 10 times that of cholesterol after a few hours of incubation with [^{14}C]acetate, seemingly confirming squalene as a cholesterol precursor. However, it was quickly realized (Nicolaides and Rothman, 1955) that in skin, squalene is synthesized in the sebaceous glands, where it is a metabolic dead end, whereas cholesterol is mostly synthesized in epidermis.

Patterson and Griesemer (1959) and Griesemer and Thomas (1965b) confirmed that squalene is synthesized in sebaceous glands and cholesterol is synthesized mostly in epidermis. Dermis from sites rich in sebaceous glands was found to incorporate more [^{14}C]acetate into squalene than did dermis from sites with sparse sebaceous glands. In contrast, incorporation of radioactivity into cholesterol was similar in epidermis from all sites. Psoriatic skin made little squalene compared to skin from an uninvolved area of the same subject. Presumably this was because the sebaceous glands in psoriatic areas of skin are often atrophic (Headington *et al.*, 1989). Homogenization of skin was found to destroy its lipogenic activity (Griesemer and Thomas, 1963a).

In the studies of the 1950s and 1960s, the first step in the analysis of labeled lipid from *in vitro* incubations was hydrolysis of the total extract. This procedure destroys information by mixing fatty acids from several lipid classes (e.g., triglycerides, wax esters, and phospholipids). Kellum *et al.* (1973) first applied thin-layer chromatography to analysis of the labeled products. Sebaceous glands were isolated by microdissection and incubated with [^{14}C]glucose. Intact lipid classes were separated by TLC and the radioactivity was measured. Most of the label was recovered in triglycerides, with little in wax esters and squalene. The reason for the low level of incorporation into squalene and wax esters was not clear, although tissue damage during the dissection process seems a possibility.

Cooper *et al.* (1976) reported that dermal lipogenesis from [^{14}C]glucose was higher in skin from subjects with acne or a past history of acne than in skin from normal subjects. The distribution of radioactivity in different lipid classes changed with age. In young adults, triglycerides contained about 50% of the radioactivity, squalene about 20%, and wax esters about 10%. (This distribution does not reflect the composition of sebum, which is about 25% wax esters and about 12% squalene. Presumably the low level of labeling of wax esters resulted from the indirect synthesis of some or all of this lipid class.) With increasing age of the skin donors, the percentage of label in squalene decreased while that in triglycerides increased. The reason for this finding is not clear, since there is no evidence that the composition of sebum changes significantly after puberty.

Kealey *et al.* (1986) reported that intact sebaceous glands can be isolated by mincing human skin with scissors under a buffer solution. It was postulated that the mincing process created shear forces that acted on weak connections between the basement membrane surrounding the glands and the connective tissue of the dermis. Using glands isolated by the shearing method, Cassidy *et al.* (1986) found that lipogenesis was stimulated by preincubation of the glands in tissue culture medium containing fetal calf serum. Following preincubation, glands were transferred to buffer, and branched-chain amino acids (valine, leucine, and isoleucine) were compared with glucose as lipid precursors. It was expected that if the branched-chain amino acids are precursors of the terminal portions of the branched-chain fatty acids in sebum, as has been proposed (Wheatley, 1974), their radioactivity would be incorporated preferentially into fatty acid-containing lipid classes. Actually, the distribution of radioactivity turned out to be similar in the various classes.

On the other hand, Middleton *et al.* (1988) compared acetate, lactate, glucose, and isoleucine as lipid precursors in glands isolated by shearing and preincubated. Acetate was the best precursor as regards total incorporation into lipids and was preferentially incorporated into squalene. Isoleucine was a poor substrate but gave the highest percentage incorporation into wax esters. Other evidence concerning the origins of the terminally branched fatty acids of sebum is discussed in the sections below.

2. *Animal Skin and Specialized Glands*

Both skin and specialized scent glands have been used for *in vitro* experiments in animals. Wheatley *et al.* (1970) compared lipogenesis in slices of guinea pig supracaudal gland, rat preputial gland, gerbil ventral gland, and hamster dorsal gland with ear slices from the same species. The ear slices were found to give higher incorporation of substrate and better reproducibility. Another advantage of the ear slices was that adipose tissue was reported to be absent from this tissue.

In further experiments with guinea pig ear slices, Wheatley *et al.* (1971) examined the rates of incorporation of various substrates into lipid. Because the first step in the lipid analysis was hydrolysis, followed by separation into fatty acids and nonsaponifiables, their experiments did not measure sebaceous lipogenesis specifically. Various precursors, including glucose, acetate, pyruvate, lactate, and branched-chain amino acids, were found to be incorporated into lipid. The presence of unlabeled glucose in the incubation medium stimulated the incorporation of all precursors other than glucose. Glucose also stimulated the incorporation of tritiated water into lipid whereas acetate did not. The results were interpreted as indicating that glucose is the major exogenous precursor of cutaneous lipid. The logic of this conclusion is not clear: if glucose stimulates lipogenesis by providing acetyl-CoA, it would be expected that unlabeled acetyl-CoA from glucose would dilute labeled acetyl-CoA from acetate and interfere with its incorporation rather than stimulate it. It seems more likely that glucose is necessary to provide ATP and NADPH, whereas the source of carbon is circulating acetate or other substrates that can provide acetyl-CoA without passing through pyruvate.

A number of studies have been reported on specialized sebaceous complexes. In hamster flank organ (Lutsky *et al.*, 1974; Bedord *et al.*, 1986), [^{14}C]acetate was incorporated mainly into triglycerides and phospholipids. Triglycerides do not appear in the secretion of the gland (Nicolaides *et al.*, 1968) and perhaps are a precursor to the lipids that are secreted, as is probably also the case for phospholipids. Triglycerides and polar lipids were also found to be the main lipids labeled in incubations of dissociated rat preputial gland cells with [^{14}C]glucose (Alves *et al.*, 1986). Therefore, in both flank organ and preputial gland, labeling patterns appear to be different from the composition of secreted lipid. Just how different is not clear, as definitive lipid analyses of the secretions of these two organs have not been published.

Kolattukudy *et al.* (1985) incubated bovine meibomian glands with [^{14}C]acetate and [^{14}C]isoleucine and did an extensive analysis of the labeled products. The labeling pattern with acetate followed the composition quite closely. Most of the label was in the wax ester and sterol ester fractions, with label distributed in fatty acid, fatty alcohol, and sterol moieties. The labeling pattern in individual fatty acids within classes also followed the chemical compositions. Isoleucine (the proposed precursor for anteiso-branched fatty acids) was incorporated into fatty acids and fatty alcohols, but not into sterols. The labeled fatty acids all appeared to be of the anteiso-branched type. The triglyceride fraction contained a series of short-chain (C_7–C_{15}) anteiso-branched acids not previously reported in meibomian glands.

Bovine meibomian glands were the source of a microsomal preparation that could catalyze fatty acid chain elongation from appropriate precursors (Anderson and Kolattuckudy, 1985). Exogenous C_{18}-CoA was elongated to products up to C_{28}. The very-long-chain fatty acids of meibomian lipids were thus shown to be products of a separate elongating system, rather than of an unusual fatty acid synthase. This work appears to be the only example of the successful application of homogenate fractionation techniques to mammalian sebaceous tissue. However, such techniques have been applied extensively to avian uropygial gland tissue (Buckner and Kolattukudy, 1976).

C. Deductions from Studies of Sebum Composition

Although the composition of sebum is unusual and highly variable, many of the lipids can be visualized as arising by established pathways of biosynthesis. This was first recognized for squalene and the unusual sterols in sebum. All the sebaceous sterols are precursors of cholesterol, as is squalene (Fig. 5). Therefore, new pathways need not be postulated for their synthesis, but simply blocks at appropriate points in the known pathway (e.g., after squalene in human sebaceous glands). The mechanism by which such blocks occur has not been determined.

Fatty acid biosynthesis in most tissues proceeds by addition of two-carbon units derived from malonyl-CoA to an acetyl-CoA primer. The odd-numbered and branched-chain fatty acids of sebum could be generated by changing the primer and/or the extender unit (Nicolaides, 1974). For example, changing the primer to propionyl-CoA would generate straight-chain fatty acids with odd numbers of carbons. Iso- or anteiso-branched primers would generate the corresponding terminally branched fatty acids, whereas the incorporation of one or more methylmalonyl extenders would generate singly or multiply internally methyl-branched acids.

The branched primers are assumed to be isobutyrate (for iso-even fatty acids), isovalerate (for iso-odd fatty acids), and 2-methylbutyrate (for anteiso-odd fatty acids). A possible source for these primers is the catabolism of the essential amino acids valine, leucine, and isoleucine, respectively. However, isoleucine is not necessarily the only source of an anteiso primer *in vivo*. Addition of a methylmalonyl extender unit to an acetate primer would also yield an anteiso fatty acid (Nicolaides, 1974). No alternative pathways have been suggested for the iso-branched fatty acids.

If amino acids are the source of the branched-chain primers, they could be derived either from the circulation or from the breakdown of protein at the end of the sebaceous cell's life. However, breakdown of cell protein would provide only a very limited supply of precursor molecules, which would be available for only part of the cell's life. The sebaceous cells of the domestic horse, which make 100% iso-branched lactones, obviously must have iso-branched precursors available throughout their lipid-synthesizing phase. Also, as discussed above, the synthesis of iso-branched chains is genetically variable in the genus *Equus* and does not occur at all in several species. On the other hand, in the human there are indications that the supply of branched primers is limited. Children, who make little sebum, tend to have a high concentration of terminally branched fatty acids in wax esters (Stewart and Downing, 1985b). With increasing sebum secretion rates, the concentration of terminally branched fatty acids declines, as if the increased amount of sebum that the cells are making *de novo* is diluting a limited amount made from the recycling of cell protein.

However, even in humans genetics plays a role in the synthesis of iso-branched fatty acids, at least of those with even numbers of carbons. Some adults retain

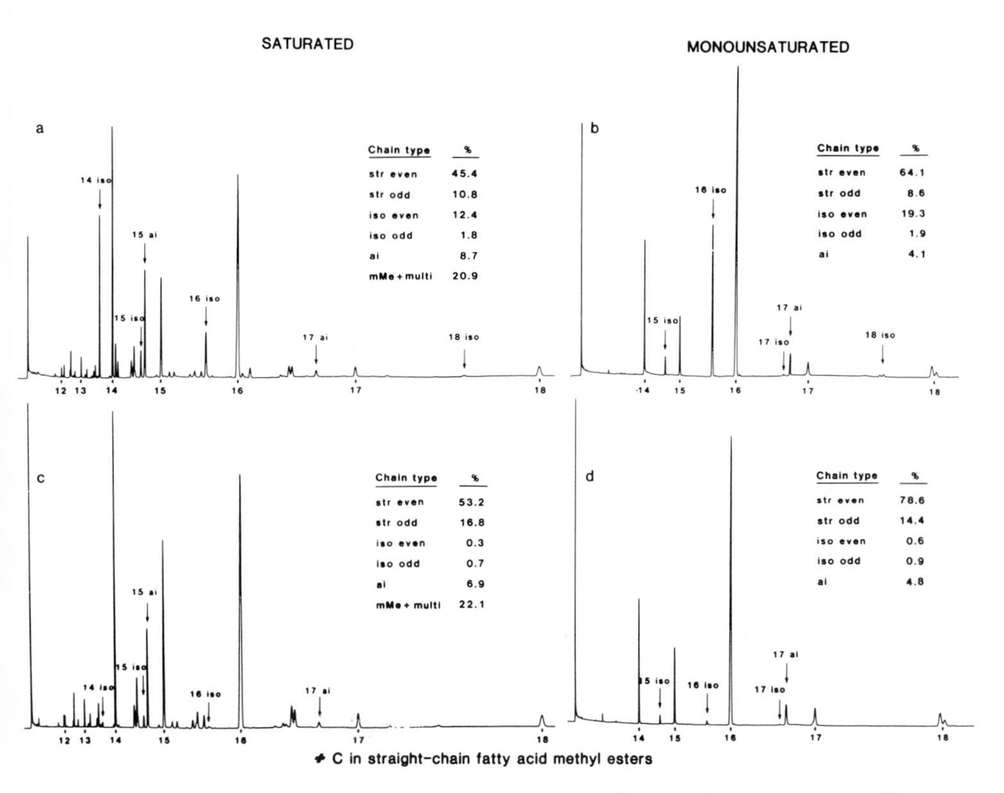

FIG. 8. Isothermal (160°C) capillary gas chromatograms of saturated and monounsaturated fatty acid methyl esters from the wax esters of two subjects. The column was a 0.2-mm x 50-m vitreous silica capillary column, wall-coated with OV 101. Note the differences in composition, especially the fairly high percentages of iso-branched fatty acids with even numbers of carbons in one subject (a and b) and the very low percentages in the other subject (c and d). Abbreviations: str, straight; ai, anteiso; mMe, internally methyl branched; multi, multiply methyl branched.

quite high levels of iso-even fatty acids (Green *et al.*, 1984), whereas others have almost none (Figs. 8 and 9). Moreover, in identical twins, intrapair differences in levels of iso-even fatty acids are very small (Stewart *et al.*, 1986a). Possibly there is a genetically variable mechanism for uptake of valine from the blood. An alternative possibility, which has not been ruled out, is an undiscovered biosynthetic pathway to isobutyrate in the sebaceous glands of some species. The latter possibility has been discounted by most workers because the iso-branched amino acids (valine and leucine) are essential.

The sebum of most species studied contains monounsaturated as well as saturated fatty acids. (Notable exceptions are sheep and cows.) In some species desaturation appears to occur at the Δ^9 position of palmitate. Subsequent elongation yields a series of ω^7 fatty acids. Whether the desaturase is the same enzyme found

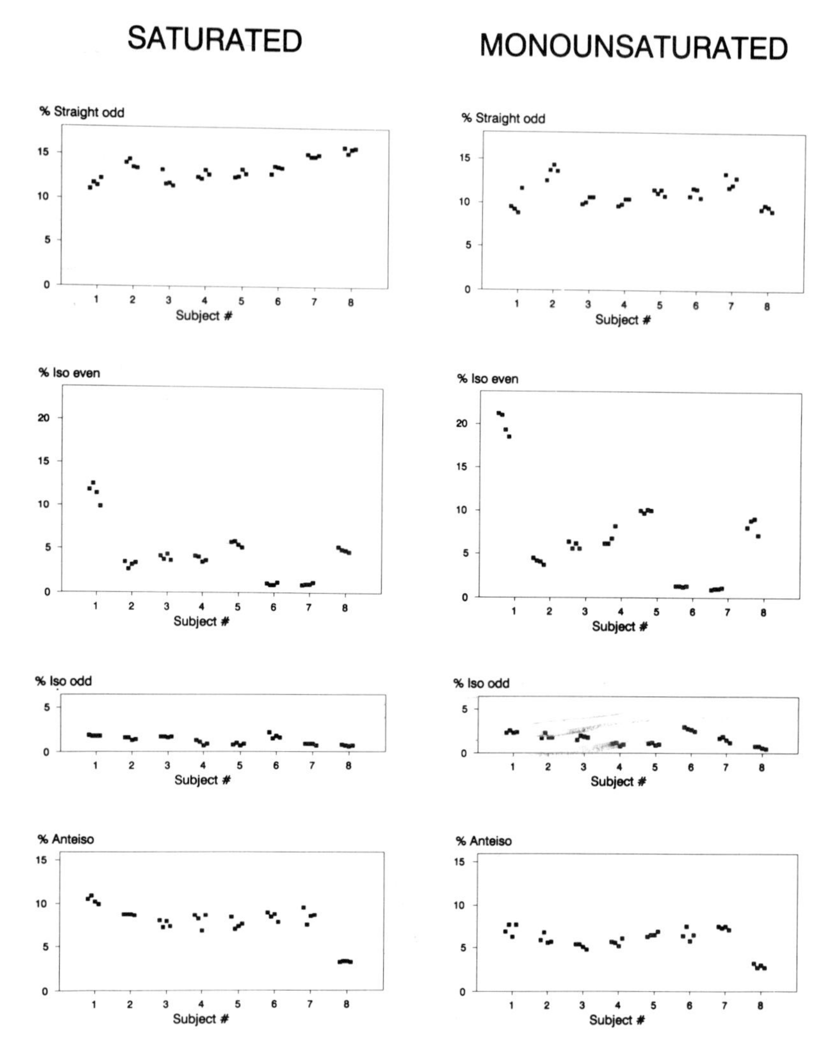

FIG. 9. Variation in wax ester fatty acid composition in five men (subjects 1–5) and three women (subjects 6–8), aged 22–29. Sebum was collected from each subject four times at 2-week intervals. Wax esters were isolated and their constituent fatty acids analyzed by capillary gas chromatography as in Fig. 8. (Redrawn from Green *et al.*, 1984.)

in other tissues has not been determined. The unique Δ^6-desaturase of humans acts on straight chains 14–17 carbons in length, with the $C_{16:1}\Delta^6$ product being the most important quantitatively (Nicolaides *et al.*, 1972). Desaturation of branched saturated fatty acids also occurs (Fig. 8), but only if there is a straight chain at least 12 carbons long from the carboxyl end to the methyl branch, i.e., desaturation of 14-iso, 15-anteiso, and internally branched fatty acids does not occur (Green *et al.*, 1984).

Very-long-chain ω-hydroxy fatty acids occur in the sebum of some species and in epidermal lipid (Wertz and Downing, 1983) and meibomian gland secretions (Nicolaides and Ruth, 1982–1983). It can be inferred from the chain-length distribution of these fatty acids that chain extension may occur while the acids are part of a membrane (Frost *et al.*, 1984). It appears that the fatty acids are chain-extended at one surface of the membrane until the methyl end is pushed to the opposite surface and encounters an enzyme that hydroxylates it. The evidence for this mechanism is that ω-hydroxy fatty acids with methyl branching or unsaturation have more carbons than do the straight-chain ω-hydroxy fatty acids, presumably because more are needed to span the bilayer. For example, saturated straight-chain ω-hydroxy fatty acids from zebra lactones are predominantly 30 and 32 carbons in length, whereas saturated iso-branched ω-hydroxy fatty acids from horse lactones are predominantly 32 and 34 carbons in length (Colton and Downing, 1983).

VII. Function of Sebum

Numerous speculations have been advanced to explain the biological value of sebum, including its having bacteriocidal activity, for lubricating and waterproofing the fur, and for minimizing water loss through the epidermis. Kligman (1963) has proposed that sebum has none of these effects, and in fact has no biological value, at least in man. He pointed out that the skin of prepubertal children suffers no defects from producing virtually no sebum. Likewise, dry, scaly skin in elderly subjects was not correlated with low sebum secretion in this age group (Frantz *et al.*, 1986).

A possible function of sebum is its ability to act as a dynamic seal around the hair shaft. In the absence of any such obstruction, a greater variety of bacteria could more readily enter and persist in the hair canal, and water could escape. Even in children, sufficient sebum is produced to provide for this function. A similar action in sealing the feather follicle in birds could be provided by the copious amounts of nonpolar epidermal lipids that are produced, whether or not the particular avian species possesses a uropygial gland.

In addition to the intrinsic interest of revealing the variety of sebum composition in the host of uninvestigated species, the patterns that remain to be elucidated might provide sufficient insight to understand the biological role of sebum and its bizarre variation in composition.

Acknowledgment

This work was supported in part by U.S. Public Health Service Grant AM 22083.

References

Ahern, D. G., and Downing, D. T. (1974). *Lipids* **9,** 8–18.

Alves, A. M., Thody, A. J., Fisher, C., and Shuster, S. (1986). *J. Endocrinol.* **109,** 1–7.

Anderson, G. J., and Kolattukudy, P. E. (1985). *Arch. Biochem. Biophys.* **237,** 177–185.

Bedord, C. J., Ruddle, D. L., and Wagener, B. M. (1986). *Arch. Dermatol. Res.* **278,** 486–490.

Benfenati, A., and Brillanti, F. (1939). *Arch. Ital. Dermatol.* **15,** 33–42.

Bortz, J. T., Wertz, P. W., and Downing, D. T. (1989). *J. Invest. Dermatol.* **93,** 723–727.

Buckner, J. S., and Kolattukudy, P. E. (1976). *In* "Chemistry and Biochemistry of Natural Waxes" (P. E. Kolattududy, ed.), pp. 148–200. Elsevier, Amsterdam.

Burken, R. R., Wertz, P. W., and Downing, D. T. (1985). *Comp. Biochem. Physiol. B* **81B,** 315–318.

Carey, F. A., and Sundberg, R. J. (1983). *In* "Advanced Organic Chemistry" (F. A. Carey and R. J. Sundberg, eds.), 2nd Ed., pp. 481–538. Plenum Press, New York.

Cassidy, D. M., Lee, C. M., Laker, M. F., and Kealey, T. (1986). *FEBS Lett.* **200,** 173–176.

Collison, D. W., Burns, T. L., Stewart, M. E., Downing, D. T., and Strauss, J. S. (1987). *Arch Dermatol. Res.* **279,** 266–269.

Colton, S. W., 6th, and Downing, D. T. (1983). *Comp. Biochem. Biophys. B* **75B,** 429–433.

Colton, S. W., 6th, and Downing, D. T. (1985a). *Biochim. Biophys. Acta* **835,** 98–103.

Colton, S. W., 6th, and Downing, D. T. (1985b). *Biochim. Biophys. Acta* **836,** 306–311.

Colton, S. W., 6th, and Downing, D. T. (1985c). *Biochim. Biophys. Acta* **837,** 190–196.

Colton, S. W., 6th, Lindholm, J. S., Abraham, W., and Downing, D. T. (1986). *Comp. Biochem. Physiol. B* **84B,** 369–371.

Cooper, M. F., McGrath, H., and Shuster, S. (1976). *Br. J. Dermatol.* **94,** 165–172.

Downing, D. T. (1968). *J. Chromatogr.* **38,** 91–99.

Downing, D. T. (1976). *In* "Chemistry and Biochemistry of Natural Waxes" (P. E. Kolattukudy, ed.), pp. 17–48. Elsevier, Amsterdam.

Downing, D. T. (1986). *In* "Biology of the Integument. Vol. 2: Vertebrates" (J. Bereiter-Hahn, A. G. Matoltsy, and K. S. Richards, eds.), pp. 833–840. Springer-Verlag, Berlin.

Downing, D. T., and Colton, S. W., 6th (1980). *Lipids* **15,** 323–327.

Downing, D. T., and Greene, R. S. (1968a). *Anal. Chem.* **40,** 827–828.

Downing, D. T., and Greene, R. S. (1968b). *J. Invest. Dermatol.* **50,** 380–386.

Downing, D. T., and Lindholm, J. S. (1982). *Comp. Biochem. Physiol. B* **73B,** 327–330.

Downing, D. T., and Sharaf, D. M. (1976). *Biochim. Biophys. Acta* **431,** 378–388.

Downing, D. T., and Stewart, M. E. (1985). *In* "Methods in Skin Research" (D. Skerrow and C. Skerrow, eds.), pp. 349–379. Wiley, New York.

Downing, D. T., and Stewart, M. E. (1987). *Comp. Biochem. Physiol. B* **86B,** 667–670.

Downing, D. T., and Stranieri, A. M. (1980). *J. Chromatogr.* **192,** 208–211.

Downing, D. T., and Strauss, J. S. (1982). *Arch. Dermatol. Res.* **272,** 343–349.

Downing, D. T., Kranz, S. H., and Murray, K. E. (1960). *Aust. J. Chem.* **13,** 80–94.

Downing, D. T., Strauss, J. S., and Pochi, P. E. (1969). *J. Invest. Dermatol.* **53,** 322–327.

Downing, D. T., Strauss, J. S., and Pochi, P. E. (1972). *Am. J. Clin. Nutr.* **25,** 365–367.

Downing, D. T., Strauss, J. S., Ramasastry, P., Abel, M., Lees, C. E., and Pochi, P. E. (1975). *J. Invest. Dermatol.* **64,** 215–219.

Downing, D. T., Strauss, J. S., Norton, L. A., Pochi, P. E., and Stewart, M. E. (1977). *J. Invest. Dermatol.* **69,** 407–412.

Downing, D. T., Stewart, M. E., and Strauss, J. S. (1981). *J. Invest. Dermatol.* **77,** 358–360.

Downing, D. T., Stranieri, A. M., and Strauss, J. S. (1982). *J. Invest. Dermatol.* **79,** 226–228.

Downing, D. T., Stewart, M. E., Wertz, P. W., and Strauss, J. S. (1986). *J. Am. Acad. Dermatol.* **14,** 221–225.

Drucker, W. D., Blumberg, J. M., Gandy, H. M., David, R. R., and Verde, A. L. (1972). *J. Clin. Endocrinol. Metab.* **35,** 48–54.
Ebling, F. J. (1974). *J. Invest. Dermatol.* **62,** 161–171.
Eckstein, H. C. (1927). *J. Biol. Chem.* **73,** 363–369.
Frantz, R. A., Kinney, C. K., and Downing, D. T. (1986). *Nurs. Res.* **35,** 98–100.
Frost, M. L., Colton, S. W., 6th, Wertz, P. W., and Downing, D. T. (1984). *Comp. Biochem. Physiol. B* **78B,** 549–552.
Gaul, B. L., Stewart, M. E., and Downing, D. T. (1985). *Comp. Biochem. Physiol. B* **80B,** 431–435.
Gray, G. M., White, R. J., Williams, R. H., and Yardley, H. J. (1982). *Br. J. Dermatol.* **106,** 59–63.
Green, S. G., Stewart, M. E., and Downing, D. T. (1984). *J. Invest. Dermatol.* **83,** 114–117.
Greene, R. S., Downing, D. T., Pochi, P. E., and Strauss, J. S. (1970). *J. Invest. Dermatol.* **54,** 240–247.
Griesemer, R. D., and Thomas R. W. (1963a). *J. Invest. Dermatol.* **41,** 95–98.
Griesemer, R. D., and Thomas, R. W. (1963b). *J. Invest. Dermatol.* **41,** 235–238.
Harris, H. H., Downing, D. T., Stewart, M. E., and Strauss, J. S. (1983). *J. Am. Acad. Dermatol.* **8,** 200–203.
Headington, J. T., Gupta, A. K., Goldfarb, M. T., Nickoloff, B. J., Hamilton, T. A., Ellis C. N., and Voorhees, J. J. (1989). *Arch. Dermatol.* **125,** 639–642.
Hedberg, C. L., Wertz, P. W., and Downing, D. T. (1988). *J. Invest. Dermatol.* **91,** 169–174.
Holick, M. F. (1986). *In* "Nutrition and the Skin" (D. A. Roe, ed.), pp. 15–43. Alan R. Liss, New York.
Ito, M., Suzuki, M., Motoyoshi, K., Maruyama, T., and Sato, Y. (1984). *J. Invest. Dermatol.* **82,** 381–385.
Jacob, J. (1976). *In* "Chemistry and Biochemistry of Natural Waxes" (P. E. Kolattukudy, ed.), pp. 94–146. Elsevier, Amsterdam.
Jacobsen, E., Billings, J. K., Frantz, R. A., Kinney, C. K., Stewart, M. E., and Downing, D. T. (1985). *J. Invest. Dermatol.* **85,** 483–485.
James, A. T., and Wheatley, V. R. (1956). *Biochem. J.* **63,** 269–273.
Jarrett, A. (1955). *Br. J. Dermatol.* **67,** 165–179.
Kandutsch, A. A. (1964). *In* "The Epidermis" (W. Montagna and W. C. Lobitz, Jr., eds.), pp. 493–510. Academic Press, New York.
Kealey, T., Lee, C. M., Thody, A. J., and Coaker, T. (1986). *Br. J. Dermatol.* **114,** 181–188.
Kellum, R. E., Toshitani, S., and Strangfeld, K. (1973). *J. Invest. Dermatol.* **60,** 53–57.
Kligman, A. M. (1963). *Adv. Biol. Skin* **4,** 110–124.
Kligman, A. M., and Shelley, W. B. (1958). *J. Invest. Dermatol.* **30,** 99–125.
Kolattukudy, P. E., Rogers, L. M., and Nicolaides, N. (1985). *Lipids* **20,** 468–474.
Kooyman, D. J. (1932). *Arch. Dermatol. Syphilol.* **25,** 444–450.
Lindholm, J. S., and Downing, D. T. (1980). *Lipids* **15,** 1062–1063.
Lindholm, J. S., McCormick, J. M., Colton, S. W., VI, and Downing, D. T. (1981). *Comp. Biochem. Physiol. B* **69B,** 75–78.
Lucas, A. M. (1972). *U.S. Dep. Agric., Agric. Handb.* No. 362, Part 2.
Lutsky, B. N., Casmer, C., and Koziol, P. (1974). *Lipids* **9,** 43–48.
Mason, R. T., Fales, H. M., Jones, T. H., Pannell, L. K., Chinn, J. W., and Crews, D. (1989). *Science* **245,** 290–293.
Menon, G. K., Aggarwal, S. K., and Lucas, A. M. (1981). *J. Morphol.* **167,** 185–199.
Middleton, B., Birdi, I., Heffron, M., and Marsden, J. R. (1988). *FEBS Lett.* **231,** 59–61.
Morello, A. M., Downing, D. T., and Strauss, J. S. (1976). *J. Invest. Dermatol.* **66,** 319–323.
Motiuk, K. (1979a). *J. Am. Oil Chem. Soc.* **56,** 91–97.
Motiuk, K. (1979b). *J. Am. Oil Chem. Soc.* **56,** 651–658.
Nazzaro-Porro, M., Passi, S., Boniforti, L., and Belsito, F. (1979). *J. Invest. Dermatol.* **73,** 112–117.

Nicolaides, N. (1967). *Lipids* **2,** 266–275.
Nicolaides, N. (1971). *Lipids* **6,** 901–905.
Nicolaides, N. (1974). *Science* **186,** 19–26.
Nicolaides, N., and Ansari, M. N. A. (1969). *Lipids* **4,** 79–81.
Nicolaides, N., and Apon, J. M. B. (1976). *Lipids* **11,** 781–790.
Nicolaides, N., and Apon, J. M. B. (1977). *Biomed. Mass Spectrom.* **4,** 337–347.
Nicolaides, N., and Rothman, S. (1955). *J. Invest. Dermatol.* **24,** 125–129.
Nicolaides, N., and Ruth, E. C. (1982–1983). *Curr. Eye Res.* **2,** 93–98.
Nicolaides, N., Reiss, O. K., and Langdon, R. G. (1955). *J. Am. Chem. Soc.* **77,** 1535–1538.
Nicolaides, N., Kellum, R. E., and Woolley, P. V., III (1964). *Arch. Biochem. Biophys.* **105,** 634–639.
Nicolaides, N., Fu, H. C., and Rice, G. R. (1968). *J. Invest. Dermatol.* **51,** 83–89.
Nicolaides, N., Fu, H. C., Ansari, M. N. A., and Rice, G. R. (1972). *Lipids* **7,** 506–517.
Nikkari, T. (1965). *Scand. J. Clin. Lab. Invest., Suppl.* No. 85, 1–140.
Noble, R. C., Crouchman, M. L., and Moore, J. H. (1974). *Res. Vet. Sci.* **17,** 372–376.
Patterson, J. F., and Griesemer, R. D. (1959). *J. Invest. Dermatol.* **33,** 281–285.
Plewig, G., and Christophers, E. (1974). *Acta Derm. Venereol.* **54,** 177–182.
Pochi, P. E., Downing, D. T., and Strauss, J. S. (1970). *J. Invest. Dermatol.* **55,** 303–309.
Pochi, P. E., Strauss, J. S., and Downing, D. T. (1977). *J. Invest. Dermatol.* **69,** 485–489.
Quay, W. B. (1986). *In* "Biology of the Integument. Vol. 2: Vertebrates" (J. Bereiter-Hahn, A. G. Matoltsy, and K. S. Richards, eds.), pp. 188–193. Springer-Verlag, Berlin.
Ramasastry, P., Downing, D. T., Pochi, P. E., and Strauss, J. S. (1970). *J. Invest. Dermatol.* **54,** 139–144.
Reinertson, R. P., and Wheatley, V. R. (1959). *J. Invest. Dermatol.* **32,** 49–59.
Rony, H. R., and Zakon, S. J. (1943). *Arch. Dermatol. Sypilol.* **48,** 601–604.
Sansone-Bazzano, G., Cummings, B., Seeler, A. K., and Reisner, R. M. (1980). *Br. J. Dermatol.* **103,** 131–137.
Sharaf, D. M., Clark, S. J., and Downing, D. T. (1977). *Lipids* **12,** 786–790.
Stewart, M. E., and Downing, D. T. (1981). *Lipids* **16,** 355–359.
Stewart, M. E., and Downing, D. T. (1985a). *J. Invest. Dermatol.* **84,** 59–61.
Stewart, M. E., and Downing, D. T. (1985b). *J. Invest. Dermatol.* **84,** 501–503.
Stewart, M. E., and Downing, D. T. (1990). *J. Invest. Dermatol.* **95,** 603–606.
Stewart, M. E., Downing, D. T., Pochi, P. E., and Strauss, J. S. (1978). *Biochim. Biophys. Acta* **529,** 380–386.
Stewart, M. E., Downing, D. T., and Strauss, J. S. (1983). *Dermatol. Clin.* **1,** 335–344.
Stewart, M. E., Benoit, A. M., Downing, D. T., and Strauss, J. S. (1984). *J. Invest. Dermatol.* **82,** 74–78.
Stewart, M. E., McDonnell, M. W., and Downing, D. T. (1986a). *J. Invest. Dermatol.* **86,** 706–708.
Stewart, M. E., Grahek, M. O., Cambier, L. S., Wertz, P. W., and Downing, D. T. (1986b). *J. Invest. Dermatol.* **87,** 733–736.
Stewart, M. E., Steele, W. A., and Downing, D. T. (1989). *J. Invest. Dermatol.* **92,** 371–378.
Strauss, J. S., and Pochi, P. E. (1961). *J. Invest. Dermatol.* **36,** 293–298.
Strauss, J. S., and Pochi, P. E. (1963). *Adv. Biol. Skin* **4,** 220–254.
Tiffany, J. M. (1987). *Adv. Lipid Res.* **22,** 1–62.
Tosti, A. (1974). *J. Invest. Dermatol.* **62,** 147–152.
Weitkamp, A. W. (1945). *J. Am. Chem. Soc.* **67,** 445–454.
Weitkamp, A. W., Smiljanic, A. M., and Rothman, S. (1947). *J. Am. Chem. Soc.* **69,** 1936–1939.
Wertz, P. W., and Downing, D. T. (1983). *J. Lipid Res.* **24,** 753–758.
Wertz, P. W., Colton, S. W., 6th, and Downing, D. T. (1983). *Comp. Biochem. Physiol. B* **75B,** 217–220.
Wertz, P. W., Stover, P. M., Abraham, W., and Downing, D. T. (1986). *J. Lipid Res.* **27,** 427–435.

Wertz, P. W., Swartzendruber, D. C., Madison, K. C., and Downing, D. T. (1987). *J. Invest. Dermatol.* **89,** 419–425.

Wheatley, V. R. (1974). *J. Invest. Dermatol.* **62,** 245–256.

Wheatley, V. R. (1986). *In* "The Physiology and Pathophysiology of the Skin" (A. Jarrett, ed.), Vol. 9, pp. 2705–2971. Academic Press, New York.

Wheatley, V. R., and James, A. T. (1957). *Biochem. J.* **65,** 36–42.

Wheatley, V. R., and Reinertson, R. P. (1958). *J. Invest. Dermatol.* **31,** 51–54.

Wheatley, V. R., Hodgins, L. T., and Coon, W. M. (1970). *J. Invest. Dermatol.* **54,** 288–297.

Wheatley, V. R., Hodgins, L. T., Coon, W. M., Kumarasiri, M., Berenzweig, H., and Feinstein, J. M. (1971). *J. Lipid Res.* **12,** 347–360.

Wilkinson, D. I., and Karasek, M. A. (1966). *J. Invest. Dermatol.* **47,** 449–455.

Yamada, K. (1932). *Folia Anat. Jpn.* **10,** 721–752.

Yamzaki, T., Seyama, Y., Otsuka, H., Ogawa, H., and Yamakawa, T. (1981). *J. Biochem. (Tokyo)* **89,** 683–691.

Yeung, D., Nacht, S., and Cover, R. E. (1981). *Biochim. Biophys. Acta* **663,** 524–535.

ADVANCES IN LIPID RESEARCH, VOL. 24

Integumental Lipids of Plants and Animals: Comparative Function and Biochemistry

NEIL F. HADLEY

Department of Zoology
Arizona State University
Tempe, Arizona 85287

I. Introduction

All organisms, in adapting to their environment, depend on a highly structured and functional external covering, or "integument." In single-celled organisms, the integument consists of a thin, fragile barrier, the plasma membrane, which separates the cell from its watery environment. In both plants and arthropods, the outer surface is known as the "cuticle," a noncellular, multilayered membrane that lies over the epidermal cells. In most arthropod species, the outer portion of the cuticle ultimately becomes stiffened or "sclerotized," thus providing the exoskeleton that characterizes insects and arachnids (Hadley, 1986). The integument of vertebrates is also multilayered; however, the layers are composed of different cell types as opposed to being noncellular. Despite having a well-defined internal skeleton, vertebrates have also evolved a variety of keratinized epidermal appendages (e.g., scales, shells, feathers, and hair), which provide additional structural integrity (Matoltsy and Bereiter-Hahn, 1986).

The integument can serve an organism in many ways. It provides mechanical protection against injury and serves as a physical barrier to predators and invasion by microorganisms. The integument regulates the movement of materials, especially water, in and out of the body. In many species, the integument also participates in temperature regulation, respiration, locomotion, chemical and visual communication, and environmental sensing (Hadley, 1984, 1985). These and other functions depend upon the unique interaction of specific structural, biochemical, and physiological properties of the integument. In virtually every case, lipids, deposited on the surface and/or impregnated within the various layers, are a key element. The functional importance of integumental lipids, recognized initially in plants and arthropods, is now well documented for vertebrates, especially mammals. This issue of *Advances in Lipid Research* highlights some of the recent developments in this latter group.

The following discussions provide a general overview of the chemical composition, physical structure, and arrangement of integumental lipids, with emphasis on lipids that are involved in waterproofing the skin or cuticle. Coverage includes a brief discussion of the physicochemical features of surface lipids, a survey of the principal lipid classes and molecular types that comprise the water barrier of different plant and animal groups, and a comparative summary of how these lipids are synthesized and transported to the surface. Although the focus is on plants, arthropods, and lower vertebrates, basic comparative information on lipid patterns and processes in the integument of mammals is included for the purpose of enabling the reader to more fully appreciate the parallel evolution that has occurred between widely divergent groups.

II. General Design of Surface Lipids

The physical and chemical features of lipids make them well suited to be an effective waterproofing component on the surface of organisms (Hadley, 1989). As a group, lipids are typically nonpolar; the electrons of the carbon–hydrogen bonds that dominate these molecules are equally attracted by the atoms forming the bond so that no asymmetry or charge distribution occurs (see Fig. 1). As a result, lipids tend to repel water when the two compounds come in contact. The degree of "nonpolarity" varies among the different lipid classes. Hydrocarbons, which have no oxygen-containing functional groups, and wax esters are very nonpolar and thus are ideal waterproofing constituents. It is not surprising that they are typically the most abundant lipid class present on the cuticle of plants and arthropods. Phospholipids, in contrast, are much more polar because many of the polar head groups of these compounds are asymmetrical or ionized. Although this feature allows phospholipids to electrostatically interact with water to form micelles or lipid bilayers, phospholipids are not readily dispersed by water and thus can also serve as barriers in plasma membranes and to a lesser extent in the stratum corneum of reptiles, birds, and mammals. Other lipid classes with intermediate polarity include triacylglycerols, free fatty acids, aliphatic alcohols, and free sterols (particularly cholesterol). As will be shown, these groups are invariably present in the integument, although the extent to which they influence transcuticular water flux in plants and arthropods is uncertain.

Many of the features responsible for a lipid's nonpolarity also contribute to its ability to function under varying and potentially adverse environmental conditions. Highly saturated, long-chain hydrocarbons have low vapor pressures and, hence, are not easily lost through volatilization. Despite this fact, in some insects, surface hydrocarbons and their oxygenated derivatives apparently function as both water barrier constituents and as chemical releasers or pheromones (Howard and Blomquist, 1982). Long-chain, saturated lipids also help protect these molecules against atmospheric oxidation and degradation by microorganisms (Kolattukudy, 1976). They are also thermally stable, having melting points that are higher than temperatures encountered by plants and animals under most natural environmental conditions. For example, *n*-alkanes (saturated, straight-chain hydrocarbons) containing 23 to 35 carbon atoms have melting points that range from 47.6 to 75°C.

Analyses of surface lipids often reveal a greater quantity of branched versus straight-chain hydrocarbons. The relative abundance of the former remains an anomaly, as branched hydrocarbons melt at lower temperatures than do *n*-alkanes of the same chain length and are also thought to be more permeable to water because they cannot pack together as closely. A permeability difference, however, has not been experimentally verified for the two hydrocarbon types when they are deposited as bulk layers on a surface. Although the following functions are largely speculative, branched hydrocarbons may help prevent the premature release of long-chain molecules during synthesis as well as increase the immunity of these molecules to enzymatic degradation (Jackson and Blomquist, 1976). Branched hydrocarbons may also facilitate the secretion and transport of lipid to the surface and help orient surface lipid molecules for optimal waterproofing properties (Hadley, 1985).

III. Composition and Structure of Integumental Lipids

Lipids associated with the integument of organisms consist of a complex mixture of long-chain and cyclic compounds (see Fig. 2). For purposes of discussion, they can be separated into four major categories: (1) hydrocarbons, (2) oxygenated derivatives of hydrocarbons, (3) cyclic compounds, and (4) polar lipids (phospholipids and sphingolipids). Not all groups are found in all organisms. In some cases, their absence may reflect the extraction solvent used or simply the fact that the investigator chose to concentrate on a more abundant class.

A. Plants

Plant surface lipids have been reviewed on several occasions (Baker, 1982; Kolattukudy, 1975; Kolattukudy and Walton, 1972; Tulloch, 1976). These surveys indicate that hydrocarbons are invariably present, although their percentage may be

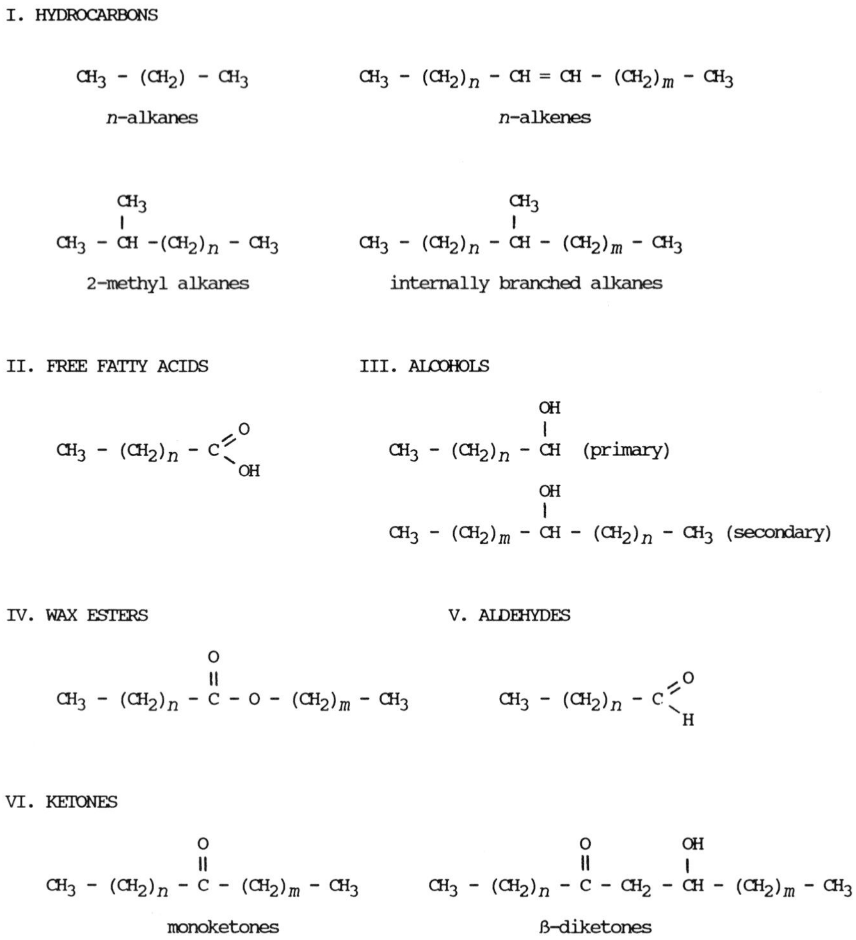

I. HYDROCARBONS

$CH_3 - (CH_2) - CH_3$

n-alkanes

$CH_3 - (CH_2)_n - CH = CH - (CH_2)_m - CH_3$

n-alkenes

$CH_3 - CH(CH_3) - (CH_2)_n - CH_3$

2-methyl alkanes

$CH_3 - (CH_2)_n - CH(CH_3) - (CH_2)_m - CH_3$

internally branched alkanes

II. FREE FATTY ACIDS

$CH_3 - (CH_2)_n - C(=O)OH$

III. ALCOHOLS

$CH_3 - (CH_2)_n - CH(OH)$ (primary)

$CH_3 - (CH_2)_m - CH(OH) - (CH_2)_n - CH_3$ (secondary)

IV. WAX ESTERS

$CH_3 - (CH_2)_n - C(=O) - O - (CH_2)_m - CH_3$

V. ALDEHYDES

$CH_3 - (CH_2)_n - C(=O)H$

VI. KETONES

$CH_3 - (CH_2)_n - C(=O) - (CH_2)_m - CH_3$

monoketones

$CH_3 - (CH_2)_n - C(=O) - CH_2 - CH(OH) - (CH_2)_m - CH_3$

ß-diketones

FIG. 1. Chemical structures of some integumental lipids. See text for discussion of their distribution among plants and animals.

very small in some species. *n*-Alkanes with odd-numbered chains containing between 21 and 37 carbon atoms predominate (especially C_{27}, C_{29}, C_{31}, and C_{33}). For example, hentriacontane (C_{31}) is the major alkane in the epicuticular wax hydrocarbons of 12 of 17 species of Ericaceae, with nonacosane (C_{29}) next in abundance (Salasco, 1989). Small quantities of branched alkanes (2-methyl, 3-methyl, dimethyl, and internally branched) and *n*-alkenes have been reported in several species, but generally they are far less abundant than *n*-alkanes. In lower plants such as algae, however, unsaturated hydrocarbons containing one, two, and even three double bonds constitute a major fraction of the total surface hydrocarbons.

VII. STEROLS

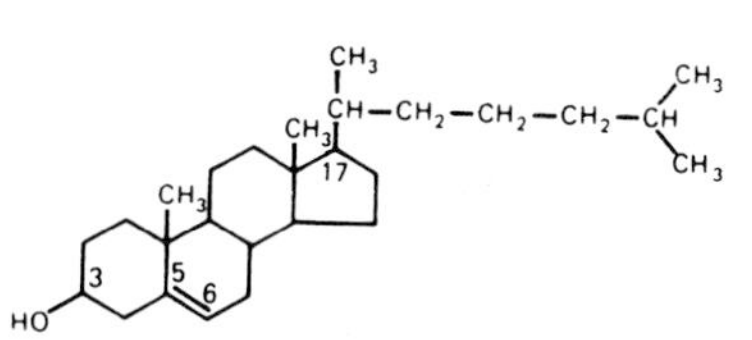

cholesterol

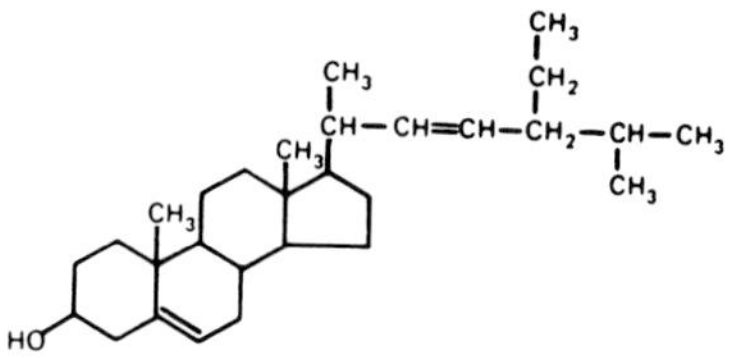

stigmasterol

VIII. TERPENES

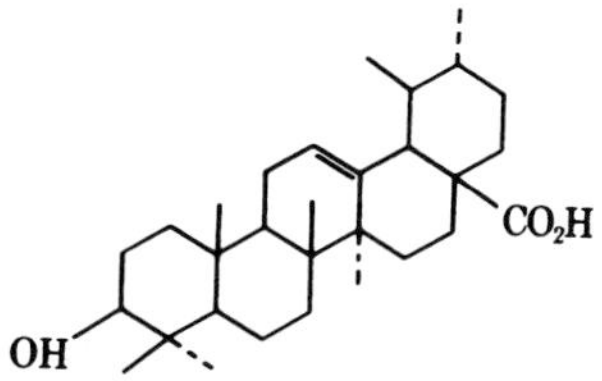

oleanolic acid (triterpene)

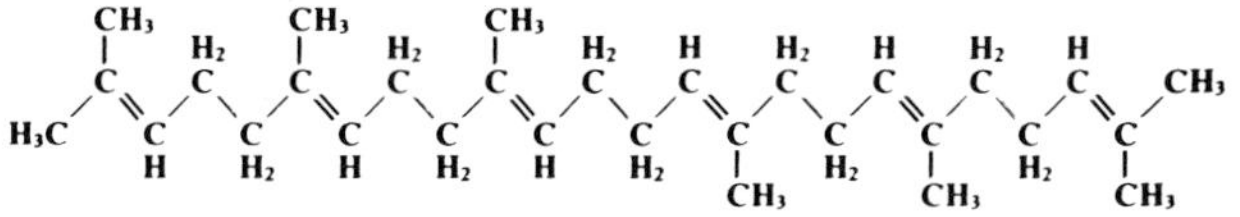

squalene

IX. CERAMIDES

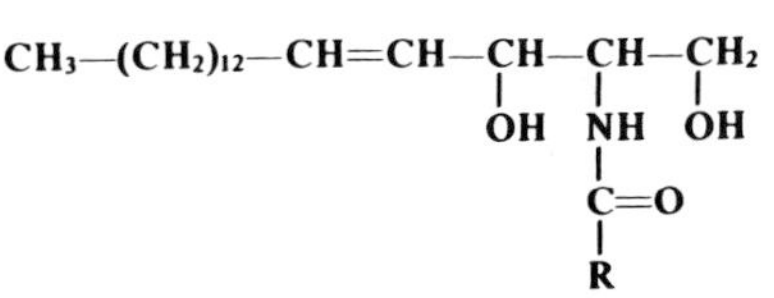

$$CH_3-(CH_2)_{12}-CH{=}CH-\underset{\underset{OH}{|}}{CH}-\underset{\underset{\underset{\underset{\underset{R}{|}}{C=O}}{|}}{NH}}{\overset{}{CH}}-\underset{\underset{OH}{|}}{CH_2}$$

FIG. 1. (*continued*)

Oxygenated derivatives of alkanes comprise a second important group of plant cuticular lipids. The most common among these are wax esters and ketones. Wax esters usually consist of fatty acids (C_{20}–C_{24}) esterified to alcohols having an even number of carbon atoms in the range C_{12}–C_{32}. A variety of ketones (monoketones, β-diketones, hydroxy β-diketones, and methyl ketones) have been extracted from plant cuticle. These typically have chain lengths similar to the

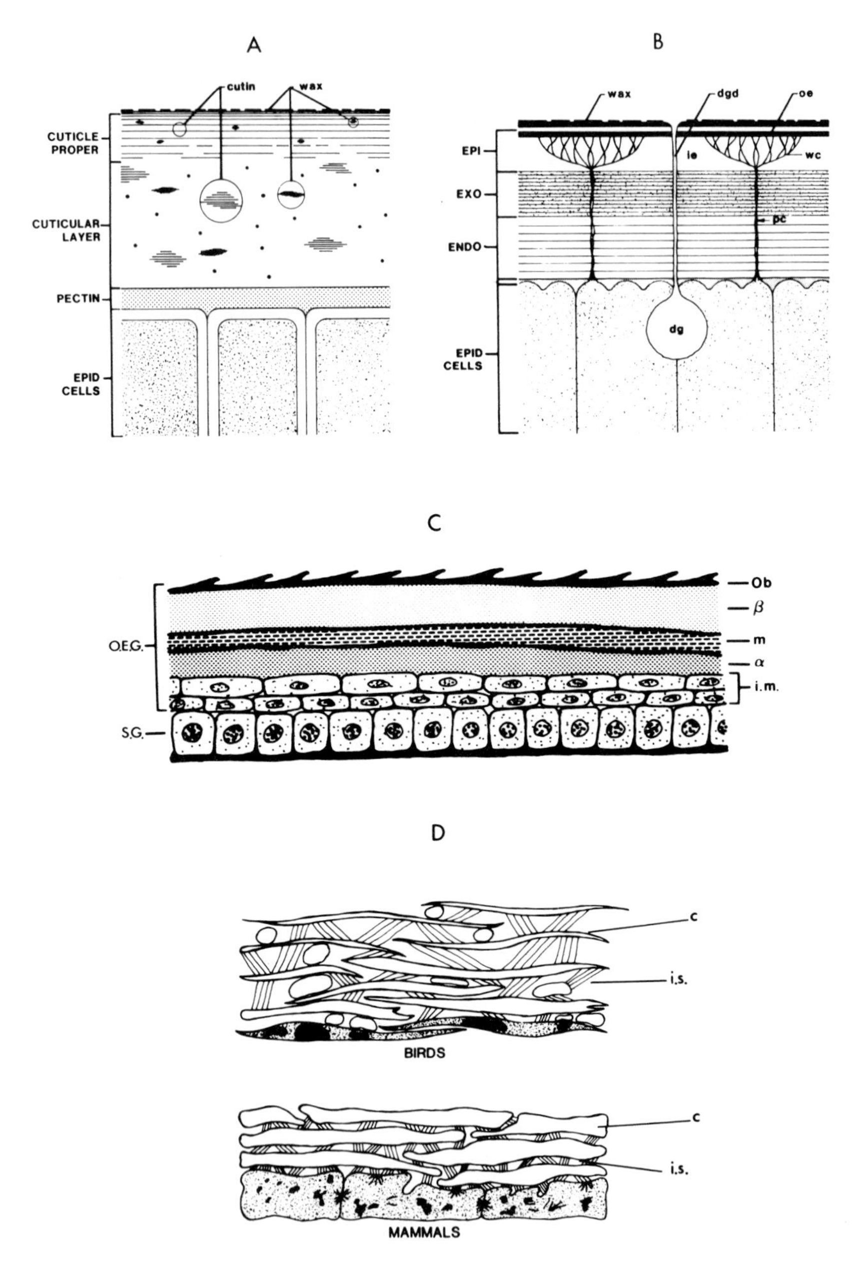

FIG. 2. Diagrammatic representations of the integument of various organisms. (A) Plant cuticle. Epicuticular wax overlies the outermost laminate cuticle proper, which is composed primarily of cutin. Some wax and cutin are also present in the underlying cuticular layer. (B) Arthropod cuticle.

alkanes present. Alcohols are usually straight-chain compounds containing an even number of carbon atoms, whereas fatty acids may be saturated, unsaturated, or even branched. Alcohols and fatty acids usually occur in only minor or trace amounts; however, primary alcohols are the major component in the leaves of beech trees, accounting for 34.8% of the crude wax (Gülz *et al.*, 1989).

Cyclic plant lipids include the pentacyclic triterpenes (β-amyrins, taraxerone), tetracyclic phytosterols (sistosterol, campesterol), flavones, and aromatic hydrocarbons. The earlier literature suggested that these compounds, although often abundant in fruits and berries, were not major constitutents in leaf waxes. More recent studies, however, have demonstrated that they occur more frequently than previously thought (Manheim and Mulroy, 1978; Barthlott and Wollenweber, 1981). In the perennial weed *Euphorbia cyparissias*, pentacyclic triterpenoids are the major component (34%) of the epicuticular wax. Because of their unique micromorphological structure, terpenoids and flavenoids can be identified on the surface of plants with the scanning electron microscope (Barthlott, 1981).

B. Arthropods

Comprehensive accounts of the lipids associated with arthropod cuticle can be found in Jackson and Blomquist (1976), Hadley (1981), Blomquist and Dillwith (1985), Blomquist *et al.* (1987), Lockey (1985, 1988), and de Renobales *et al.* (1991). The compositional theme that emerges from these reviews is one of complexity, with extracts typically including straight-chain and methyl-branched hydrocarbons (saturated and unsaturated), wax and sterol esters, acetate esters of keto alcohols, ketones, alcohols, aldehydes, and acids. As in plants, phospholipids and acylglycerols are usually absent from cuticular lipid extracts, although they are found internally.

Although *n*-alkanes have been found in almost every arthropod species studied (see, however, Malinski *et al.*, 1986), their percentage of the total lipid fraction is

---

Waxes cover the surface of the epicuticle and are also abundant in the outer epicuticle (oe) and inner epicuticle (ie). Secretions from a dermal gland (dg) reach the cuticle surface via the dermal gland duct (dgd). The latter is flanked by two smaller diameter pore canals (pc) that divide into still smaller diameter wax canals (wc) as they approach the epicuticle. (C) Reptilian epidermis covering the outer scale surface. The outer epidermal generation (O.E.G.) is divided into an outermost serrated Oberhäutchen (Ob), a beta layer (β), a mesos layer (m), an alpha layer (α), and two or more layers of immature cells (i.m.) that overlie the stratum germinativum (S.G.). (D) Stratum corneum of birds and mammals. Lipids extruded into the intercellular spaces (i.s.) of the stratum corneum are believed to provide the principal barrier to water efflux in both groups. Avian corneocytes (c) contain few keratin filaments relative to mammals and thus are more attentuated. In mammals, the corneocytes are substantial structures due to an abundance of keratin filaments. A, B, and D are from Hadley (1989); C is from Hadley (1985).

highly variable, ranging from trace amounts in tsetse flies to nearly 98% in tenebrionid beetles. Unsaturated hydrocarbons (primarily *n*-alkenes) have been reported in approximately 50% of the insects studied thus far; in some species, they are the dominant hydrocarbon. Terminally and internally methyl-branched hydrocarbons, which are much more common in arthropods than in plants, constitute the hydrocarbons with the longest chain lengths. Molecular species of branched hydrocarbons include monomethyl-, dimethyl-, and, more rarely, trimethylalkanes. Recently, novel tetramethylalkanes were identified in both male and female tsetse flies (Nelson *et al*, 1988).

Oxygenated derivatives are usually present on the surface of arthropods; however, with few exceptions they have not received the attention given hydrocarbons. Wax (alkyl) esters vary from rather simple to complex molecules, depending on the combination of acids and alcohols. They are the dominant surface lipid in black widow spiders, sand cockroaches, and scale insects. The wax esters of some grasshoppers contain secondary alcohols, which are rare in insects and not reported for plants. Mixtures of saturated and unsaturated fatty acids having even carbon numbers from 10 to 32 are common constituents, whereas free alcohols have been reported in less than half of the species investigated to date. Most alcohols are saturated and unbranched with even carbon numbers. Aldehydes and ketones have been reported in some insects; however, the quantity of each seldom exceeds 10% of the total surface lipid.

Whereas plant surface lipids often possess a relatively rich complement of cyclic compounds, the arthropod cuticle apparently contains only simple sterols, primarily cholesterol, and in rather small amounts (<5%). The presence of cholesterol, nonetheless, is noteworthy, because neither insects nor arachnids can synthesize the sterane nucleus and thus must obtain cholesterol or its precursors from their diet. It is not known if cholesterol participates in the cuticular water barrier. Other sterols reported include stigmasterol in the grasshopper genus *Melanoplus*, and sitosterol, which occurs along with stigmasterol in two species of ants.

C. Amphibians and Reptiles

The integumental lipids of amphibians and reptiles have received little study, especially when compared with plants and arthropods. There are several reasons for this. Most, if not all, amphibians require a moist or aquatic environment to survive and reproduce. With few exceptions, their integument is highly permeable to the bidirectional movement of water. Without an apparent epidermal water barrier, there has been no reason to expect lipids to be a significant component of amphibian skin. Reptiles, in contrast, have made a complete transition to a terrestrial existence. Their integumental water loss rates are among the lowest reported for terrestrial animals. Until fairly recently, however, the low permeability was attributed to a thickened, heavily keratinized epidermis combined with the presence of scales. In both groups, integumental lipids have now been shown to contribute to the waterproofing process or are suspected of doing so.

The most detailed information on amphibian skin lipids comes from studies of South American phyllomedusine frogs. Several species secrete a lipid material onto the surface and then spread it over the integument. When dry, the lipid covering helps prevent evaporative water loss, thereby allowing the frogs to bask in the sun. Analyses of this secretion show wax esters to be dominant, with smaller quantities of triacylglycerols, hydrocarbons, cholesterol, cholesterol esters, and free fatty acids also present (McClanahan *et al.*, 1978). The wax esters average about 46 carbons in length and usually contain a double bond. A variety of neutral and polar lipids (e.g., cholesterol, cholesteryl esters, fatty acids, triacylglycerols, and phospholipids) can be extracted from *Hyperolius viridiflavus*, another xeric-adapted species that does not secrete lipids onto the surface (Giese and Linsemair, 1986). Finally, phospholipid and neutral lipids are reported to be present in definite stoichiometric proportions in the skin of *Rana pipiens*, in an arrangement that may govern the normal permeability of the skin (Abhinender Naidu *et al.*, 1984).

Information on the lipid composition of the reptilian integument is unfortunately restricted to a few species of snakes and lizards. In most cases, the data were obtained from whole bodies or shed skins extracted with chloroform:methanol to determine if epidermal lipids have an integral role in the waterproofing mechanism. In snakes, lipids account for approximately 1.4–14% of the weight of the shed skin and usually consist of a mixture of hydrocarbons, wax and sterol esters, di- and triacylglycerols, free fatty acids, alcohols, cholesterol, ceramides, and a variety of phospholipids (Roberts and Lillywhite, 1983; Burken *et al.*, 1985; Weldon and Bagnall, 1987). On the basis of results obtained from frozen sections of skin stained with oil red O, it appears that lipids are present in all epidermal layers (alpha, beta, and mesos), but are perhaps most abundant in the mesos layer (Roberts and Lillywhite, 1980). The lipid histochemistry of the scale and hinge epidermis of the chequered water snake, *Natrix piscator*, throughout the sloughing cycle has recently been described by Singh and Mittal (1989). They found comparatively high concentrations of phospholipids in the mesos and alpha layers. In addition, free fatty acids were present in almost all epidermal layers and in eosinophilic granular cells.

D. Birds and Mammals

The integumental lipids of higher vertebrates have been extensively studied, especially in mammals, because of the medical interest in the human skin and its cosmetic function. Only a brief summary is presented here, as the structure and function of mammalian epidermal lipids are described in great detail in several other articles in this volume.

Most reports of avian integumental lipids are actually descriptions of the secretions of the uropygial gland, which is located on the dorsum of the tail. Uropygial secretions, or "preen waxes," contain a complex mixture of lipids in which wax esters usually predominate. It is not uncommon for the acids and alcohols that comprise these esters to be largely branched-chain compounds (Downing, 1985). Smaller quantities of saturated and unsaturated hydrocarbons and triacylglycerols may also be present in some avian species. Preen waxes, when spread over the feathers, effectively prevent water penetration and may also help protect against bacteria and fungi (Jacob and Ziswiler, 1982). The most conclusive information to date on the origin and composition of lipids secreted by the avian epidermis is provided by Menon *et al.* (1986). They found that neutral lipids (triacylglycerols, sterol esters, sterols, and free fatty acids) and hydrocarbons accounted for about 67% of the lipids in the stratum corneum of the domestic pigeon. Sphingolipids, comprising a mixture of ceramides and glycosphingolipids, were next in abundance (25%), followed by phospholipids (<8%).

As in birds, the integumental lipids of mammals are derived from glandular secretions as well as from epidermal cells. Sebaceous glands are unique to mammals. Except for the palms and soles, they are distributed throughout the dermis, where they are generally connected via a duct to a hair follicle. Sebaceous glands produce an oily secretion called sebum. Although its composition varies among different species of mammals, it tends to consist of nonpolar, non-membrane-forming classes of lipids (e.g., sterols, sterol esters, wax esters, triacylglycerols, and squalene) (Stewart, 1985). Human sebum, which has received the most study, is unique in that triacyglycerols (55%) are the major component. Human sebum also contains some unusual methyl-branched fatty acids along with a more standard complement of saturated and unsaturated straight-chain acids. Approximately 12% of human sebum consists of squalene, a 30-carbon terpene molecule that is an intermediate in the synthesis of cholesterol. In other mammals, almost all of the sebum is squalene (Lindholm and Downing, 1980). The chemistry and function of sebaceous lipids are discussed more thoroughly by Stewart and Downing (this volume).

The source of epidermally derived lipids is believed to be special organelles called lamellar bodies, which are present in the stratum granulosum. These bodies ultimately migrate to the apical surface of the keratinocytes where they empty their lipid contents into the intercellular spaces. Here they are transformed from a relatively polar lipid mixture containing substantial amounts of phospholipids into a predominately nonpolar mixture that includes neutral lipids (free sterols, free fatty acids, triacylglycerols, and nonpolar *n*-alkanes) and sphingolipids (mostly ceramides) (Elias *et al.*, 1977; Lampe *et al.*, 1983). Despite the absence of phospholipids, there is evidence that these intercellular lipids can arrange themselves in bilayers that can exhibit thermal transitions similar to those described for artificial and biomembranes (Gray and White, 1979; Golden *et al.*, 1987). The composition and biochemistry of these stratum corneum lipids, plus their role in providing a barrier against transcutaneous water loss, are examined in depth by Schurer and Elias (this volume).

IV. Synthesis and Transport of Integumental Lipids

Whereas the composition and structure of integmental lipids are reasonably well understood, there is less certainty as to exactly where lipids are synthesized or how they are transported to their point of deposition. Similarly, little is known regarding possible biochemical changes that occur once lipids have reached their final destination. The following discussion is a brief summary of the principal sites and events involved in the synthesis and transport of integumental lipids. Again, the emphasis is on the cuticular lipids of plants and arthropods. In many cases, more comprehensive information is provided in several of the reviews previously cited.

A. Lipid Origins

In plants, epidermal cells, which lie directly beneath the leaf cuticle, appear to be the principal site for the synthesis of cuticular waxes. Supporting experimental evidence comes largely from studies demonstrating *de novo* synthesis of long-chain lipid compounds from radiolabeled precursors by excised epidermis of leaves and, conversely, the absence of synthesis when cell-free preparations are used. The epidermis is also the source of cuticular lipids in arthropods; however, here the manufacture of lipids and perhaps their precursors is associated with specific cell types. The most important of these are oenocytes, which are derived from and remain associated with the epidermis. There is considerable evidence (reviewed in Lockey, 1988) that both structural lipid and free lipid are synthesized by the oenocytes. Ultimately, this lipid is transferred to the epidermal cells. In the bloodsucking bug *Rhodnius*, this transfer is facilitated by cytoplasmic extensions produced by the epidermal cells that connect them to the oenocytes (Wigglesworth, 1988). Peripheral fat body cells, which are also closely associated with oenocytes and which can also become attached to the epidermis, may also participate in the synthesis of specific lipid classes. Some of the synthesized lipid is incorporated directly into the overlying cuticle, and another portion can enter the hemolymph, where it combines with lipophorin carrier molecules and is distributed throughout the body (Katase and Chino, 1982, 1984).

Whereas both plants and arthropods can synthesize cuticular lipids *de novo*, arthropods are also able to incorporate dietary lipids into their cuticle. This can be tested by feeding labeled alkanes to insects and, after an appropriate incubation period, verifying the presence of the labeled molecule in lipids extracted from the cuticle or by identifying a specific alkane on the surface that is present in the diet, but which is not normally present in the consumer (Blomquist and Jackson, 1973; Hall and Hadley, 1982). Although the dietary contribution to cuticular lipids is generally believed to be secondary to those manufactured directly by the arthropod, Espelie and Bernays (1989) provide convincing data showing that the surface chemistry of *Manduca sexta* larvae varies with their food source. Obviously, this is an area that warrants further study.

B. Principal Synthetic Pathways

The biosynthesis of cuticular lipids has received fairly extensive study, especially by Blomquist, Kolattukudy, and their co-workers. As in compositional analyses, the focus has been on cuticular hydrocarbons rather than their oxygenated derivatives. With few exceptions, the primary pathways and enzymatic reactions appear to be identical in plants and arthropods. In fact, it is likely that similar metabolic pathways occur in vertebrates (see Feingold, this volume). Because very little is known about the synthesis of sterols in plant and arthropod cuticle, the following discussion will focus on hydrocarbon biosynthetic pathways.

The hydrocarbon components of surface lipids were, for a long time, thought to be made by condensation and decarboxylation reactions. According to this scheme, two molecules of C_{15} might condense head to head, followed by the decarboxylation of one of the molecules to give a C_{29} ketone. This ketone, upon reduction, would produce a C_{29} alcohol. Further dehydration followed by reduction would give rise to a C_{29} alkane (Kolattukudy and Walton, 1972). Despite its attractive simplicity, the existing experimental evidence strongly argues against this metabolic route. Instead, hydrocarbon biosynthesis is believed to occur by the elongation of a fatty acid precursor until a decarboxylation reaction yields an alkane that is one carbon shorter than the substrate. In this sequence, the C_{29} alkane (nonacosane) would be produced by successively adding two carbon units to palmitic acid (C_{16}) until a C_{30} chain is created, followed by a decarboxylation reaction to release the hydrocarbon. The enzyme systems that carry out the elongation of the fatty acyl chain are known as elongases. In plants, these elongases and the decarboxylase system are believed to be located in the epidermal cells, where they are affiliated with or part of the microsomal membranes (von Wettstein-Knowles, 1987). Incubation of epidermal-enriched microsomal preparations have also resulted in the production of a series of long-chain, even-numbered carbon acyl moieties in several insect species, suggesting the presence of elongase enzymes (de Renobales *et al.*, 1991).

The *de novo* synthesis of the unsaturated and branched hydrocarbons also occurs by the elongation–decarboxylation pathway, but requires a different starter acid. Oleic acid ($C_{18:1}$) and linoleic acid ($C_{18:2}$) can serve as the precursors for the synthesis of long-chain monoenes and dienes. The precursors for the biosynthesis of branched molecules vary according to the position of the methyl branch. For 2-methylalkanes having an even number of carbon atoms, the methyl branch arises from the carbon skeleton of the amino acid valine, whereas leucine serves as the precursor for odd-numbered 2-methylalkanes. Two pathways have been suggested for the biosynthesis of 3-methylalkanes (Blomquist and Dillwith, 1985). The first, which occurs in plants, involves the conversion of isoleucine to 2-methylbutyric acid, which is then incorporated into 3-methylalkanes. The second, noted in cockroaches, involves the conversion of propionate to methylmalonyl-CoA, which becomes the precursor for the branching methyl group. Internally branched alkanes could originate by substituting methylmalonyl-CoA for malonyl-CoA during the elongation process or by methylating a preformed chain.

As indicated above, the biosynthesis of oxygenated derivatives of hydrocarbons has not been as extensively studied. It is commonly presumed that the long-chain fatty acids that result from the elongation reactions and that give rise to hydrocarbons upon reduction are also the source of the primary alcohols and aldehydes present in cuticular extracts. In plants, the chain length distribution of aldehydes is often very similar to that of alcohols, suggesting a precursor–product relationship. Both short- and long-chain fatty acids can also join with fatty alcohols to produce wax esters. Formation of the latter requires the presence of an esterase-type enzyme. The chain length of the fatty alcohols in the wax esters often closely resembles that of the free alcohols, indicating that they originate from a common metabolic pool.

The biosynthesis of cyclic compounds such as terpenoids and sterols is well known because of their presence in other plant and animal tissues. The early steps, which are common to all organisms, involve the formation of mevalonic acid from three molecules of acetyl-CoA. Mevalonic acid then gives rise, through a series of three phosphorylations, to isopentenyl pyrophosphate, which possesses the five-carbon isopentenoid structure characteristic of terpenoids. A detailed summary of these plus the remaining steps can be found in Harwood and Russell (1984). In mammals, the skin is a major site of total body sterol synthesis, with the epidermis accounting for approximately 30% of this total (Feingold *et al.*, 1983). Recent studies have demonstrated that within the epidermis the biosynthetic activity of the stratum granulosum layer is comparable to or greater than that of the combined basal/spinous layers (Monger *et al.*, 1988).

C. Transport to the Surface

1. Plants

Lipids synthesized by the epidermis or associated cells must pass through the cuticle before being deposited on or near the surface. How this transport occurs is not well understood. Plants are particularly troublesome in that most species have neither pores nor pits on the surface, nor a system of vertical canals that links the epidermis to the surface. [Miller (1986), however, claims that discrete, naturally occurring cuticular pores and transcuticular canals are randomly distributed throughout the cuticular matrix of all plant species.] It is possible that channels or pores are present in living hydrated tissue, but that they cannot be demonstrated in the dry cuticular membranes required for electron microscopy (Hadley, 1981). Whatever the nature of the conducting pathway, it is generally agreed that the lipid is carried to the surface in a solution that subsequently evaporates, causing the dissolved lipids to crystallize on the surface (Jeffree *et al.*, 1975; Chambers *et al.*, 1976). Some specific physical possibilities for transport include the lipid being solubilized in a volatile organic solvent, present as micelles in an aqueous system, packaged as an oil/water emulsion particle, or secreted as lipoproteins.

2. *Arthropods*

Lipid transport in arthropods is routinely associated with a complex system of canals that traverse the cuticle. The largest diameter ducts in this system are the pore canals, which extend from the apical membrane of the epidermal cells to approximately the lower boundary of the epicuticle. Here each pore canal connects with several smaller epicuticular channels, which perforate the outer layers of the epicuticle. Smaller diameter tubular filaments are often visible within the lumen of the pore canals and epicuticular channels. Despite this elaborate interconnecting system, no satisfactory explanation has been put forth to explain how lipids are actually transported or in what form. There is a good possibility that a protein or lipoprotein carrier molecule is involved. The filaments likely contain or are themselves composed of some protein. Moreover, protein is often associated with epicuticular lipids (Hadley, 1979; Wigglesworth, 1985). The lipid, in combination with a protein carrier molecule, could be transmitted to the surface inside the tubular filaments or in the lumen between the filaments and the inside wall of the pore canals. It should be noted that filamentous wax blooms that characterize the surface of several species of tenebrionid beetles reach the surface via dermal gland ducts rather than the pore canal complex described above (Hanrahan *et al.*, 1984).

3. *Birds and Mammals*

Lipids synthesized in the viable layers of the epidermis are ultimately sequestered into either multigranular bodies (MGBs) (birds) or lamellar bodies (LBs) (mammals). These membrane-bound structures contain stacks of membrane bilayer disks, the content and arrangement of which vary among the two vertebrate groups (Elias *et al.*, 1987). The fate of he MGB is uncertain. In the pigeon, the MGB appears to disintegrate, resulting in the formation of large, non-membrane-bound lipid droplets that appear in the cytoplasm of the lowermost cells in the stratum corneum. In this species, little if any of the content of the MGB is secreted into the intercellular spaces; however, in the zebra finch, some of the MGB disks do appear to be secreted into these spaces (Menon *et al.*, 1989). In mammals, the disks inside the lamellar bodies become exocytosed at the interface of the stratum granulosum and stratum corneum. They subsequently fuse to form multilamellar sheets in the intercellular space of the stratum corneum.

V. Water Barrier Function: Working Models

There is strong evidence, both direct and indirect, that lipids associated with the integument provide the principal barrier to water efflux for most terrestrial organisms. Transintegumentary water loss usually increases significantly when these lipids are removed by solvents, mechanically disrupted, absorbed by dusts applied to the surface, or structurally altered by exposure to high temperature (Hadley, 1981, 1989). Schönherr (1976) reported a 350- to 500-fold increase in diffusion across isolated cuticle segments of citrus leaves following extraction of the soluble lipids; Hadley and Quinlan (1989) observed a 100-fold increase in the permeability of abdominal cuticle of the black widow spider following gentle rubbing with chloroform:methanol. In reptiles, lipid extraction increased transepidermal permeation as much as 35- to 175-fold in seven snake species (Burken *et al.*, 1985). Treatment of mammalian skin with lipid extractants resulted in water loss rates that approach the 100-fold increase observed when the stratum corneum is removed (Smith *et al.*, 1982). Investigators have also demonstrated inverse relationships between the amount or surface density of lipids and transpiration as well as correlated barrier effectiveness with surface wax morphology, arrangement, and chemical composition. Finally, both plants and animals are capable of modifying lipid amounts, composition, and/or structure in ways that theoretically at least should improve waterproofing effectiveness during times when there is a greater need for water conservation (see Geyer and Schönherr, 1990; Hadley, 1985, 1989, for references).

Several models have been proposed to explain the functional organization of barrier lipids and how the waterproofing mechanism operates. In plants, each of the cuticle layers is rich in lipids; however, only the "soluble cuticular lipids (SCLs)," which can be easily extracted from the cutin matrix with nonpolar solvents, are believed to contribute substantially to the barrier. The SCLs are thought to consist of crystalline and amorphous regions that are arranged in one or several layers that are more or less parallel to the surface of the cuticle (Riederer and Schneider, 1990). The crystalline region consists of the hydrocarbon chains of the various lipid constituents packed in an orthorhombic lattice. The amorphous zones consist of the flexible chain ends and any substituent groups. Together, the crystalline and amorphous regions comprise SCL "flakes" that are embedded in the cutin matrix. Water moving from the epidermis to the cuticle surface would be able to pass only through the matrix regions surrounding the SCL flakes. The effectiveness of the SCL flakes in reducing permeability would depend on their volume fraction, which, in turn, depends on the number and size of the flakes and their lateral and vertical distances (Cussler *et al.*, 1988).

The "stacked-flake" model described above has several features that are consonant with the observed structure and functioning of the cuticular waxes in plants. An orthorhombic perpendicular arrangement of the hydrocarbon chains, which has been proposed for the cuticular waxes of the carnauba palm, is the probable packing pattern for plant waxes in general (Basson and Reynhardt, 1988). Moreover, it has been shown by polarizing microscopy that the long axes of the molecules of crystalline SCL are indeed oriented normal to the plane of the cuticle (Sitte and Rennier, 1963). The model also helps explain some of the reported failures to establish correlations between the measured permeability of plant cuticles to water and the chemical composition of the SCL. According to this model, what really matters in building a barrier to the efflux of water is the effect of packed hydrocarbon chains and not so much that of any polar substitution among these hydrocarbons (Riederer and Schneider, 1990). The specific packing of aliphatic hydrocarbon chains below the melting temperature of a pure crystal or a solid solution can provide a layer that is inaccessible and thus impermeable to water molecules. Although the general properties of these aliphatic solids are modulated to varying degrees by functional groups substituted to the hydrocarbon chain, these effects diminish with increasing chain length.

In the arthropod cuticle, lipids are most abundant in the outer layer, or "epicuticle." Here they are incorporated into various horizontal subdivisions as well as being superficially deposited on the surface. It is these epicuticular lipids that are primarily responsible for restricting transcuticular water loss; however, their organization is uncertain. There is increasing evidence that they may simply reside as a heterogeneous mixture in layers that parallel the surface. Temperature-induced permeability changes in the cuticle of the black widow spider suggest that there may be two components to the lipid barrier: a primary outer barrier, nonpolar in composition, which is easily removed by solvents, and a secondary, more resistant inner barrier containing more polar lipids that are perhaps bound to protein (Hadley and Quinlan, 1989). It is not known if water that does escape through the cuticle does so by diffusing through the general epicuticular surface or is restricted to fixed points such as the terminations of the epicuticular filaments.

In higher vertebrates, the epidermal permeability barrier consists of multiple lipid bilayers that form in the intercellular space of the stratum corneum (see Elias and Menon, this volume). These disks ultimately fuse to form uninterrupted sheetlike lamellae that are arranged parallel to the plasma membranes of the adjacent keratinocytes. The resultant lipid matrix prevents both transcellular and paracellular water flux. Current and comprehensive information on the origin, structure, and composition of these lipids and the general properties of the barrier is contained elsewhere in this volume. The reader will note that although many features of the system in birds and mammals are similar to those described for plants and arthropods, the currently accepted "lipid model" represents a unique development in the evolution of integumentary water barriers.

References

Abhinender Naidu, K., Ramamurthi, R., Akhilender Naidu, K., and Hanke, W. (1984). *Comp. Biochem. Physiol. A* **79A,** 49–52.

Baker, E. A. (1982). *In* "The Plant Cuticle" (D. F. Cutler, K. L. Alvin, and C. E. Price, eds.), pp. 139–165. Academic Press, New York.

Barthlott, W. (1981). *Nord. J. Bot.* **1,** 345–355.

Barthlott, W., and Wollenweber, E. (1981). *Trop. Subtrop. Pflanzenwelt* **32,** 1–67.

Basson, I., and Reynhardt, E. C. (1988). *J. Phys. D* **21,** 1429–1433.

Blomquist, G. J., and Dillwith, J. W. (1985). *In* "Comprehensive Insect Physiology, Biochemistry and Pharmacology" (G. A. Kerkut and L. I. Gilbert, eds.), pp. 117–154. Pergamon, Oxford.

Blomquist, G. J., and Jackson, L. L. (1973). *J. Insect Physiol.* **19,** 1639–1647.

Blomquist, G. J., Nelson, D. R., and de Renobales, M. (1987). *Arch. Insect Biochem. Physiol.* **6,** 227–265.

Burken, R. R., Wertz, P. W., and Downing, D. T. (1985). *Comp. Biochem. Physiol. A* **81A,** 213–216.

Chambers, T. C., Ritchie, I. M., and Booth, M. A. (1976). *New Phytol.* **77,** 43–49.

Cussler, E. L., Hughes, S. E., Ward, W. J., and Aris, R. (1988). *J. Membr. Sci.* **38,** 161–174.

de Renobales, M., Nelson, D. R., and Blomquist, G. J. (1991). *In* "The Physiology of the Insect Epidermis" (A. Retnakaran and K. Binnington, eds.). Inkata Press, North Clayton, Australia. In press.

Downing, D. T. (1985). *In* "Biology of the Integument. Vol. 2: Vertebrates" (J. Bereiter-Hahn, A. G. Matoltsy, and K. S. Richards, eds.), pp. 833–840. Springer-Verlag, Berlin.

Elias, P. M., Goerke, J., and Friend, D. S. (1977). *J. Invest. Dermatol.* **69,** 535–546.

Elias, P. M., Menon, G. K., Grayson, S., Brown, B. E., and Rehfeld, S. J. (1987). *Am. J. Anat.* **180,** 161–177.

Espelie, K. E., and Bernays, E. A. (1989). *J. Chem. Ecol.* **15,** 2003–2017.

Feingold, K. R., Brown, B. E., Lear, S. R., Moser, A. H., and Elias, P. M. (1983). *J. Invest. Dermatol.* **81,** 365–369.

Geise, W., and Linsenmair, K. E. (1986). *Oecologia* **68,** 542–548.

Geyer, U., and Schönherr, J. (1990). *Planta* **180,** 147–153.

Golden, G. M., Guzek, D. B., Kennedy, A. H., McKie, J. E., and Potts, R. O. (1987). *Biochemistry* **26,** 2382–2388.

Gray, G. M., and White, R. J. (1979). *Biochem. Soc. Trans.* **7,** 1129–1131.

Gülz, P. G., Müller, E., and Prasad, R. B. N. (1989). *Z. Naturforsch., C* **44C,** 731–734.

Hadley, N. F. (1979). *Science* **203,** 367–369.

Hadley, N. F. (1981). *Biol. Rev. Cambridge Philos. Soc.* **56,** 23–47.

Hadley, N. F. (1984). *In* "Biology of the Integument. Vol. 1: Invertebrates" (J. Bereiter-Hahn, A. G. Matoltsy, and K. S. Richards, eds.), pp. 685–693. Springer-Verlag, Berlin.

Hadley, N. F. (1985). "The Adaptive Role of Lipids in Biological Systems." Wiley, New York.

Hadley, N. F. (1986). *Sci. Am.* **254,** 104–112.

Hadley, N. F. (1989). *Prog. Lipid Res.* **28,** 1–33.
Hadley, N. F., and Quinlan, M. C. (1989). *J. Comp. Physiol.* **159,** 243–248.
Hall, R. L., and Hadley, N. F. (1982). *J. Exp. Zool.* **224,** 195–203.
Hanrahan, S. A., McClain, E., and Gernecke, D. (1984). *S. Afr. J. Sci.* **80,** 176–181.
Harwood, J. L., and Russell, N. J. (1984). "Lipids in Plants and Microbes." Allen Unwin, London.
Howard, R. W., and Blomquist, G. J. (1982). *Annu. Rev. Entomol.* **27,** 149–172.
Jackson, L. L., and Blomquist, G. J. (1976). *In* "Chemistry and Biochemistry of Natural Waxes" (P. E. Kolattukudy, ed.), pp. 201–233. Elsevier, Amsterdam.
Jacob, J., and Ziswiler, V. (1982). *In* "Avian Biology" (D. S. Farner, J. R. King, and K. C. Parkes, eds.), Vol. 6, pp. 199–324. Academic Press, New York.
Jeffree, C. E., Baker, E. A., and Holloway, P. J. (1975). *New Phytol.* **75,** 539–549.
Katase, H., and Chino, H. (1982). *Biochim. Biophys. Acta* **710,** 341–348.
Katase, H., and Chino, H. (1984). *Insect Biochem.***14,** 1–6.
Kolattukudy, P. E. (1975). *In* "Recent Advances in the Chemistry and Biochemistry of Plant Lipids" (T. Gallard and E. I. Mercer, eds.), pp. 203–246. Academic Press, New York.
Kolattukudy, P. E. (1976). *In* "Chemistry and Biochemistry of Natural Waxes" (P. E. Kolattukudy, ed.), pp. 1–15. Elsevier, Amsterdam.
Kolattukudy, P. E., and Walton, T. J. (1972). *Prog. Chem. Fats Other Lipids* **13,** 121–175.
Lampe, M. A., Burlingame, A. L., Whitney, J. A., Williams, M. L., Brown, B. E., Roitman, E., and Elias, P. M. (1983). *J. Lipid Res.* **24,** 120–130.
Lindholm, J. S., and Downing, D. T. (1980). *Lipids* **15,** 1062–1063.
Lickey, K. H. (1985). *Comp. Biochem. Physiol. B* **81B,** 263–273.
Lockey, K. H. (1988). *Comp. Biochem. Physiol. B* **89B,** 595–645.
Malinski, E., Kusmierz, J., Szafranek, J., Dubis, E., Poplawski, J., Wrobel, J. T., and Konig, W. A. (1986). *Z. Naturforsch., B* **41B,** 567–574.
Manheim, M. S., and Mulroy, T. W. (1978). *Phytochemistry* **17,** 1799–1800.
Matoltsy, A. G., and Bereiter-Hahn, J. (1986). *In* "Biology of the Integument. Vol. 2: Vertebrates" (J. Bereiter-Hahn, A. G. Matoltsy, and K. S. Richards, eds.), pp. 1–7. Springer-Verlag, Berlin.
McClanahan, L. L., Stinner, J. L., and Shoemaker, V. H. (1978). *Physiol. Zool.* **51,** 179–187.
Menon, G. K., Brown, B. E., and Elias, P. M. (1986). *Tissue Cell* **18,** 71–82.
Menon, G. K., Baptista, L. F., Brown, B. E., and Elias, P. M. (1989). *Tissue Cell* **21,** 83–92.
Miller, R. H. (1986). *Ann. Bot.* **58,** 407–416.
Monger, D. J., Williams, M. L., Feingold, K. R., Brown, B. E., and Elias, P. M. (1988). *J. Lipid Res.* **29,** 603–612.
Nelson, D. R., Carlson, D. A., and Fatland, C. L. (1988). *J. Chem. Ecol.* **14,** 963–987.
Riederer, M., and Schneider, G. (1990). *Planta* **180,** 154–165.
Roberts, J. B., and Lillywhite, H. B. (1980). *Science* **207,** 1077–1079.
Roberts, J. B., and Lillywhite, H. B. (1983). *J. Exp. Zool.* **228,** 1–9.
Salasoo, I. (1989). *Biochem. Syst. Ecol.* **17,** 381–384.
Schönherr, J. (1976). *Planta* **131,** 159–164.
Singh, J. P. N., and Mittal, A. K. (1989). *J. Zool.* **218,** 39–50.
Sitte, P., and Rennier, R. (1963). *Planta* **60,** 19–40.
Smith, W. P., Christensen, M. S., and Nacht, S. J. (1982). *Invest. Dermatol.* **78,** 7–11.
Stewart, M. E. (1985). *In* "Biology of the Integument. Vol. 2: Vertebrates" (J. Bereiter-Hahn, A. G. Matoltsy, and K. S. Richards, eds.), pp. 824–832. Springer-Verlag, Berlin.
Tulloch, A. P. (1976). *In* "Chemistry and Biochemistry of Natural Waxes" (P. E. Kolattukudy, ed.), pp. 235–287. Elsevier, Amsterdam.
von Wettstein-Knowles, P. (1987). *In* "Plant Molecular Biology" (D. von Wettstein and N.-H. Chua, eds.), pp. 305–314. Plenum, New York.
Weldon, P. J., and Bagnall, D. (1987). *Comp. Biochem. Physiol. B* **87B,** 345–349.
Wigglesworth, V. B. (1985). *Tissue Cell* **17,** 227–248.
Wigglesworth, V. B. (1988). *Tissue Cell* **20,** 919–932.

Epidermal Vitamin D Metabolism, Function, and Regulation

SREEKUMAR PILLAI AND DANIEL D. BIKLE

Departments of Dermatology and Medicine
University of California School of Medicine
San Francisco, California 94143
and
Dermatology Service and Endocrinology Section
Veterans Administration Medical Center
San Francisco, California 94121

I. Introduction

The secosteroid hormone, 1,25-dihydroxy vitamin D_3 [1,25-$(OH)_2D_3$], the major biologically active form of vitamin D, plays a central role in calcium homeostasis. Substantial advances have been made in our understanding about both the production and the metabolism of vitamin D over the last two decades. For example, the unique role of the skin in the production of vitamin D and its hydroxylation by the liver and kidney to form the biologically active hormone are now well appreciated. Likewise, the mode of action of vitamin D on the calcium flux across intestinal mucosa and renal tubule, and its role in the mobilization of calcium from bone, have been well elucidated. The reader is referred to several excellent reviews that have appeared recently on these aspects of vitamin D metabolism (P. A. Bell, 1978; DeLuca and Schnoes, 1983; Henry and Norman, 1984; Audran, 1985; MacLaughlin and Holick, 1983; N. H. Bell, 1985; Stumpf, 1988; Holick, 1988; Reichel *et al.*, 1989). There is growing evidence that the vitamin D endocrine system is important not only for the regulation of bone and mineral metabolism, but also in the modulation of function and differentiation in other tissues as well. The recent findings that 1,25-$(OH)_2D_3$ receptors are present in tissues other than the gut, kidney, and bone suggest other functions for this hormone. This review will focus on the possible autocrine role of vitamin D in the epidermis, i.e., production of the hormonally active form of vitamin D in the epidermis, and its modulation of epidermal cell differentiation.

II. Historical Perspective

Rickets was recognized as early as the seventeenth century as a disease of the bone in children living in the crowded, polluted cities of Great Britain (Holick, 1986). By the early twentieth century, the cause of the disease was attributed to nutritional deficiency, and the curative effect of fish liver oils was demonstrated by Mellanby (1918). However, nutrition alone failed to explain the occurrence of rickets only in industrialized towns. Children who had poor nutrition and lived in the countryside or in underdeveloped countries of the world did not develop the dreaded disease (Sniadecki, 1931; Owen, 1889). These findings led Palm (1890) to hypothesize that the cause of rickets in children was the lack of exposure to sunlight. Proof for the curative effect of UV radiation was obtained with the demonstration that exposure to sunlight (Hess, 1921) or to a mercury vapor arc lamp (Huldschinsky, 1919) could cure rickets in children. The action of UV radiation was initially attributed to mobilization of an antirachitic factor from the skin (Holick *et al.*, 1982b). The demonstration by Hess and Weinstock (1924) and Steenbock and Black (1924) that irradiation imparted antirachitic activity to foods indicated the presence of a nutrient with antirachitic activity, which they deduced would also be produced by UV radiation of the skin. The isolation and identification of vitamin D_2 from irradiated mixtures of plant sterols was achieved by Askew *et al.* (1931) and by Windus and collaborators (1931, 1932), who also isolated and identified the precursor of vitamin D_2, ergosterol. However, vitamin D_2 was less effective as an antirachitic factor in chickens than the antirachitic factor produced by the skin. Subsequently, Windus and Bock (1937) structurally characterized the provitamin D of the human skin as 7-dehydrocholesterol (7-DHC). In the mid 1970s, Petrova *et al.* (1976) and Holick *et al.* (1977), using tritiated 7-DHC, demonstrated the conversion of 7-DHC to pre-D_3. Finally, using high-performance liquid chromatography systems and radiolabeled precursors, Holick and colleagues (1979, 1980) localized the various metabolites of 7-DHC to specific epidermal layers and elucidated the pathways of vitamin D_3 synthesis in the skin.

III. Chemistry of Vitamin D

Vitamin D and its metabolites (Fig. 1) are secosteroids, or cholesterol-like compounds in which one of the rings has undergone fission. In vitamin D, the

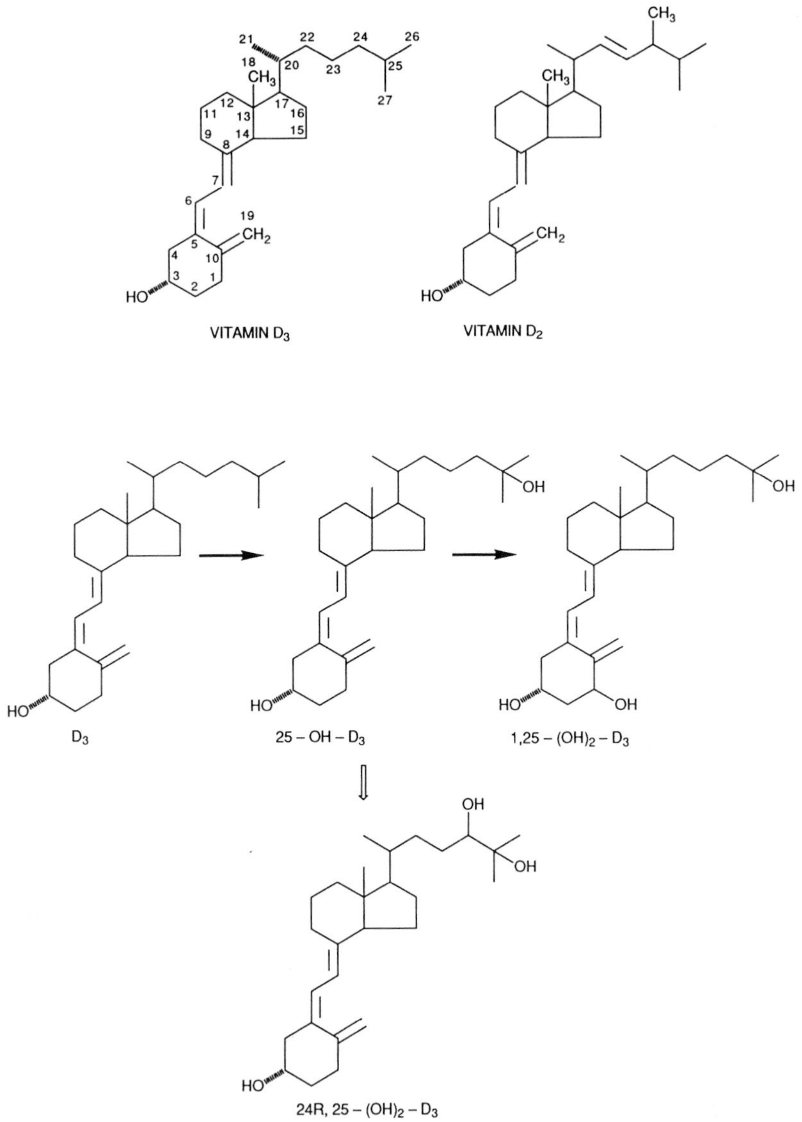

FIG. 1. Structure of Vitamin D and its major biologically active metabolites.

9,10 carbon bond of ring B is broken. Because of the seco nature of vitamin D, the A ring is inverted, with rotation occurring around the bond between C-7 and C-8. The product obtained from irradiation of ergosterol is termed vitamin D_2. Vitamin D_3 is the product obtained from 7-dehydrocholesterol, the 5,7-diene sterol containing a cholesterol-like side chain. Vitamin D also exists in nature in the

D_4–D_8 forms, which arise from a similar 5,7-diene sterol containing different side chains (Fieser and Fieser, 1959). Vitamin D_3 is hydroxylated at the C-25 position in the liver to form 25-(OH)D_3, the major circulating form of vitamin D. The kidney and other tissues function as the site of two further hydroxylations to form 1,25-$(OH)_2D_3$ and 24,25-$(OH)_2D_3$ (Fig. 1) (Fraser and Kodicek, 1970; Gray *et al.*, 1979). Several trihydroxy and lactone metabolites have also been isolated and may represent catabolic products of vitamin D_3, although several have unique biologic activities. Bile represents the major excretory route for vitamin D catabolic products (Kumar *et al.*, 1980).

IV. Role of Vitamin D in Calcium Homeostasis

1,25-$(OH)_2D_3$ is important for the maintenance of calcium homeostasis. In concert with parathyroid hormone and calcitonins, 1,25-$(OH)_2D_3$ exerts actions on the classic target tissues—bone, intestine, and kidney—to regulate the circulating levels of calcium.

The concentration of calcium in the blood is maintained between 2.2 and 2.5 m*M*. About 30% of the calcium is bound to proteins, primarily serum albumin; another 10% is complexed with various chelators, such as citrate, and the remaining 60% is free or ionized. The latter form is important in biological activities, and its concentration is tightly regulated. The blood and extracellular fluid calcium make up only 0.1% of the total body calcium. About 1% is stored intracellularly and the bulk (99%) resides in bone and teeth. Therefore, bone tissue constitutes a major organ in the regulation of calcium homeostasis in the body.

Bone undergoes constant remodeling under normal circumstances. Osteoclasts mediate bone resorption, and osteoblasts mediate new bone formation (Canalis *et al.*, 1988; Raiw and Kream, 1983). 1,25-$(OH)_2D_3$ regulates both of these processes (Stern, 1980). 1,25-$(OH)_2D_3$ exerts its effects on osteoblasts through 1,25-$(OH)_2D_3$ receptors present in these cells (Kurihara *et al.*, 1986). These effects include stimulation of alkaline phosphatase (Kurihara *et al.*, 1986), stimulation of synthesis of bone G1a protein (osteocalcin) (Pan and Price, 1984) and matrix G1a protein (Fraser *et al.*, 1988), increase in receptors for epidermal growth factor (EGF) and transforming growth factor β (TGF β) (Petkovich *et al.*, 1987), and down-regulation of type I collagen synthesis (Rowe and Kream, 1982). Thus, the role of vitamin D on bone mineralization appears to be mediated through its direct action on osteoblasts. On the other hand, the effect of vitamin D on bone resorption may be more indirect. Mature osteoclasts do not contain vitamin D receptors. Exposure to 1,25-$(OH)_2D_3$ for several weeks increases the number of bone osteoclasts (Holtrop *et al.*, 1981; Roodman *et al.*, 1985). Osteoclasts originate from a hematopoietic cell of early macrophage lineage (Burger *et al.*, 1982). A role for 1,25-$(OH)_2D_3$ on the differentiation and maturation of myeloid hematopoietic precursors cells into multinucleated osteoclasts is now generally accepted (Burger *et al.*, 1982). Therefore, the effect of 1,25-$(OH)_2D_3$ on bone resorption may be closely linked to the effect of this hormone in induction of cell differentiation.

Another mode of action of Vitamin D in the maintenance of serum calcium and bone mineralization is mediated by its effect on calcium absorption from the intestine. 1,25-$(OH)_2D_3$ stimulates the transport of calcium and phosphate from the intestinal lumen into plasma (Haussler and McCain, 1977). This effect requires both genomic and nongenomic actions of 1,25-$(OH)_2D_3$. The best studied genomic action is the induction of an intestinal calcium-binding protein known as calbindin (Dupret *et al.*, 1987). Although the amount of this protein correlates with the rate of calcium absorption in the intestine, the exact role of calbindin in calcium transport has not been delineated. In addition to its genomic action, vitamin D also stimulates intestinal calcium transport by increasing the calmodulin content of the brush border membrane (Nemere *et al.*, 1984; Bikle and Chafouleas, 1984). A 105-kDa protein present in the brush border membrane binds calmodulin in a manner stimulated by 1,25-$(OH)_2D_3$ (Bikle and Munson, 1985). This binding increases calcium transport across the brush border membrane, suggesting that it may regulate a calcium channel in the brush border membrane.

The effect of 1,25-$(OH)_2D_3$ on calcium transport in the kidney is not well established. The direction of vitamin D-mediated renal calcium transport (excretion versus resorption) may depend on the plasma calcium level (Agus, 1983). Under hypocalcemic conditions, 1,25-$(OH)_2D_3$ appears to reduce the excretion of calcium in the kidney. Phosphate reabsorption appears to be enhanced by 1,25-$(OH)_2D_3$, an action facilitated by parathyroid hormone (Lyles and Drezner, 1982). However, the detailed mechanisms of this regulation are still unclear.

Abnormalities in the vitamin D endocrine system occur in many conditions. Diseases involving the liver, kidney, or intestine are sometimes associated with defects in the metabolism or action of this hormone. These conditions result in disorders of bone mineralization, such as osteoporosis, osteomalacia, and rickets. Therapy with vitamin D often improves calcium balance, increases calcium absorption, increases bone volume, and reduces bone fractures. Two rare syndromes, vitamin D-dependent rickets type II (VDDR II) and hereditary vitamin D-resistant rickets (HVDRR), involve defective or absent 1,25-$(OH)_2D_3$ receptors (Brooks *et al.*, 1978), respectively. These individuals develop rickets that is not responsive or is poorly responsive to 1,25-$(OH)_2D_3$. Of relevance to the cutaneous actions of 1,25-$(OH)_2D_3$ is the observation that many of these subjects have alopecia, suggesting a role for 1,25-$(OH)_2D_3$ in hair growth (Bell *et al.*, 1978).

V. Other Biological Actions of 1,25-$(OH)_2D_3$

In addition to its role in calcium homeostasis, 1,25-$(OH)_2D_3$ has been shown to influence other cellular functions as well. In analogy to the vitamin A metabolite, retinoic acid, 1,25-$(OH)_2D_3$ regulates proliferation and differentiation of

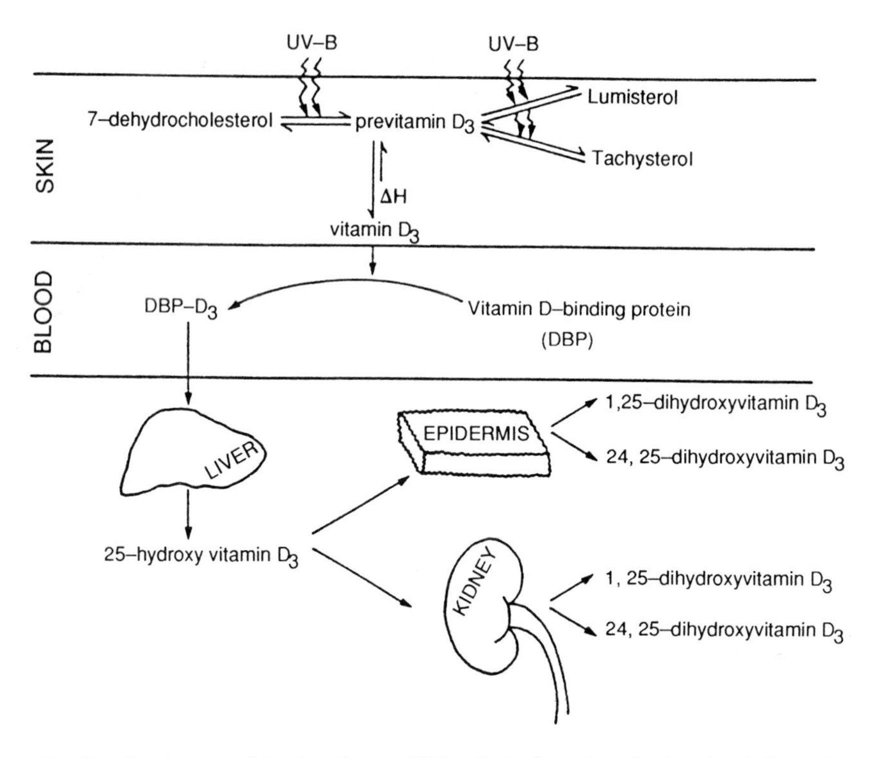

FIG. 2. Involvement of the skin, liver, and kidney in the formation of active vitamin D metabolites. Pre-vitamin D_3 is formed in the skin from 7-dehydrocholesterol during exposure to UV light from the sun. Thermal isomerization of previtamin D_3 to vitamin D_3 occurs in the basal layers of the epidermis. The vitamin D_3 formed is specifically translocated by the vitamin D-binding protein (DBP) from skin into the circulation. Vitamin D_3 then undergoes sequential hydroxylation in the liver and kidney to generate the hormonally active forms of vitamin D_3. Recent studies indicate that skin and other nonrenal tissues may participate in the production of 1,25-$(OH)_2D_3$. (From Pillai *et al.*, 1988b.)

because the levels of 7-DHC were minimal in cultured cells in comparison to the epidermis, cultured cells required treatment with an inhibitor of 7-DHC reductase, AY-9944, in order to increase the pool of available 7-DHC for vitamin D_3 production. Further conversion of pre-D_3 to D_3 also occurred in these cells. Keratinocytes were a more prolific source of the vitamin than were fibroblasts.

VIII. Production of 1,25-$(OH)_2D_3$ and Its Regulation

Vitamin D_3 is hydroxylated at the C-25 position by vitamin D_3 25-hydroxylase to form 25-$(OH)D_3$, primarily in the liver (Ponchon *et al.*, 1969) and to some extent in other tissues, including lung, intestine, and kidney (Ichikawa *et al.*, 1983). 25-(OH)D_3 is then further hydroxylated to other metabolites, the most important of which is 1,25-$(OH)_2D_3$. The enzyme responsible for 1,25-$(OH)_2D_3$ production is the 25-(OH)D_3 1α-hydroxylase, found primarily in the mitochondria of the proximal tubule of the kidney (Fraser and Kodicek, 1970). In contrast, the 25-hydroxylase is a cytochrome *P*-450-dependent multienzyme complex distributed between microsomes and mitochondria (Bhattacharyya and DeLuca, 1974; Bjorkhem and Holmberg, 1978). The 25-hydroxylase activity of the liver does not appear to be as strictly regulated as the 1α-hydroxylase of the kidney. Both the 25-(OH)D_3 1α-hydroxylase and the 25-(OH)D_3 24*R*-hydroxylase of the kidney belong to a class of enzymes known as mitochondrial mixed-function oxidases. They are composed of three proteins in the inner mitochondrial membrane: a nonheme iron protein, ferrodoxin; ferrodoxin reductase; and cytochrome *P*-450 (Ghazarian and DeLuca, 1974; Ghazarian *et al.*, 1974; Pederson *et al.*, 1976). The activity of this complex is strictly regulated by the 1,25-$(OH)_2D_3$ status of the cell. In the vitamin D-deficient state, the production of 1,25-$(OH)_2D_3$ is maximal and 24,25-$(OH)_2D_3$ production is minimal or undetectable (Tanaka and DeLuca, 1974; Henry *et al.*, 1974). The situation is reversed in the presence of 1,25-$(OH)_2D_3$. This regulation is seen both *in vivo* and *in vitro* (Stern *et al.*, 1981; Bikle *et al.*, 1986a). The effect of 1,25-$(OH)_2D_3$ on both enzymes is inhibited by cycloheximide and actinomycin D, suggesting genomic regulation. Other hormonal factors, such as parathyroid hormone, calcitonin, estrogens, and pituitary hormones, as well as plasma levels of calcium and phosphorus, also regulate the production of 1,25-$(OH)_2D_3$ (Henry and Norman, 1984).

There is increasing evidence in recent years for extrarenal production of 1,25-$(OH)_2D_3$. Anephric humans and pigs demonstrate low but detectable circulating levels of 1,25-$(OH)_2D_3$, which increase after systemic vitamin D administration (Lambert *et al.*, 1982; Littledike and Horst, 1982). Examples of nonrenal cells that produce 1,25-$(OH)_2D_3$ include macrophages (Koeffler *et al.*, 1985; Reichel *et al.*, 1987), bone cells (Howard *et al.*, 1981; Turner *et al.*, 1980), melanomas (Frankel *et al.*, 1983), sarcoid tissue (Barbour *et al.*, 1981), placenta (Gray *et al.*, 1979; Whitsett *et al.*, 1981), and human foreskin keratinocytes (Bikle *et al.*, 1986a,b). The regulation of 1,25-$(OH)_2D_3$ production by these nonrenal cells is similar but not identical to that by renal cells.

IX. Production of 1,25-$(OH)_2D_3$ by Epidermal Cells in Culture

Human foreskin keratinocytes in culture produce 1,25-$(OH)_2D_3$ from its precursor 25-(OH)D_3 (Bikle *et al.*, 1986b). The addition of [^{3}H]25-(OH)D_3 to keratinocytes in culture results in rapid uptake of the radiolabel, reaching a maximum by 1 hour (Bikle *et al.*, 1986a). Within 1 hour, several new metabolites are

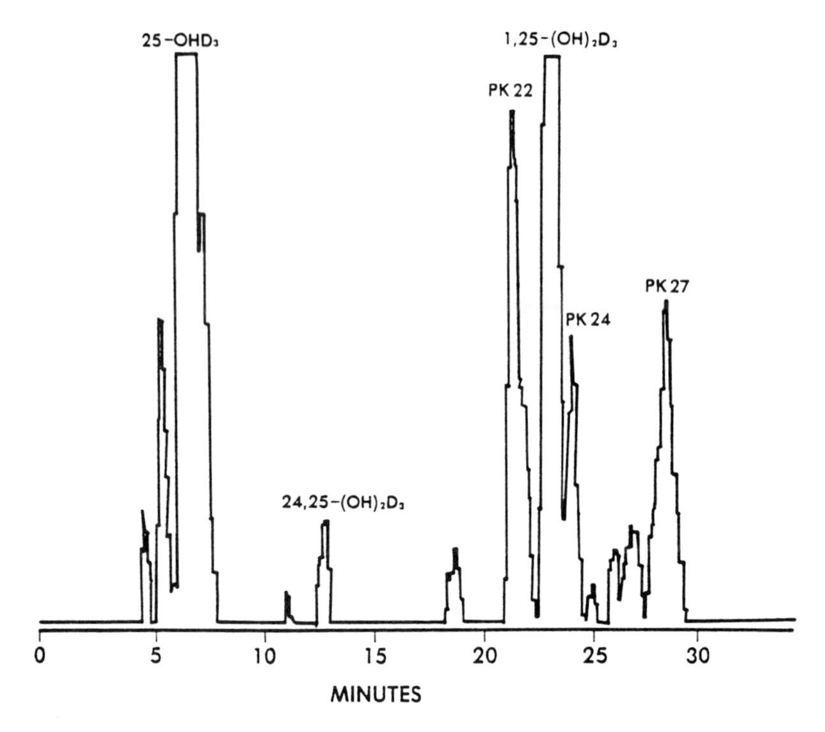

FIG. 3. A representative chromatogram of the lipid extract of a keratinocyte culture incubated with [^{3}H]25-(OH)D_3. The sample was incubated with 1.5 x 10^{-8} *M* 25-(OH)D_3 for 1 hour. The peaks eluting in the positions of the standards 25-(OH)D_3, 24,25-$(OH)_2D_3$, and 1,25-$(OH)_2D_3$ are labeled. (From Bikle *et al.*, 1986b.)

formed (Fig. 3). One of the metabolites cochromatographs with chemically synthesized 1,25-$(OH)_2D_3$ in multiple HPLC solvent systems (both normal and reverse phase), and displaces [^{3}H]1,25-$(OH)_2D_3$ in the chicken gut cytosol receptor assay with a potency equivalent to that of authentic 1,25-$(OH)_2D_3$ (Bikle *et al.*, 1986a). Finally, the mass spectrum of this putative 1,25-$(OH)_2D_3$ peak is identical to that of authentic 1,25-$(OH)_2D_3$ (Bikle *et al.*, 1986b). In addition to 1,25-$(OH)_2D_3$, keratinocytes also produce lower but variable amounts of the other major dihydroxy metabolite, 24,25-$(OH)_2D_3$, and a number of other metabolites, as well as the trihydroxy metabolite 1,24,25-$(OH)_3D_3$.

The production of these vitamin D metabolites by human keratinocytes is a regulatable process. Exogenous 1,25-$(OH)_2D_3$ at physiologically relevant concentrations (10^{-9} to 10^{-12} *M*) inhibits 1,25-$(OH)_2D_3$ production while increasing the 24,25-$(OH)_2D_3$ production by these cells (Bikle *et al.*, 1986a). The cells have to be exposed to exogenous 1,25-$(OH)_2D_3$ for at least 4 hours to observe the inhibitory effect on 1,25-$(OH)_2D_3$ production, and even longer incubations are required for maximal stimulation of 24,25-$(OH)_2D_3$ production. The half-maximal effect is at 10^{-11} *M* of the free 1,25-$(OH)_2D_3$ (not the total concentration) (Bikle and Gee, 1989). Actinomycin D blocks both these effects, indicating the genomic mechanism of action of this hormone on the regulation of its own production. Parathyroid hormone stimulates the production of 1,25-$(OH)_2D_3$, and the phosphodiesterase inhibitor, isobutylmethylxanthine, increases the 1,25-$(OH)_2D_3$ levels further, primarily by inhibiting its catabolism (Bikle *et al.*, 1986a; Bikle and Pillai, 1988). Finally, factors that induce keratinocyte differentiation, as well as the differentiation state of the cells themselves, regulate the propensity of these cells to generate 1,25$(OH)_2D_3$ or 24,25-$(OH)_2D_3$ (Pillai *et al.*, 1988a,b). For example, keratinocytes grown in 0.1 m*M* calcium (conditions favoring proliferation) make more 1,25-$(OH)_2D_3$ than do keratinocytes grown in 1.2 m*M* calcium (conditions favoring differentiation) (Pillai *et al.*, 1988b). Moreover, cytokines that induce keratinocyte differentiation (e.g., tumor necrosis factor-α and interferon-γ) also stimulate 1,25-$(OH)_2D_3$ production in subconfluent cultures (Pillai *et al.*, 1987; Bikle *et al.*, 1989) but inhibit production of the 1,25-$(OH)_2D_3$ metabolite in more differentiated cultures. Finally, undifferentiated preconfluent cells make more 1,25-$(OH)_2D_3$ and less 24,25-$(OH)_2D_3$, whereas postconfluent, differentiated keratinocytes make less 1,25-$(OH)_2D_3$ and progressively more 24,25-$(OH)_2D_3$ (Pillai *et al.*, 1988a).

X. Effects of 1,25-$(OH)_2D_3$ on Growth and Differentiation of Keratinocytes

The skin is not only important in the synthesis of vitamin D_3 and 1,25-$(OH)_2D_3$, but is also a prominent target organ for this hormone. Histochemical analysis of 1,25-$(OH)_2D_3$ receptors in various rat tissues indicates a high concentration of receptors in the epithelium of the outer hair sheaths, followed by the basal layers of the epidermis (Stumpf *et al.*, 1982). Various epidermal layers and dermal cells also contain specific 1,25-$(OH)_2D_3$ receptors (Pillai *et al.*, 1988a,b; Hosomi *et al.*, 1983), the concentrations of which appear to vary with the degree of differentiation of the cells *in vitro* (Pillai *et al.*, 1988a) and with development from fetus to adult (Horiuchi *et al.*, 1985). Differentiation of keratinocytes is associated with a decrease in the number of, but not the affinity of, the receptors (Pillai *et al.*, 1988a). The number of receptors increases markedly in the neonatal stage and reaches a maximum on day 10 after birth in mouse skin (Horiuchi *et al.*, 1985). During the neonatal period, when the receptor number is increased, the epidermal and dermal layers thicken rapidly and hair grows, suggesting a role for 1,25-$(OH)_2D_3$ in this process (Horiuchi *et al.*, 1985). Thus, changes in receptor number may be associated with changes in hormonal responsiveness at different developmental stages of cell growth and differentiation.

Table I

EFFECTS OF 1,25-$(OH)_2D_3$ ON KERATINOCYTE DIFFERENTIATION

Acute effects (within minutes to hours)	Long-term effects (within hours to days)
Inhibition of endogenous 1,25-$(OH)_2D_3$ production	Decrease in basal cells
Stimulation of 1,25-$(OH)_2D_3$ production	Increase in squamous cells
	Decrease in DNA synthesis
Stimulation of phosphoinositide turnover	Increase in transglutaminase activity
Acute increase of intracellular calcium levels	Slow and sustained increase of intracellular calcium
Translocation and activation of protein kinase C	Increase in cornified envelope formation
	Decrease in high-affinity EGF receptors
	Decrease in c-*myc* mRNA level

Several recent studies indicate a role for 1,25-$(OH)_2D_3$ in the growth and differentiation of epidermal cells (Pillai *et al.*, 1988a, 1990; Hosomi *et al.*, 1983; Smith *et al.*, 1986). A summary of the effects of 1,25-$(OH)_2D_3$ on the growth and differentiation of keratinocytes is given in Table I. Terminal differentiation of mouse epidermal cells grown in serum-containing medium is stimulated in a dose-dependent fashion by 1,25-$(OH)_2D_3$ (Hosomi *et al.*, 1983). Specifically, DNA synthesis is inhibited, the number of cornified cells is increased, and the number of basal cells is decreased as a result of 1,25-$(OH)_2D_3$ treatment. Only 1,25-$(OH)_2D_3$, not other metabolites of vitamin D, has this activity, suggesting the involvement of high-affinity 1,25-$(OH)_2D_3$ receptors in this process. Likewise, Smith and colleagues (1986) have reported the prodifferentiating effect of 1,25-$(OH)_2D_3$ on human keratinocytes grown in a serum-free, defined medium using several morphological and biochemical criteria. Inhibition of DNA synthesis, activation of transglutaminase, and an increase in the cornified envelopes were specifically attributed to the 1α form of 1,25-$(OH)_2D_3$, the 1β form being inactive. Studies using deepidermized dermis to induce a still greater degree of differentiation indicate that 1,25-$(OH)_2D_3$ dramatically increases the number of cornified layers and decreases the number of viable, basallike layers of keratinocytes (Regnier and Darmon, 1988). The effects of 1,25-$(OH)_2D_3$ on both the inhibition of growth and the induction of cornified envelopes on preconfluent cells grown in a serum-free medium are independent of extracellular calcium concentrations from 0.25 to 3.0 m*M* (McLane *et al.*, 1990). 1,25-$(OH)_2D_3$ decreased the number of high-affinity receptors for EGF without affecting the total receptor number (Matsumoto *et al.*, 1990b). Inhibition of DNA synthesis by 1,25-$(OH)_2D_3$ is accompanied by a decrease in c-*myc* mRNA expression within 3 hours of treatment, suggesting that decreased c-*myc* expression is one of the primary effects of this hormone on the growth inhibition of keratinocytes (Matsumoto *et al.*, 1990b).

Recently, several synthetic analogs of 1,25-$(OH)_2D_3$ have been found to be equally or more effective in inhibition of keratinocyte growth. 1,24*R*-$(OH)_2D_3$ binds to keratinocyte 1,25-$(OH)_2D_3$ receptors with a higher affinity than 1,25-$(OH)_2D_3$ and inhibits their growth to a greater extent (Matsumoto *et al.*, 1990a). Another 1,25-$(OH)_2D_3$ analog, calcipotriol (MC 903), inhibits keratinocyte growth and induces their terminal differentiation (Kragballe and Wildfang, 1990). In contrast, this analog was over 100-fold less potent than 1,25-$(OH)_2D_3$ in its effect on bone resorption, calciuria, and hypercalcemia. Therefore, it may be possible to separate the effects of vitamin D on growth from those on calcium metabolism. These studies should also point to potential new agents for effective treatment of psoriasis and other disorders of epidermal differentiation while minimizing the risk of hypercalcemia.

Because keratinocytes produce 1,25-$(OH)_2D_3$, and because exogenous 1,25-$(OH)_2D_3$ alters their differentiation, we explored the role of endogenously produced 1,25-$(OH)_2D_3$ on keratinocyte differentiation. The production of 1,25-$(OH)_2D_3$ in keratinocyte cultures peaks at the time of confluence and then falls off slowly (Pillai *et al.*, 1988a). Two weeks after confluence, keratinocytes make little 1,25-$(OH)_2D_3$. In contrast, the production of 24,25-$(OH)_2D_3$ is minimal in preconfluent cultures but increases after confluence is achieved (Fig. 4). 1,25-$(OH)_2D_3$ production parallels transglutaminase activity and involucrin content (markers of early stages of keratinocyte differentiation), whereas the subsequent production of 24,25-$(OH)_2D_3$ parallels the appearance of an important marker of terminal differentiation, cornified envelopes (Fig. 4). Thus, the correlation of 1,25-$(OH)_2D_3$ production with the early stages of differentiation and of 24,25-$(OH)_2D_3$ production with the later stages of differentiation support a role for these two vitamin D metabolites in the regulation of distinct stages in epidermal differentiation. The effect of high concentrations (10^{-6} *M*) of 25-$(OH)D_3$ on keratinocyte differentiation observed in one study (Smith *et al.*, 1986) may be attributed to the conversion of 25-$(OH)D_3$ to 1,25-$(OH)_2D_3$ rather than to the binding of 25-$(OH)D_3$ to the 1,25-$(OH)_2D_3$ receptor, because binding affinity of 25-$(OH)D_3$ to 1,25-$(OH)_2D_3$ receptor is typically at least three orders of magnitude less than that of 1,25-$(OH)_2D_3$. In preliminary studies, we observed that 25-$(OH)D_3$ displayed prodifferentiating activity comparable in potency to that of 1,25-$(OH)_2D_3$. Whether this activity is inhibitable in the presence of a specific 1α-hydroxylase inhibitor, such as ketoconazole, remains to be seen.

The effects of vitamin D deficiency on rat skin include (1) diminution in the number of granular cells, (2) change in the polarity of basal cells from cuboidal to longitudinal, and (3) increased cell proliferation (Pillai *et al.*, 1988b). Vitamin D-deficient rats maintained in the dark on an isocaloric diet supplemented with essential fatty acids demonstrate striking changes in gross appearance, including moist and a matted hair, patchy alopecia, scaling and erythema, and a variable but significant defect in the barrier to transcutaneous water loss (Pillai *et al.*, 1988b).

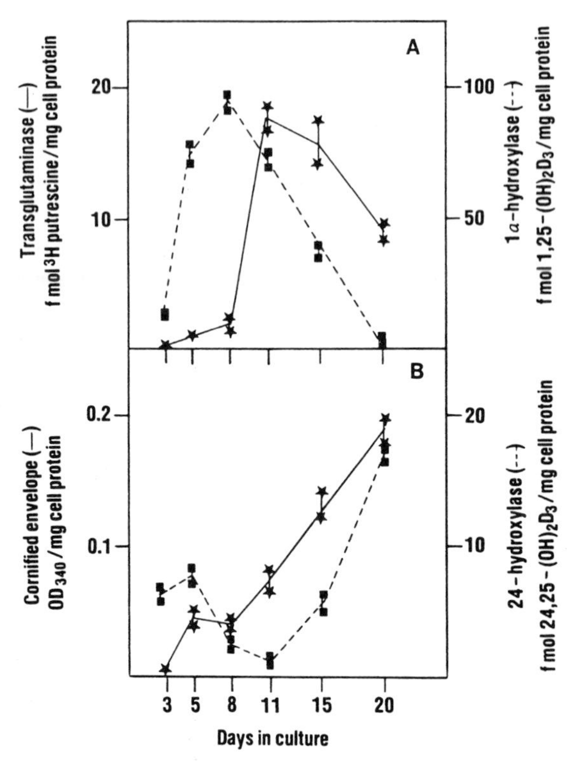

FIG. 4. Production of 1,25-$(OH)_2D_3$ and 24,25-$(OH)_2D_3$ in relation to keratinocyte differentiation markers. (A) The specific activity of keratinocyte 1α-hydroxylase paralleled the early differentiation marker, transglutaminase activity; (B) the activity of 24-hydroxylase paralleled the late differentiation marker, cornified envelope formation of keratinocytes. (From Pillai *et al.*, 1988a.)

These systemic effects support the concept that vitamin D and/or its metabolites play an important modulatory role in keratinocyte differentiation.

XI. Mechanism of Action of 1,25-$(OH)_2D_3$ in the Regulation of Epidermal Differentiation

One potential mechanism by which 1,25-$(OH)_2D_3$ may regulate epidermal differentiation involves calcium, because physiological levels of extracellular calcium are required for terminal differentiation of keratinocytes (Hennings *et al.*, 1980; Pillai *et al.*, 1990). Extracellular calcium increases intracellular calcium levels of keratinocytes and induces their differentiation (Pillai and Bikle, 1989;

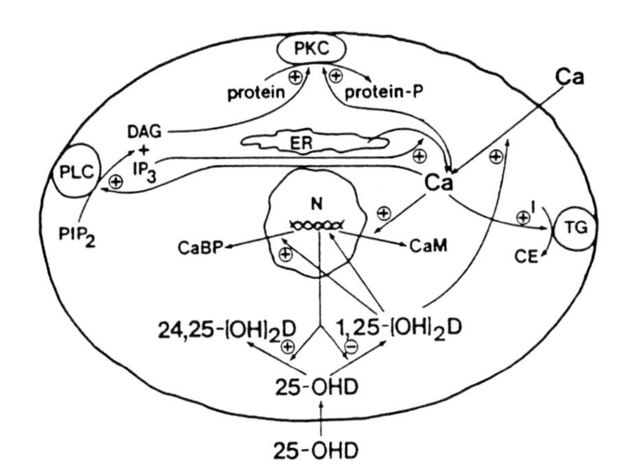

FIG. 5. A model for the role of 1,25-$(OH)_2D_3$ in keratinocyte differentiation. 1,25-$(OH)_2D_3$ produced within the cell may regulate calcium entry into cells and stimulate cornified envelope formation (CE) from precursors such as involucrin (I) by activating membrane-bound transglutaminase (TG) activity. In addition, 1,25-$(OH)_2D_3$ may translocate protein kinase C (PKC) to the membrane and Ca_i may stimulate PKC and polyphosphoinositide (PIP_2) turnover. Long-term effects of 1,25-$(OH)_2D_3$ may include nuclear regulation of the synthesis of calcium-binding protein (CaBP) and calmodulin (CaM), and the regulation of 1- and 24-hydroxylase activity. (From Pillai *et al.*, 1988b.)

Sharpe *et al.*, 1989; Hennings *et al.*, 1989). 1,25-$(OH)_2D_3$ also increases the intracellular calcium levels of keratinocytes (MacLaughlin *et al.*, 1990; Yada *et al.*, 1989; Tang *et al.*, 1987), possibly by regulating calcium flux across the plasma membrane (Fig. 5). An increase in intracellular free calcium may activate several enzyme systems, such as protein kinases (Schulman and Greengard, 1978), phosphodiesterases (Schulman and Greengard, 1978), adenylyl cyclases (Cheung *et al.*, 1978), and phospholipases (Daniel, 1985). These enzymes in turn may serve signaling mechanisms involved in proliferation and differentiation. Other calcium-requiring enzymes, such as thymidylate synthetase (Whitfield *et al.*, 1976) and transglutaminase (Rice *et al.*, 1988; Rubin and Rice, 1986), also are clearly involved in cellular proliferation and differentiation. The latter enzyme, transglutaminase, is essential for the formation of γ-glutamyl ϵ-lysyl cross-links during cornified envelope formation. An increase in intracellular calcium may also stimulate the redistribution of desmosomal proteins to the membranes and help in the formation of intercellular connections that may be associated with the differentiation of keratinocytes (Hennings and Holbrook, 1983).

One mechanism by which 1,25-$(OH)_2D_3$ may increase the intracellular calcium of keratinocytes is via increased phosphoinositide turnover. This results in a rise in inositol trisphosphates, which rapidly release calcium from intracellular stores (MacLaughlin *et al.*, 1990; Yada *et al.*, 1989; Tang *et al.*, 1987). This increase in

Ca_i, along with the diacyl glycerol produced by phosphoinositide hydrolysis, may induce the translocation and stimulation of protein kinase C (Yada *et al.*, 1989) leading to the phosphorylation of critical cellular proteins important in the regulation of growth and differentiation.

1,25-$(OH)_2D_3$ may induce new protein synthesis to either increase Ca_i levels or stimulate differentiation. Some studies indicate that longer incubations of keratinocytes with 1,25-$(OH)_2D_3$ (hours or days) are required to increase Ca_i levels and induce differentiation of keratinocytes (Pillai and Bikle, 1991). This is consistent with findings in other cell types, such as HL-60, which require hours or days of exposure to 1,25-$(OH)_2D_3$ before changes in Ca_i and differentiation are observed (Hruska *et al.*, 1988). Moreover, in other classical systems of 1,25-$(OH)_2D_3$ action, such as intestinal calcium absorption, the effect of 1,25-$(OH)_2D_3$ takes hours (Bikle, 1990). It is possible that just as in intestinal epithelial cells, 1,25-$(OH)_2D_3$ may regulate calcium transport across the keratinocyte cell membrane by both genomic and nongenomic mechanisms. It is conceivable that a vitamin D-dependent calcium-binding protein, which has been described in the skin (Laouari *et al.*, 1980), could participate in the regulation of Ca_i in keratinocytes. Involucrin, the γ-carboxyglutamic acid-rich protein precursor of cornified envelopes, is another potential 1,25-$(OH)_2D_3$-inducible protein. This would be analogous to the bone wherein 1,25-$(OH)_2D_3$ induces a similar γ-carboxyglutamic acid-rich protein, bone G1a protein (Price and Baukol, 1981) (see above). Thus, several possible mechanism(s) involving calcium may account for the regulation of keratinocyte differentiation by 1,25-$(OH)_2D_3$.

XII. Effect of 1,25-$(OH)_2D_3$ on Other Types of Skin Cells

Two other major skin cells, dermal fibroblasts and epidermal melanocytes, also contain high-affinity 1,25-$(OH)_2D_3$ receptors (Stumpf *et al.*, 1982; Hosomi *et al.*, 1983; Colston *et al.*, 1981). 1,25-$(OH)_2D_3$ causes a dose-dependent inhibition of growth of receptor-positive dermal fibroblasts, whereas the growth of receptor-defective VDDR II fibroblasts is not affected by 1,25-$(OH)_2D_3$ (Clements *et al.*, 1983). Dermal fibroblasts of psoriatic patients have been reported to respond abnormally to the growth inhibitory effects of 1,25-$(OH)_2D_3$, despite having normal numbers of high-affinity receptors (MacLaughlin *et al.*, 1985), suggesting the possibility of defective genomic mechanisms in their response to 1,25-$(OH)_2D_3$.

Epidermal melanocytes, in addition to possessing high-affinity 1,25-$(OH)_2D_3$ receptors, also appear to have the ability to make 1,25-$(OH)_2D_3$ from 25-$(OH)D_3$ (Frankel *et al.*, 1983). Two melanin-producing melanoma cell lines produced a metabolite that comigrates with authentic 1,25-$(OH)_2D_3$ in several chromatographic systems (Frankel *et al.*, 1983). Treatment of melanocyte cultures with 1,25-$(OH)_2D_3$ has been reported to have variable effects on their growth (Colston *et al.*, 1981; Abdel-Malek *et al.*, 1988). 1,25-$(OH)_2D_3$ may also increase melanin synthesis by stimulating tyrosinase activity at very high concentrations (10^{-5} *M*) (Oikawa and Nakayasu, 1974) or inhibit melanin synthesis by suppressing tyrosinase activity at lower concentrations (10^{-8} *M*) (Abdel-Malek *et al.*, 1988). If this occurs in epidermal melanocytes *in vivo*, 1,25-$(OH)_2D_3$ may act as a negative feedback mechanism for the production of vitamin D_3 by the skin by increasing melanin levels, thereby reducing penetration of sunlight into the epidermis. But some studies suggest a positive, rather than a negative, feedback regulation of vitamin D_3 production by 1,25-$(OH)_2D_3$ in the skin. Administration of 1,25-$(OH)_2D_3$ to vitamin D-deficient rats increases the cutaneous provitamin D_3 concentration (Esvelt *et al.*, 1980), and incubation of human keratinocytes with 1,25-$(OH)_2D_3$ results in a slight elevation of the amount of 7-DHC, possibly by inhibiting conversion of 7-DHC to cholesterol (Holick *et al.*, 1982a).

XIII. Concluding Remarks

In the last few years, it has become clear that the vitamin D endocrine system is important not only in the regulation of calcium homeostasis but also in the growth and differentiation of cells of the immune system and skin. The effect of 1,25-$(OH)_2D_3$ on epidermal cell differentiation may be mediated by the Ca_i levels of these cells. A better understanding of the basic mechanism of action of 1,25-$(OH)_2D_3$ on the differentiation of keratinocytes may have important pharmacological and therapeutic implications. Manipulation of epidermal 1,25-$(OH)_2D_3$ production could provide a novel approach to the therapy of certain skin diseases characterized by aberrant proliferation and differentiation, such as psoriasis or carcinomas.

Acknowledgments

We appreciate the support of Dr. Peter Elias for helpful discussions and NIH Grants AR-383860 and AR-39448 and a merit review from the Veterans Administration.

References

Abdel-Malek, Z. A., Ross, R., Trinkle, L., Swope, V., Pike, J. W., and Nordlund, J. J. (1988). *J. Cell. Physiol.* **136,** 273–280.

Agus, Z. S. (1983). *Kidney Int.* **24,** 113–123.

Askew, F. A., Bourdillon, R. B., Bruce H. M. Jenkins, R. G. C., and Webster, T. A. (1931). *Proc. R. Soc. London, Set. B* **107,** 76–90.

Audran, M. (1985). *Mayo Clin. Proc.* **60,** 851–866.

Baker, A. R., McDonnell, D. P., and Hughes, M. (1988). *Proc. Natl. Acad. Sci. U.S.A.* **85,** 3294–3298.

Baran, D. T., and Milne, M. L. (1986). *J. Clin. Invest.* **77,** 1622–1626.

Barbour, G. L., Coburn, J. W., Slatopolsky, E., Normal, A. W., and Horst, R. L. (1981). *N. Engl. J. Med.* **305,** 440–443.

Barsony, J., and Marx, S. (1988). *Proc. Natl. Acad. Sci. U.S.A.* **85,** 1223–1226.

Bell, N. H. (1985). *J. Clin. Invest.* **76,** 1–5.

Bell, N. H., Hamstra, A. J., and De Luca, H. F. (1978). *N. Engl. J. Med.* **298,** 996–999.

Bell, P. A. (1978). *In* "Vitamin D" (D. E. M. Lawson, ed.), pp. 1–23. Academic Press, New York.

Bhattacharyya, M. H., and DeLuca, H. F. (1974). *Arch. Biochem. Biophys.* **160,** 58–62.

Bikle, D. D. (1990). *In* "Membrane Transport and Information Storage" (R. C. Aloia, C. C. Curtain, and L. M. Gordon, eds.), pp. 191–219. Wiley-Less, New York.

Bikle, D. D., and Chafouleas, J. G. (1984). *In* "Endocrine Control of Bone and Calcium Metabolism" (D. V. Cohn, T. Fujita, J. T. Potts, and R. V. Talmage, eds.), pp. 320–323. Elsevier, Amsterdam.

Bikle, D. D., and Gee, E. (1989). *Endocrinology (Baltimore)* **124,** 649–654.

Bikle, D. D., and Munson, S. (1985). *J. Clin. Invest.* **76,** 2312–2316.

Bikle, D. D., and Pillai, S. (1988). *Ann. N.Y. Acad. Sci.* **548,** 27–44.

Bikle, D. D., Nemanic, M. K., Gee, E., and Elias, P. M. (1986a). *J. Clin. Invest.* **78,** 557–566.

Bikle, D. D., Nemanic, M. K., Whitney, J. O., and Elias, P. M. (1986b). *Biochemistry* **25,** 1545–1548.

Bikle, D. D., Pillai, S., Gee, E., and Hincenbergs, M. (1989). *Endocrinology (Baltimore)* **124,** 655–660.

Bjorkhem, I., and Holmberg. (1978). *J. Biol. Chem.* **253,** 842–849.

Brooks, M. H., Bell, N. H., Love, L., Stern, P. H., Orfei, E., Wueener, S. F., Hamstra, A. J., and DeLuca, H. F. (1978). *N. Eng. J. Med.* **298,** 996–1001.

Brumbaugh, P. F., and Haussler, M. R. (1974). *J. Biol. Chem.* **249,** 1251–1262.

Burger, E. H., Van der Meer, J. W. M., van de Gevel, J. S., Gribnau, J. C., Thesingh, C. W., and van Furth, R. (1982). *J. Exp. Med.* **156,** 1604–1614.

Canalis, E., McCarthy, T., and Centrella, M. (1988). *J. Clin. Invest.* **81,** 277–281.

Cheung, W. Y., Lynch, T. J., and Wallace, R. W. (1978). *Adv. Cyclic Nucleotide Res.* **9,** 233–251.

Clemens, T. L., Adams, J. S., Horiuchi, N., Gilchrest, B. A., Cho, H., Tsuchiya, Y., Matsuo, N., Suda, T., and Holick, M. F. (1983). *J. Clin. Endocrinol. Metab.* **56,** 824–830.

Colston, K., Colston, M. J., and Feldman, D. (1981). *Endocrinology (Baltimore)* **108,** 1083–1086.

Daniel, L. W. (1985). *In* "Biochemistry of Arachidonic Acid Metabolism" (W. E. M. Lands, ed.), pp. 175–189. Nijhoff, The Hague.

DeLuca, H. F., and Schnoes, H. K. (1983). *Annu. Rev. Biochem.* **52,** 411–439.

Dodd, R. C., Cohen, M. S., Newman, S. L., and Gray, T. K. (1983). *Proc. Natl. Acad. Sci. U.S.A.* **80,** 7538–7541.

Dupret, J. M., Brun, P., Perret, C., Lomri, N., Thomasset, M., and Cuisinier-Gleizes, P. (1987). *J. Biol. Chem.* **262,** 16553–16557.

Esvelt, R. P., De Luca, H. F., Wichmann, J. K., Yoshizua, S., Zurcher, J., Sar, M., and Stumpf, W. E. (1980). *Biochemistry* **19,** 6158–6163.

Fieser, L. D., and Fieser, M. (1959). "Steroids," pp. 90–168. Reinhold, New York.

Frankel, T. L., Mason, R. S., Hersey, P., Murray, E., and Posen, S. (1983). *J. Clin. Endocrinol. Metab.* **57,** 627–631.

Fraser, D. R., and Kodicek, E. (1970). *Nature (London)* **228,** 764–766.

Fraser, J. D., Otawara, Y., and Price, P. A. (1988). *J. Biol. Chem.* **263** 911–916.

Ghazarian, J. G., and DeLuca, H. F. (1974). *Arch. Biom. Biophys.* **160,** 63–72.

Ghazarian, J. G., Jefcoate, C. R., Knutson, J. C., Orme-Johnson, W. H., and DeLuca, H. F. (1974). *J. Biol. Chem.* **449,** 3026–3033.

Gray, T. K., Lester, E. G., and Lorenc, R. S. (1979). *Science* **204,** 1311–1313.

Haussler, M. R., and McCain, T. A. (1977). *N. Engl. J. Med.* **297,** 974–983, 1041–1050.

Hennings, H., and Holbrook, K. (1983). *Exp. Cell Res.* **143,** 127–142.

Hennings, H., Michael, D., Cheng, C., Steinert, P., Holbrook, K., and Yuspa, S. H. (1980). *Cell* **19,** 245–254.

Hennings, H., Kruszewski, F. H., Yuspa, S. H., and Tucker, R. W. (1989). *Carcinogenesis* **4,** 777–780.

Henry, H. L., and Norman, A. W. (1984). *Annu. Rev. Nutr.* **4,** 493–520.

Henry, H. L., Midget, R. J., and Norman, A. W. (1974). *J. Biol. Chem.* **249,** 7584–7592.

Hess, A. F. (1921). *J. Am. Med. Assoc.* **77,** 39.

Hess, A. F., and M. Weinstock, (1924). *J. Biol. Chem.* **62,** 301–313.

Holick, M. F. (1986). *In* "Nutrition and the Skin" (D. Roe, ed.), pp. 15–43. Alan R. Liss, New York.

Holick, M. F. (1988). *In* "Cutaneous Aging" (A. Klingman and Y. Takase, Eds.), pp. 223–246. Univ. of Tokyo Press, Tokyo.

Holick, M. F., Frommer, J. E., McNeill, S. C., Richtand, N. M., Henley, J. F., and Potts, J. T., Jr. (1977). *Biochem. Biophys. Res. Commun.* **76,** 107–114.

Holick, M. F., Richtand, N. M., McNeill, S. C., Holick, S. A., Frommer, J. E., Henley, J. W., and Potts, J. T., Jr. (1979). *Biochemistry* **18,** 1003–1008.

Holick, M. F., MacLaughlin, J. A., Clark, M. B., Holick, S. A., Potts, J. T., Jr., Anderson, R. R., Blank, I. H., Parrish, J. A., and Elias, P. M. (1980). *Science* **210,** 203–205.

Holick, M. F., MacLaughlin, J. A., and Doppelt, S. H. (1981). *Science* **211,** 590–593.

Holick, M. F., Adams, J. T., Clemens, T. L., MacLaughlin, J. A., Horiuchi, N., Smith, E., Holick, S. A., Nolan, J., and Hannifan, N. (1982a). *In* "Vitamin D, Chemical, Biochemical and Clinical Endocrinology of Calcium Metabolism" (A. W. Norman, K. Schaefer, D. V. Herrath, and H. G. Grigoleit, eds.), pp. 1151–1156. de Gruyter, Berlin.

Holick, M. F., MacLaughlin, J. A., Parrish, J. A., and Anderson, R. R. (1982b). *In* "The Science of Photomedicine" (J. D. Regan and J. A. Parrish, eds.) pp. 195–218. Plenum, New York.

Holtrop, M. E., Cox, K. A., Clark, M. B., Holick, M. F., and Anast, C. S. (1981). *Endocrinology (Baltimore)* **108,** 2293–2301.

Horiuchi, N., Clemens, T. L., Schiller, A. L., and Holick, M. F. (1985). *J. Invest. Dermatol.* **84,** 461–464.

Hosomi, J., Hosoi, J., Abe, E., Suda, T., and Kuroki, T. (1983). *Endocrinology (Baltimore)* **113,** 1950–1957.

Howard, G. A., Turner, R. T., Sharrad, D. J., and Baylink, D. J. (1981). *J. Biol. Chem.* **256,** 7738–7740.

Hruska, K. A., Bar-Shavit, Z., Malone, J. D., and Teitelbaum, S. (1988). *J. Biol. Chem.* **263,** 16039–16044.

Huldschinsky, K. (1919). *Dtsch. Med. Wochenschr.* **14,** 712–713.

Ichikawa, Y., Hiwatashi, A., and Mishii, Y. (1983). *Comp. Biochem. Physiol. B* **75D,** 479–488.

Koeffler, H. P., Reichel, H., Bishop, J. E., and Norman, A. W. (1985). *Biochem. Biophys. Res. Commun.* **127,** 596–603.

Kragballe, K., and Wildfang, I. L. (1990). *Arch. Dermatol. Res.* **282,** 164–167.

Kumar, R., Bagubandi, S., Mattox, V. R., and Londowski, J. M. (1980). *J. Clin. Invest.* **65,** 277–284.

Kurihara, M., Ishizuka, S., Kiyoki, M., Haketa, Y., Ikeda, K., and Kumegawa, M. (1986). *Endocrinology (Baltimore)* **118,** 940–947.

Lambert, P. W., Stern, P. H., and Avioli, R. C. (1982). *J. Clin. Invest.* **69,** 722–725.

Laouari, D., Pavlovitch, H., and Deceneux, G. (1980). *FEBS Lett.* **111,** 285–289.

Lieberhen, M. (1987). *J. Biol. Chem.* **262,** 13168–13171.

Littledike, E. T., and Horst, R. L. (1982). *Endocrinology (Baltimore)* **111,** 2008–2013.

Lyles, K. W., and Drezner, M. K. (1982). *J. Clin. Endocrinol. Metab.* **54,** 638–644.

MacLaughlin, J. A., and Holick, M. F. (1983) *In* "Biochemistry and Physiology of the Skin" (L. A. Goldsmith, ed.). pp. 734–754. Oxford Univ. Press, London.

MacLaughlin, J. A., Gnage, W., Taylor, D., Smith, E., and Holick, M. F. (1985). *Proc. Natl. Acad. Sci. U.S.A.* **83,** 5409–5414.

MacLaughlin, J. A., Cantley, L. C., and Holick, M. F. (1990). *J. Nutr. Biochem.* **1,** 81–87.
Matsumoto, K., Hashimoto, K., Kiyoki, M., Yamamoto, M., and Yoshikawa, K. (1990a). *J. Dermatol.* **17,** 97–103.
Matsumoto, K., Hashimoto, K., Nishiida, Y., Hashiro, M., and Yoshikawa, K. (1990b). *Biochem. Biophys. Res. Commun.* **166,** 916–923.
McCarthy, D. M., San Miguel, J. F., Freake, H. C., Green, P. M., Zola, H., Catovsky, D., and Goldman, J. M. (1983). *Leuk. Res.* **7,** 51–55.
McDonnell, D. P., Mangelsdorf, D. J., Pike, J. W., Haussler, M. R., and O'Malley, B. W. (1987). *Science* **235,** 1214–1217.
McLane, J. A., Katz, M., and Abdelkader, N. (1990). *In Vitro Cell Dev. Biol.* **26,** 379–387.
Mellanby, E. J. (1918). *J. Physiol. (London)* **52,** 11–14.
Monolaguas, S. L., Burton, D. W., and Deflos, L. J. (1981). *J. Biol. Chem.* **256,** 7115–7117.
Nemanic, M. K., Whitney, J., Arnaud, S., and Elias, P. M. (1983). *Biochem. Biophys. Res. Commun.* **115,** 444–449.
Nemanic, M. K., Whitney, J., and Elias, P. M. (1985). *Biochim. Biophys. Acta* **841,** 267–277.
Nemere, I., Yoshimoto, Y., and Norman, A. W. (1984). *Endocrinology (Baltimore)* **115,** 1476–1483.
Oikawa, A., and Nakayasu, M. (1974). *FEBS Lett.* **42,** 32–35.
Olsson, I., Gullberg, U., Ivhed, I., and Nilsson, K. (1984). *Cancer Res.* **43,** 5862–5867.
Owen, I. (1889). *Br. Med. J.* **1,** 113–116.
Palm, T. A. (1890). *Practitioner* **45,** 270–279, 321–342.
Pan, I. C., and Price, P. A. (1984). *J. Biol. Chem.* **259,** 5844–5847.
Pederson, J. I., Gazarian, J. G., Orme-Johnson, N. R., and DeLuca, H. F. (1976). *J. Biol. Chem.* **251,** 3933–3941.
Petkovich, P. M., Wrana, J. L., Grigoriadis, A. E., Heersche, J. N. M., and Sodek, J. (1987). *J. Biol. Chem.* **262,** 13424–13428.
Petrova, E. A., Nikulicheva, S. I., and Lazareva, N. P. (1976). *Vopr. Pitan.* **5,** 50–52.
Pike, J. W., Marion, S. L., Donaldson, C. A., and Haussler, M. R. (1983). *J. Biol. Chem.* **258,** 1289–1296.
Pillai, S., and Bikle, D. D. (1989). *J. Invest. Dermatol.* **92,** 500a.
Pillai, S., and Bikle, D. D. (1991). *J. Cell. Physiol.* **146,** 94–100.
Pillai, S., Bikle, D. D., Essalu, T. E., Aggarwal, B., and Elias, P. M. (1987). *Clin. Res.* **35,** 194A.
Pillai, S., Bikle, D. D., and Elias, P. M. (1988a). *J. Biol. Chem.* **263,** 5390–5395.
Pillai, S., Bikle, D. D., and Elias, P. M. (1988b). *Skin Pharmacol.* **1,** 149–160.
Pillai, S., Bikle, D. D., Mancianti, M.-L., Cline, P., and Hincenbergs, M. (1990). *J. Cell. Physiol.* **143,** 294–302.
Ponchon, G., Kennan, A. L., and DeLuca, H. F. (1969). *J. Clin. Invest.* **48,** 2032–2037.
Price, P. A., and Baukol, S. A. (1980). *J. Biol. Chem.* **255,** 11660–11663.
Price, P. A., and Baukol, S. A. (1981). *Biochem. Biophys. Res. Commun.* **99,** 928–935.
Pryke, A. M., Duggan, C., White, C. P., Posen, S., and Mason, R. S. (1990). *J. Cell. Physiol.* **142,** 652–656.
Raiw, K. G., and Kream, B. E. (1983). *N. Engl. J. Med.* **29,** 83–89.
Regnier, M., and Darmon, M. Y. (1988). *J. Invest. Dermatol.* **90,** 600.
Reichel, H., Koeffler, H. P., and Norman, A. W. (1987). *J. Biol. Chem.* **262,** 10931–10937.
Reichel, H., Koeffler, P., and Norman, A. W. (1989). *N. Engl. J. Med.* **320,** 980–991.
Rice, R. H., Chakravarty, R., Chen, J., O'Callahan, W., and Rubin, A. L. (1988). *In* "Advances in Post-Translational Modification of Proteins and Aging" (V. Zappia and P. Galletti, eds.), pp. 51–61. Plenum, New York.
Roodman, G. D., Ibbotromn, K. J., MacDonald, B. R., Kuchi, T. J., and Mundy, G. R. (1985). *Proc. Natl. Acad. Sci. U.S.A.* **82,** 8213–8217.
Rowe, D. W., and Kream, B. E. (1982). *J. Biol. Chem.* **257,** 8009–8015.

Rubin, A. L., and Rice, R. H. (1986). *Cancer Res.* **46,** 2356–2361.

Schulman, H., and Greengard, P. (1978). *Proc. Natl. Acad. Sci. U.S.A.* **75,** 5432–5436.

Sharpe, G. R., Gillespie, J. I., and Greenwell, J. R. (1989). *FEBS Lett.* **354,** 25–28.

Smith, E. L., Walworth, N. C., and Holick, M. F. (1986). *J. Invest. Dermatol.* **86,** 709–714.

Sniadecki, J. (1931). *Nature (London)* **143,** 121.

Spencer, R., Charman, M., Emtage, J. S., and Lawson, D. E. M. (1976). *Eur. J. Biochem.* **71,** 399–409.

Spencer, R., Charman, M., and Lawson, D. E. M. (1978). *Biochem. J.* **175,** 1089–1094.

Steenbock, H., and Black, A. (1924). *J. Biol. Chem.* **61,** 408–422.

Stern, P. H. (1980). *Pharmacol. Rev.* **32,** 47–80.

Stern, P. H., Taylor, A. B., Bell, N. H., and Epstein, S. (1981). *J. Clin. Invest.* **68,** 1374–1379.

Stumpf, W. E. (1988). *Histochemistry,* **89,** 209–219.

Stumpf, W. E., Sar, M., Reid, F. A., Tanaka, Y., and De Luca, H. F. (1982). *J. Clin. Endocrinol. Metab.* **54,** 638–646.

Suda, T., Miyaura, C., Abe, E., and Kuroki, T. (1986). *Bone Miner. Res.* **4,** 1–48.

Tanaka, H., Abe, E., Miyaura, C., Kuribayashi, T., Konno, K., Nishii, Y., and Suda, T. (1982). *Biochem. J.* **204,** 713–719.

Tanaka, Y., and DeLuca, H. F. (1974). *Science* **183,** 1198–1200.

Tang, W., Ziboh, V., Isseroff, R. R., and Martinez, D. (1987). *J. Cell. Physiol.* **132,** 131–136.

Turner, R. R., Puzan, J. E., Frote, M. D., Lester, G. E., Gray, T. K., Howard, G. A., and Baylink, D. J. (1980). *Proc. Natl. Acad. Sci. U.S.A.* **77,** 5720–5724.

Wali, R. K., Baum, C. L., Sitrin, M. D., and Brasitus, T. A. (1990). *J. Clin. Invest.* **85,** 1296–1303.

Whitfield, J. F., MacManus, J. P., Rixon, R. H., Baynton, A. L., Youdale, T., and Swierenga, S. (1976). *In Vitro* **12,** 1–18.

Whitsett, J. A., Ho, M. T., Sang, R. C., Norman, E. J., and Adams, K. G. (1981). *J. Clin. Endocrinol. Metab.* **53,** 484–488.

Windus, A., and Bock, F. (1937). *Hoppe-Seyler's Z. Physiol. Chem.* **245,** 168.

Windus, A., Luttringhaus, A., and Deppe, M. (1931). *Justus Liebig's Ann. Chem.* **489,** 252–269.

Windus, A., von Werder, F., and Luttringhaus, A. (1932). *Justus Liebig's Ann. Chem.* **499,** 188.

Yada, Y., Ozeki, T., Meguro, S., Mori, S., and Nozawa, Y. (1989). *Biochem. Biophys. Res. Commun.* **163,** 1517–1522.

Zerwekh, J. E., Haussler, M. R., and Lindell, T. J. (1974). *Proc. Natl. Acad. Sci. U.S.A.* **71,** 2337–2341.

INDEX

F

G

H

I

M

N

O

P

T

U

V

W

X

CONTENTS OF PREVIOUS VOLUMES

Volume 19

Volume 20

Volume 21

Volume 22

Volume 23